AF895799

Lecture Notes in Operations Research

Editorial Board

Ana Paula Barbosa-Povoa, University of Lisbon, Lisboa, Portugal

Adiel Teixeira de Almeida, Federal University of Pernambuco, Recife, Brazil

Noah Gans, The Wharton School, University of Pennsylvania, Philadelphia, USA

Jatinder N. D. Gupta, University of Alabama in Huntsville, Huntsville, USA

Gregory R. Heim, Mays Business School, Texas A&M University, College Station, USA

Guowei Hua, Beijing Jiaotong University, Beijing, China

Alf Kimms, University of Duisburg-Essen, Duisburg, Germany

Xiang Li, Beijing University of Chemical Technology, Beijing, China

Hatem Masri, University of Bahrain, Sakhir, Bahrain

Stefan Nickel, Karlsruhe Institute of Technology, Karlsruhe, Germany

Robin Qiu, Pennsylvania State University, Malvern, USA

Ravi Shankar, Indian Institute of Technology, New Delhi, India

Roman Slowiński, Poznań University of Technology, Poznan, Poland

Christopher S. Tang, Anderson School, University of California Los Angeles, Los Angeles, USA

Yuzhe Wu, Zhejiang University, Hangzhou, China

Joe Zhu, Foisie Business School, Worcester Polytechnic Institute, Worcester, USA

Constantin Zopounidis, Technical University of Crete, Chania, Greece

Lecture Notes in Operations Research is an interdisciplinary book series which provides a platform for the cutting-edge research and developments in both operations research and operations management field. The purview of this series is global, encompassing all nations and areas of the world.

It comprises for instance, mathematical optimization, mathematical modeling, statistical analysis, queueing theory and other stochastic-process models, Markov decision processes, econometric methods, data envelopment analysis, decision analysis, supply chain management, transportation logistics, process design, operations strategy, facilities planning, production planning and inventory control.

LNOR publishes edited conference proceedings, contributed volumes that present firsthand information on the latest research results and pioneering innovations as well as new perspectives on classical fields. The target audience of LNOR consists of students, researchers as well as industry professionals.

Ali Emrouznejad · Panagiotis D. Zervopoulos ·
Ilhan Ozturk · Dima Jamali · John Rice
Editors

Business Analytics and Decision Making in Practice

Proceedings of the International Conference on Business Analytics in Practice (ICBAP 2024), Sharjah, UAE

Springer

Editors
Ali Emrouznejad
Surrey Business School
University of Surrey
Guildford, UK

Ilhan Ozturk
College of Business Administration
University of Sharjah
Sharjah, United Arab Emirates

College of Business Administration
Nişantaşı University
İstanbul, Türkiye

John Rice
College of Business Administration
University of Sharjah
Sharjah, United Arab Emirates

Panagiotis D. Zervopoulos
College of Business Administration
University of Sharjah
Sharjah, United Arab Emirates

Dima Jamali
Canadian University in Dubai
Dubai, United Arab Emirates

ISSN 2731-040X ISSN 2731-0418 (electronic)
Lecture Notes in Operations Research
ISBN 978-3-031-61588-7 ISBN 978-3-031-61589-4 (eBook)
https://doi.org/10.1007/978-3-031-61589-4

© The Editor(s) (if applicable) and The Author(s), under exclusive license to Springer Nature Switzerland AG 2024

This work is subject to copyright. All rights are solely and exclusively licensed by the Publisher, whether the whole or part of the material is concerned, specifically the rights of translation, reprinting, reuse of illustrations, recitation, broadcasting, reproduction on microfilms or in any other physical way, and transmission or information storage and retrieval, electronic adaptation, computer software, or by similar or dissimilar methodology now known or hereafter developed.
The use of general descriptive names, registered names, trademarks, service marks, etc. in this publication does not imply, even in the absence of a specific statement, that such names are exempt from the relevant protective laws and regulations and therefore free for general use.
The publisher, the authors and the editors are safe to assume that the advice and information in this book are believed to be true and accurate at the date of publication. Neither the publisher nor the authors or the editors give a warranty, expressed or implied, with respect to the material contained herein or for any errors or omissions that may have been made. The publisher remains neutral with regard to jurisdictional claims in published maps and institutional affiliations.

This Springer imprint is published by the registered company Springer Nature Switzerland AG
The registered company address is: Gewerbestrasse 11, 6330 Cham, Switzerland

If disposing of this product, please recycle the paper.

Preface

In a business context where success or failure is increasingly determined by effective business analytics, the integration of data into current decision-making models has become a vital driver of sustainable growth, innovation, efficiency, and competitive advantage. Harnessing the power of data, business analytics allows organizations to generate actionable insights, identify emergent trends, and better anticipate future contextual changes. In this book, *Business Analytics and Decision Making in Practice*, we seek to investigate some practical applications of these analytical methodologies by offering readers a sample of papers that navigate illustrate the use of Business Analytics in real-world organizations and operational scenarios. Using both novel theoretical frameworks and examples from real-world practice, provides managers and researchers the tools and insights necessary to begin to unlock the potential of data-driven decision-making.

The papers in this volume were submitted at the International Conference on Business Analytics in Practice (ICBAP) 2024, held at the University of Sharjah in the UAE in 2024. Hosted in collaboration by leading institutions including the College of Business Administration at the University of Sharjah, UAE, and the Centre for Business Analytics in Practice at Surrey Business School in the University of Surrey, UK, the conference was the site of a vibrant exchange of theory, practical insights and innovative thinking for those in attendance.

The conference program included papers within the business analytics, decision support systems, and sustainability analytics fields. Recurrent themes included the interplay of institutional factors and environmental sustainability issues, in different national contexts. The construction industry is also featured in a number of papers, both in terms of sustainability and worker safety. Data envelopment analysis was employed in various contexts, with extensions to this method a highlight of a number of papers. Cutting-edge applications of machine learning were also presented, especially in financial analytics and portfolio management. At the individual level, the influence of personality traits on organizational behavior was explored in a number of submissions. Finally, novel papers were presented exploring digital supply chain logistics, the application of AI in preserving food supply chains, and the effects of circular economy practices on firm performance.

This volume comprises 29 chapters that offer a cross-section of novel applications of business analytics to real-world challenges. From the use of the Internet of Things (IoT) in the hospitality industry to the detection of anomalies in enterprise payment systems through the use of machine learning techniques, each paper provides novel conceptual ideas and applied data to illustrate practical implications of these new techniques. The book will be of interest to anyone seeking to leverage data-driven approaches to improve business and public sector performance.

We extend our appreciation to all the authors whose contributions have made this volume possible. In addition, we thank the reviewers whose feedback and constructive recommendations have enhanced the quality and rigor of each chapter. We would also like to express special thanks to Springer publisher and Springer Production, especially Dr. Christian Rauscher (Editorial Director at Springer) and Ms. Jialin Yan (Springer Editor), who supported the publication of this volume in the Springer series of Lecture Notes in Operational Research.

Finally, it is essential to clarify that the responsibility for the content, accuracy, and interpretations presented in each chapter lies solely with the respective authors. As editors, we facilitate the compilation and organization of the material, but we do not assume any responsibility for errors, inaccuracies, or omissions that may be present within individual chapters. While every effort has been made to ensure the quality and integrity of the content, readers should exercise their judgment and discretion when interpreting and utilizing the information provided.

Guildford, UK
Sharjah, United Arab Emirates
Sharjah, United Arab Emirates/
İstanbul, Türkiye
Dubai, United Arab Emirates
Sharjah, United Arab Emirates
March 2024

Ali Emrouznejad
Panagiotis D. Zervopoulos
Ilhan Ozturk
Dima Jamali
John Rice

Contents

Chapter 1
Internet of Things (IoT) a Trending Technology: Transforms the Hospitality Industry. Case Study: W Costa Navarino (Messinia, Greece)

Nikoletta Klada and Nikolaos Georgopoulos

Abstract The main study of this research is the Internet of Things (IoT) and the vital application in the Hospitality Industry. The methodology that has been used in this paper is based on literature review, which tries to find the gap in the international literature and understand the concepts which are under examination. The world community have tried to define the term "Internet of Things (IoT)", although its initial use has been attributed to Kevin Ashton, an expert on digital innovation. He is known for coining the term "The Internet of Things". Tourism is one of the economic sectors in which the rise of Internet of Things (IoT) applications is expected to greatly boost development. The Internet of Things (IoT) has the potential to transform the Hospitality Industry. Internet of Things (IoT) will help hotel owners, managers and others in Tourism Sector to understand the potential that this trending technology holds. IoT is already being applied in the hospitality industry and the future potential it has for hotel owners and managers makes it necessary for them to understand and use IoT. The survey was held in Costa Navarino (Messinia, Greece), in order to determine how important and efficient is for a hotel to use Internet of Things technology. The IoT is an integral part of tourism sector and W Costa Navarino is using IoT to automate every possible service for improving guest experience.

Keywords Internet of Things (IoT) · Hospitality industry · Tourism

Introduction

In an effort to find a unique definition for Internet of Things (IoT) it is generally accepted by academics, researchers, practitioners and innovators that there isn't one. Overall, what all of the definitions have in common is the idea that the first version

N. Klada · N. Georgopoulos (✉)
Department of Business Administration, University of Piraeus, 18534 Piraeus, Greece
e-mail: ngeorgop@unipi.gr

© The Author(s), under exclusive license to Springer Nature Switzerland AG 2024

A. Emrouznejad et al. (eds.), *Business Analytics and Decision Making in Practice*, Lecture Notes in Operations Research, https://doi.org/10.1007/978-3-031-61589-4_1

of the Internet was about data created by people, while the next version is about data created by things.

The term "Internet of Things" was first used by Kevin Ashton in 1999. IoT is a network that can connect with anything anytime and anyplace by technologies of RFID (Radio Frequency IDentification), WSN (Wireless Sensor Network) and 3G/4G/5G mobile communication, according to an agreed protocol, in order to identify, locate, track, monitor and manage smart objects [10].

The potential of the Internet of Things market is huge, where "things" interacting with each other act as active participants in the business [5]. Objects connected to the global Internet contribute both to improving the quality of life of the population and influence the processes taking place in the surrounding world. Such interconnection of devices in a single network brings huge opportunities, opening up new and broader prospects in the field of management, analytics, and security [14].

Literature Review

The basic concept of literature review is to understand the main topics and theories associated with the phenomenon, the trending Technology, Internet of Things (IoT). Some definitions of Internet of Things will be presented below, started with the most recent one.

"The Internet of Things (IoT) is defined as a paradigm in which objects equipped with sensors, actuators, and processors communicate with each other to serve a meaningful purpose [12]". U's Telecommunication Standardization Sector (ITU-T) defines IoT as "A global infrastructure for the information society enabling advanced services by interconnecting (physical and virtual) things based on existing and evolving interoperable information and communication technologies."

IoT is an "Interconnection of sensing and actuating devices providing the ability to share information across platforms through a unified framework, developing a common operating picture for enabling innovative applications" [4].

Nunberg [11] considers that the Internet of Things is a novel paradigm shift in IT arena. The phrase "Internet of Things" which is also shortly well-known as IoT is coined from the two words i.e. the first word is "Internet" and the second word is "Things". The Internet is a global system of interconnected computer networks that use the standard Internet protocol suite (TCP/IP) to serve billions of users worldwide. It is a network of networks that consists of millions of private, public, academic, business, and government networks, of local to global scope, that are linked by a broad array of electronic, wireless and optical networking technologies.

The Internet of Things can also be considered as a global network which allows the communication between human-to-human, human-to-things and things-to-things, which is anything in the world by providing unique identity to each and every object [1].

Internet of Things is maturing and continues to be the latest, most hyped concept in the IT world. Over the last decade the term Internet of Things (IoT) has attracted attention by projecting the vision of a global infrastructure of networked physical objects, enabling anytime, anyplace connectivity for anything and not only for any one [8].

"Internet of Things can also be realized in three paradigms—internet - oriented (middleware), things oriented (sensors) and semantic- oriented (knowledge)" [2].

According to Kranenburg [9] "Internet of Thing is a dynamic global network infrastructure with self-configuring capabilities based on standard and interoperable communication protocols where physical and virtual "Things" have identities, physical attributes, and virtual personalities and use intelligent interfaces, and are seamlessly integrated into the information network".

Internet of Things: Application in the Hospitality Industry via Solutions/Examples

The Hospitality Industry consists of multiple stakeholders and impacts global economy. New disruptive technologies, such as Internet of Things (IoT), Artificial Intelligence (AI), Robots, Social Media and others are crucial to achieving more efficiency and productivity in Hospitality Industry.

IoT is used in many different areas to improve efficiency and provide a better customer experience. Homes are increasingly turning into smart homes and beginning to give way to smart hotels, bringing hoteliers an opportunity to better serve guests, offer value-added services, and run the facility on the back end with far greater efficiency than was ever before possible [3].

Innovations in smart devices and IoT are driving the reform of technology used in the hospitality service platform [7]. The hotel industry can use IoT to provide integrated services such as application-driven devices and automated triggers like automated door locks, set-top-boxes, thermostats, telephones, light switches, voice-based interaction, electric blinds and other devices that are connected on a common network to enable the services that guests want. These kinds of facilities will make the guest experience more personalized and speedier, and the hotelier, more efficient. Also, hotels will be able to deliver more value to their guests at lower costs [3].

Table 1.1 illustrates several solutions and examples of IoT in the Hospitality Industry.

IoT is already being applied in the hospitality industry and the future potential it has for hotel owners makes it important for them to understand and use IoT. In addition, it can help to automate processes, improve the guest experience and help hospitality companies to save money on energy costs and maintenance. To come to closer terms with their present and future guests, hotels need to continuously monitor trends on the ever more discerning and burgeoning tourism market and adapt to the development of new technologies to remain competitive [3].

In the hospitality industry IoT applications should be embraced among businesses, and building this platform is necessary for innovators hotel owners and managers.

Table 1.1 Several solutions of IoT in the hospitality industry

Solution	Application/Example
Personalised hotel rooms	Hilton and Marriott have experimented with slightly different takes on the 'connected room' concept, where users are able to control many of the room's features from their mobile phone, or from a provided, tablet. In addition, IoT platforms could over time memorize a guest's specific comfort preferences, such as temperature, lights, TV channels & shades, and automatically set up the room for their next stay. Hotels can automatically send electronic key cards to guests' smartphones, allowing them to check-in without anyone's assistance
Voice-based interaction	*Amazon Alexa* enables the guest to use voice-control on different smart home devices. It's basically a guest's personal butler who never gets tired of fulfilling their needs. Alexa can be configured by hospitality providers to allow guests to control and adjust in-room devices such as lights, thermostats, blinds and TVs
Integration with mobile	*Fast check-in:* Starwood and Hilton, for example, already offer an option of checking in via a mobile gadget instead of spending time at the front desk. *Control device*: all thee functions performed by quite a few remote controls, key cards and switches are now available at the click of a button on a mobile device
Body area sensors	*Wireless medical sensor* technology further expands the scope of date collection by providing detailed data about organs and system within the body. For example, service providers can filter out high carbohydrate and sugary meal options for diabetic guests, high cholesterol meal options for patients with heart disease, etc.
Inventory management	IoT will change the way hotels manage their inventories. IoT implemented systems will keep record of the inventory and manage the changes in them automatically
Location-based information	For example, sending SMS messages about menu items at the restaurant when guests are close by or advertising gym services when they are near the gym. It may also mean sending up-to-date information about local transport link, or nearby attractions…
Flic—a wireless button	For example, when someone calls reception to ask for more towels or coffee. With *Flic*, guests could notify housekeepers about it with a press of a button. Ordering services and sending automated messages has never been this easy
Building automation and monitoring	For example, in-room monitoring system can be used to detect whether a room is occupied or unoccupied so as to schedule housekeeping services
Valpas—autonomous bed bug prevention	"Valpas has developed an autonomous bed bug prevention system that allows guests to stay carefree from bed bug anxiety at exciting hotels around the world. By replacing their existing bed legs with Valpas' smart legs, hotels can monitor the rooms in real-time and receive notifications of eliminated bed bugs caught inside the legs"

(continued)

Table 1.1 (continued)

Solution	Application/Example
Automation	IoT automates the business functions of travel and hotel businesses. For example, hotels can track supply chains more efficiently through sensors in shipments, enabling the to get ready for any future contingency and avoid service disruptions to guests
Augmented reality and beacon technology	For example, this technology can be used to provide guests with services such as digitally guided tours, previews of in-room environment (e.g., décor, facilities and amenities, etc.), immediate translation services for signs and other written materials, interactive restaurant menus with dish previews, critic reviews, food allergy information, etc.
Maintenance	By installing sensors, hotels can track the condition of all electronic devices and schedule the work of personnel according to whether the guest is in or out of the room. In addition, predictive maintenance will take hoteliers one-step further; it will use sensors to recognize the problems and alert before the issue hazardous
Energy saving	While IoT can enable personalization, it can also offer businesses financial benefits through automated or smart energy saving

Source Car et al. [3]

Case Study W Costa Navarino

Founded in 1997, TEMES is a leading investor, developer and operator in the high-end tourism and real estate sector in Greece. Costa Navarino, its flagship development, is one of the largest tourism investments in the Mediterranean. Based on solid financial foundations, the development will ultimately comprise four (4) resort areas with 5-star hotels, quality facilities and world-class golf courses. TEMES has fulfilled the vision of its founder, Captain Vassilis Constantakopoulos, to establish his homeland Messinia as a top international destination.

Costa Navarino is a sustainably driven destination in the Mediterranean, located in the Greek region of Messinia in the southwest Peloponnese. One of the most unspoiled and breathtaking seaside landscapes, this is a region shaped by 4,500 years of history. The Costa Navarino philosophy is driven by a genuine desire to preserve the natural beauty and heritage of Messinia.

With a total of four luxury resorts (Table 1.2), four signature golf courses, private residences, more than 40 dining and entertainment venues, and an abundance of sports and entertainment experiences across its sites, Costa Navarino has established Messinia as a place to visit as well as to live.

In this paper we will present W Costa Navarino Resort as an example, which provides the Technology IoT in many different areas, in order to improve efficiency and provide a better customer experience.

Table 1.2 Costa Navarino capacity

	Brands	Resort	Rooms
1	The luxury collection	Romanos resort	321
2	Westin	The Westin resort Costa Navarino	445
3	W hotels	W Costa Navarino	246
4	Mandarin	Mandarin oriental, Costa Navarino	99
		Total No Rooms	**1111**

W Costa Navarino marks the debut of W Hotels in Greece. This new resort opened in August 2022. Vivid and bright, with a distinctive design inspired by the region's architecture, W Costa Navarino shows 246 stylish rooms (www.costanavarino.com).

Applications of Internet of Things in W Costa Navarino

Costa Navarino invests significantly in research and development with respect to their guests' changing needs in areas such as nutrition, wellbeing, fitness and technology.

Since 2017, Costa Navarino's hotels have been using the **Marriott Guest Voice platform** to collect feedback via online Guest Satisfaction Surveys (GSS) sent to guests after their stay at Costa Navarino. In addition to the online surveys, a dedicated team monitors social media and relevant websites, collecting feedback and responding to guest inquiries in real time. The cornerstone of their philosophy is to handle any guest enquiry or request in less than 24 h. The combined analysis of the information collected covers the entire guest experience from arrival to departure, assessing the quality of activities and services, from housekeeping and staff attitude, to pools and fitness centers, Spa, golf courses, bars & restaurants, lobbies, shops, kid's facilities and all other guest areas. The use of Key Performance Indicators provides a credible and quantifiable assessment of how well they meet guests' expectations, together with valuable insights into how to make their services even better.

The Costa Navarino Mobile application (10 + thousand downloads—Google store) is a one-stop guide to plan a next visit to the destination.

The guest can use the Costa Navarino App to:

- Explore the destination and its full offering.
- Locate his room & all facilities within the map.
- Discover events and activities that are happening daily in our Calendar.
- Browse the menus of our restaurants and make dinner reservations.
- Discover the Navarino Agora, a bustling new open marketplace, with retail outlets, dining venues & pop-up events.
- Select among various sports, indoor and outdoor activities.
- Book Spa & Wellness treatments to relax and unwind.
- Order in-room dining, book a transfer or contact Costa Navarino Concierge.

Table 1.3 Marriott Bonvoy App

	Marriott Bonvoy App	W Costa Navarino
1	Mobile check in/Mobile key	
2	Manage marriot account	✓
3	Mobile chat and requests	✓
4	Hotel booking	✓
5	Mobile dining	✓
6	Shuttle services/shuttle tracker	✓

Another platform technology which Costa Navarino uses, is **Marriott Bonvoy Application** (10 + million downloads—Google store), as it is shown in Table 1.3. Marriott Bonvoy Application provides features as:

1. Mobile Check-In. The guest is able to use Mobile Check-In, when he opens the app, then selects "check in" and then sets his estimated arrival time and confirm the check-in by hitting the button on the bottom of the screen. Due to the Greek Government this feature is not allowed. Upon arrival the guest must give a passport and sign the registration form in the Reception area.
2. Manage Marriot account: the guest is able to view stay history, earn and redeem points as a Marriot Member.
3. Via Mobile chat Mobile Requests, the guest is able to:

   - Customize the room with extra blankets or pillows
   - Request services like housekeeping and luggage assistance (valet)
   - Choose a list of amenities (towel, shampoo, toothbrush, etc.).

4. Hotel booking: the guest browses, books a hotel and manages a new booking.
5. Mobile Dining: the guest can browse the hotel menu and order food and drinks.
6. Shuttle services: the guest can order a shuttle service and see the location of the hotel shuttle in real time with shuttle tracker.

Triparound's platform: is making success accessible for any travel business worldwide. Maximizing guest spending and offering personalized service. Triparound's concierge software works as an all-in-one central hub for hotels and resorts to add, manage and promote all in-house and third-party experiences (like trips, excursions, book outdoor activities, etc.).

The KNX installation: The Guest Room Management System (GRMS) shall be programmed to provide lighting and shading for the individual guest room with the possibility of future upgrades.

The system shall include, but shall not be limited to, the following for each guest room:

Room control devices and system components
Outside courtesy indicator panel
Energy saver card holder
Room call system in rooms for disabled guests (optional)
Centralized monitoring/control and interfaces.

LG application in A/C: There are different ways to optimize air conditioning in W Costa Navarino. Air conditioning and heating control must be carried out both in common areas such as lobby, living rooms, restaurants, reception, bathrooms, as well as in private areas like guest rooms.

At the air conditioning level, limiting the setpoint temperatures in public areas such as main rooms or large rooms can be very useful. It is also important to limit the setpoint temperatures to prevent hotel visitors from misusing the air conditioning system in their rooms, which otherwise leads to overconsumption.

By using monitoring and control systems for optimization of energy consumption, immediate value can be obtained. Management of consumption and real time energy control allows for identification and resolution of unnecessary energy usage. Current control systems allow monitoring of a large number of variables including for example: setpoint temperature, open window, occupancy and scenes.

Conclusions

Nunberg [11] considers that the Internet of Things is a novel paradigm shift in IT arena. The phrase "Internet of Things" which is also shortly well-known as IoT is coined from the two words i.e.the first word is "Internet" and the second word is "Things". The Internet is a global system of interconnected computer networks that use the standard Internet protocol suite (TCP/IP) to serve billions of users worldwide. It is a network of networks that consists of millions of private, public, academic, business, and government networks, of local to global scope, that are linked by a broad array of electronic, wireless and optical networking technologies.

Internet of Things (IoT) can tremendously revolutionize the hospitality industry. With IoT for hospitality industry, hotel owners, managers and others in tourism can increase guest satisfaction and overall productivity. Thanks to The Internet of Things (IoT) an essential opportunity is presented to the hospitality industry to reduce operational costs.

The boom in IoT Technology will boost the future of the hospitality industry; it will provide a competitive edge in the market and through the interconnection of devices, (sensors, actuators, identification tags, mobile, etc.) through the internet. IoT is no longer just a concept, it is very much a part of the industry and statistics are growing because IoT enables processes, data, and outcomes [13].

As the Fourth Industrial Revolution continues to run its course—especially with technologies such as IoT, which have an endless array of applications—now is the best time for a hotel owner or a manager to be an innovator. Very soon applications like personalized hotel rooms, fast check via a mobile device, etc., will become a real need in every hotel worldwide.

As a result, Costa Navarino saw this need, draw on, transforms every year and add to their capital resources to offer their assets, products and services (like IoT Technology) and achieves sustainable business growth and creating value for all their stakeholders. The IoT is an integral part of tourism sector and W Costa Navarino is using IoT to automate every possible service for improving guest experience.

Acknowledgments This work has been partly supported by the University of Piraeus Research Center.

References

1. Aggarwal, R., & Lal Das, M. (August 2012). RFID security in the context of "Internet of Things." *First International Conference on Security of Internet of Things, Kerala, 17–19*, 51–56. https://doi.org/10.1145/2490428.2490435
2. Atzori, L., Iera, A., & Morabito, G. (2010). The internet of things: A survey. *Computer Networks, 54*(15), 2787–2805.
3. Car, T., Pilepić Stifanich , L. J., & Šimunić, M. (2019). Internet of Things (IoT) IN tourism and hospitality: Opportunities and challenges, ToSEE—Tourism in Southern and Eastern Europe, Vol. 5, pp. 163–175.
4. Gubbi, J., Buyya, R., Marusic, S., & Palaniswami, M. (2013). Internet of Things (IoT): A vision, architectural elements, and future directions. *Future Generation Computer Systems, 29*(7), 1645–1660. https://doi.org/10.1016/j.future.2013.01.010
5. Hakan, Ö. (2023). Bibliometric analysis and scientific mapping of IoT. *Journal of Computer Information Systems*. https://doi.org/10.1080/08874417.2023.2167135
6. ITU. (2015). Internet of Things Global Standards Initiative. Retrieved 17 January, 2019, from https://www.itu.int/en/ITU-T/gsi/iot/Pages/default.aspx.
7. Kansakar, P., Munir, A. & Shabani, N. (2017). Technology in hospitality industry: Prospects and challenges. arXiv preprint arXiv.1709.00105
8. Kosmatos, E. A., Tselikas, N. D., & Boucouvalas, A. C. (2011). Integrating RFIDs and smart objects into a unified internet of things architecture. *Advances in Internet of Things: Scientific Research, 1*, 5–12. https://doi.org/10.4236/ait.2011.11002
9. Kranenburg, R. V. (2008). *The internet of things: A critique of ambient technology and the all-seeing network of RFID*. Insitute of Network Cultures.
10. Mingjun, W., Zhen, Y., Wei, Z., Xishang, D., Xiaofei, Y., Chenggang, S., & Jinghai, H. (2012, October). A research on experimental system for Internet of things major and application project. In 2012 3rd International Conference on System Science, Engineering Design and Manufacturing Informatization, IEEE, Vol. 1, pp. 261–263.
11. Nunberg, G. (2012). The Advent of the Internet: 12th April, Courses.
12. Sethi, P., & Sarangi, S. R. (2017). Internet of things: Architectures, protocols, and applications. *Journal of Electrical and Computer Engineering, 2017*, 1–25. https://doi.org/10.1155/2017/9324035
13. Verma, A., & Shukla, V. (2019). Analyzing the Influence of IoT in Tourism Industry," in International Conference on Sustainable Computing in Science, Technology & Management.
14. Ystgaard, K. F., Atzori, L., Palma, D., Heegaard, P. E., Bertheussen, L. E., Jensen, M. R., & De Moor, K. (2023). Review of the theory, principles, and design requirements of human-centric Internet of Things (IoT). *Journal of Ambient Intelligence and Humanized Computing*. https://doi.org/10.1007/s12652-023-04539-3

Company's Sustainability Report

15. https://www.costanavarino.com/wp-content/uploads/2022/05/sustainability-2019_20.pdf

Destination Website

16. https://www.costanavarino.com/

Hotel Websites

17. The Westin Resort Costa Navarino → https://www.marriott.com/en-us/hotels/klxwi-the-westin-resort-costa-navarino/overview/
18. The Romanos, a Luxury Collection Resort → https://www.marriott.com/en-us/hotels/klxlc-the-romanos-a-luxury-collection-resort-costa-navarino/overview/
19. W Costa Navarino → https://www.marriott.com/en-us/hotels/klxwh-w-costa-navarino/overview/

Mobile Applications

20. Marriott Bonvoy Mobile Application → https://mobile-app.marriott.com/en-us
21. Costa Navarino Mobile Application → https://apps.apple.com/us/app/costa-navarino-greece/id1458397325?l=el&ls=1

TripAround Application

22. https://www.triparound.com/
23. https://www.fortunegreece.com/article/triparound-metaschimatizontas-psifiaka-ton-touristiko-klado/

KNX installation Application

24. https://www.knx-iotech.org/

LG application

25. https://www.lg.com/us

Chapter 2
Anomaly Detection in Enterprise Payment Systems: An Ensemble Machine Learning Approach

Basem Torky, Ioannis Karamitsos, and Tariq Najar

Abstract With the exponential growth of digital transactions, ensuring the integrity and authenticity of payment systems has become imperative. This paper investigates the effectiveness of machine-learning techniques in detecting anomalous patterns in large-scale payment datasets. Ensemble methods are widely used in the field of anomaly detection in enterprise systems to improve the accuracy and robustness of these systems. Anomaly detection aims to detect abnormal patterns that deviate from the rest of the data and are referred to as anomalies or outliers. With millions of services or sub-systems to monitor such as e-commerce platforms and governmental portals, our study focuses on using forecasting methods to develop a model that can be used in these enterprise systems to avoid huge financial impacts, bad reputation, and customer dissatisfaction. Our methodology combines multiple time series methods such as Seasonal Autoregressive integrated moving average (SARIMAX) and Facebook-Prophet and SVM to create a more robust and accurate ensemble model for anomaly detection. Anomaly detection can help highlight where exactly an incident is occurring. This proactive detection greatly improves the root cause analysis of the problem and has a positive impact on business continuity. The three different types of anomalies can occur in the datasets of pointers, conditional, and collective or accumulative anomalies. The main approaches to solve anomaly detection problems are either rule-based or machine learning approaches. In this paper, we focus on the machine learning approach as it is more reliable and effective as it complements the rule-based human capabilities with the machine learning and artificial intelligence capabilities. Three widely used forecast models SARIMAX, Facebook-Prophet and SVM are compared and analysed for the payment transactions. For the evaluated performance SVM model is best performed with R squared accuracy values of 80.7%. Overall, the results demonstrated that the SVM method can provide better performance than SARIMAX and Prophet methods for payment transactions data.

B. Torky · I. Karamitsos (✉) · T. Najar
Graduate and Research Department, Rochester Institute of Technology, Dubai, UAE
e-mail: ixkcad1@rit.edu

© The Author(s), under exclusive license to Springer Nature Switzerland AG 2024
A. Emrouznejad et al. (eds.), *Business Analytics and Decision Making in Practice*, Lecture Notes in Operations Research, https://doi.org/10.1007/978-3-031-61589-4_2

Keywords Anomaly detection · ARIMA · Forecasting · Facebook-prophet · Proactive anomaly detection · SARIMA · SVM

Introduction

Anomaly detection can assist in pinpointing the exact spot of an incident amongst millions of services or sub-systems that need to be monitored. This proactive detection greatly improves the root cause investigation of the problem and has a beneficial impact on business continuity.

As we begin to solve the problem, it is critical that we identify the kind of anomaly that we are working with. The datasets may contain any one of the three different categories of anomalies listed below:

1. Pointer anomalies: these clearly visible dataset outliers require further investigation to determine whether they represent anomalies or false positives.
2. Conditional Anomalies: These are the data points that, only in a certain situation, deviate from the dataset's overall normal distribution. Seasonality in a time series data set is a common example.
3. Collective or Accumulative Anomalies: These are the data points that, when repeated consecutively, may raise an indication of an anomaly even though they appear normal on their own.

The problem of anomaly detection can be approached in a variety of ways. However, we will focus on two primary methods here:

- **Rule-based anomaly detection**: this technique requires a subject matter expert to develop the rules based on his experience and involves the programmer defining a set of rules for the anomaly detector to apply over a dataset.
- **Machine learning anomaly detection**, which defines and distributes the anomalies throughout the dataset primarily through machine learning techniques and algorithms. Due to the utilization of machine learning to enhance human capabilities, this technique is more dependable and efficient.

These days, managing and anticipating disruptions to any corporate system is extremely difficult. These disruptions, which various factors can bring on, can significantly negatively influence finances, reputation, and customer satisfaction. Determining the anomalous deviations in extensive runtime dataflow that includes useful system logs for a functional system is a complex procedure requiring profound technical and commercial knowledge. The impacted businesses are looking for irregularities such as equipment failure during operations, fraud in the finance sector, or cybersecurity.

This study aims to explore the effectiveness of machine learning-based anomaly detection in predicting anomalies and portal outages within an online services portal. In our research study is used a dataset of detailed financial transaction data for an online services portal. Using machine learning techniques, we can forecast abnormalities and portal outages based on past data.

The main contribution of this paper is:

- Developing three time series models to predict the anomalies in order to identify the next service interruption or outage that should be avoided;
- Selecting the best model in terms of the performance.
- Creating real-time dashboards (BI) in order to address the anomalies and lessen their impact by creating an alerting mechanism in the event that a predetermined error threshold is exceeded.

This remainder of the paper has been organized in the following way. Section 2.2 begins by laying out the related work on anomaly detection. Section 2.3 is concerned with the time series modelling methodology used for this study. Section 2.4 presents the findings of the study, focusing on the comparison between different forecasting models for the anomaly detection of the payment transactions. In Sect. 2.5 analyses the results of selected time-series model with real dataset implementing forecasting models. Finally, in Sect. 2.6 the conclusion gives a brief summary of the findings and areas for further work are identified.

Related Work

The literature review delves into various approaches for anomaly detection through machine learning in diverse contexts. In their study, Wang et al. [1] specifically investigated the identification of abnormal accounts in online banking. They achieved this by analyzing the geonet features of each account and identifying distinctive patterns that set outliers apart.

Hisham et al. [2] highlighted the efficacy of ensemble approaches amalgamating weak models to create more robust ones, supported by earlier studies showcasing the accuracy gains from combining ensemble techniques.

In their study, Nkongolo et al. [3] presented a new dataset for anomaly detection, incorporating cyclostationary zero-day threat behaviors. The researchers deployed Random Forest and Decision Tree algorithms together with an Ensemble Learning strategy approach to analyze the data.

Petrariu et al. [4] explored unsupervised anomaly detection algorithms on an IT department's traceability dataset, achieving up to 91.73% accuracy in identifying abnormal transitions.

Hajialian and Toma [5] developed a model using the ML Pipeline, showcasing high precision and accuracy, especially with 32 trees in the Random Forest classifier.

Zhao and Keikhosrokiani [6] integrated RFM analysis and multiple analytics algorithms for sales prediction and product recommendation, achieving a 77.82% accuracy rate.

Abdelrahman and Keikhosrokiani [7] scrutinized assembly data, utilizing KNN and ABOD algorithms to identify aberrant data points and causes without over-rejection in assembly machines.

Pavlyshenko [8] favored regression over time series analysis for sales forecasting, employing stacking to enhance accuracy.

Kotios et al. [9] established the foundation for novel financial services by proposing a hybrid transaction categorization model and an accurate cash flow prediction model.

Ibrahim et al. [10] utilized Facebook-Prophet, AutoEncoder Long Short-Term Memory, and Isolation Forest algorithms to detect anomalies in photovoltaic components. Among these algorithms, AutoEncoder Long Short-Term Memory had the highest level of accuracy.

Soro et al. [11] conducted a study on detecting anomalies in banking transactions. They are utilizing adaptive threshold algorithms and unsupervised machine learning on a dataset of over two thousand customers to provide a reliable solution for real-time or semi-real-time anomaly detection.

Edholm [12] successfully obtained an accuracy rate of 87% in detecting anomalies by utilizing Artificial Neural Networks for invoice processing in a company experiencing unidentified revenue loss.

Knutsen [13] recognized the effectiveness of Bayesian networks in managing service and availability risks. The Nova system outperformed Sensex in terms of detection and scoping time, while still preserving accuracy. This was achieved by adding the causal structure of the service. Collectively, these research provide significant contributions to the understanding of various anomaly detection methods and their practical implementations in various fields.

Research Methodology

In this study, we applied the methodology includes multiple and various steps and activities. In Fig. 2.1 is depicted the overall time series process flow.

Data Collection and Preprocessing

The collected "Payment Transactions" dataset contains 10 attributes, the time range from 2014 to 2023 but for the study purpose and due to the memory limitation, we will take the year of 2022 as a base for all the modeling and visualization (Table 2.1).

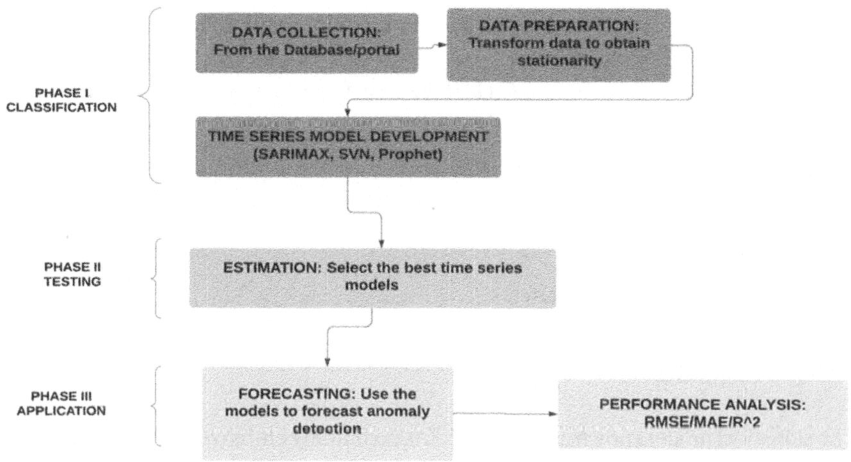

Fig. 2.1 Time series process flow

Table 2.1 Data dictionary

Variable name	Data type	Description
CREATED	Timestamp	The transaction created date
TXN_ID	String	It is the transaction unique number
ACTIVITY_TYPE	Categorical	It is the payment method type (GooglePay, ApplePay, etc.)
APP_NAME	Categorical	The application name used to complete the transaction
AMOUNT	Integer	The transaction paid amount
BENIFICIARY_TYPE	Categorical	The customer type (individual or corporate)
SERV_CODE	Categorical	The unique service code
SP_CODE	Categorical	The service provider unique code
STATUS	Categorical	The transaction status (success or failed)
SERVICENAME	String	The provided service name

Time Series Model Development

We employ a range of time series models to develop anomaly detections models. These models are trained on historical data to identify patterns indicative of anomalies.

Facebook-Prophet Model

The Facebook library called Facebook Prophet is used to forecast time series data with multiple or non-linear growth. The library fits a time series dataset to an additive model using Bayesian inference, as stated in the following formula:

$$X(t) = T(t) + S(t) + H(t) + \varepsilon_t$$

where:

- T(t): is the dataset trend
- S(t): is the time series seasonality
- H(t): holidays values
- ε_t: unpredicted/error/bias effects

SARIMAX Model

The statistical model known as SARIMAX is commonly employed in the analysis and prediction of time series data. It is an altered variant of ARIMA designed to enable time series that exhibit seasonality over the course of the dataset. Its acronym, as shown below, is ARIMA (p, d, q) × $(P, D, Q)_s$:

- (p, d, q): indicating non-seasonal properties
- (P, D, Q): indicating seasonal timeseries part
- S: indicating seasonality periodicity

Support Vector Machines Model

The Support Vector Machines (SVMs) is a highly potent and versatile machine learning model capable of performing linear or nonlinear classification, regression, and more over can identify outlier identification. Support Vector Machines (SVMs) are particularly suitable for classifying datasets that are both complex and of small or medium size. To transform our transactions time series dataset into a supervised machine learning issue, we will employ the sliding window technique. The SVM results obtained from the dataset with engineered features could potentially serve as a performance benchmark for the given challenge.

Experimental Analysis

During this part of the time series process, which is essential as it entails collecting all the information and displaying it in a visually captivating chart or graph, we will focus on delivering intriguing descriptive analytics insights from our data collection. Additionally, it is vital for presenting the business stakeholders with a clear and comprehensible visual depiction of our discoveries. We graph the temporal distribution of payment transactions using a dataset of transactions. The timeseries graph presented below illustrates a rising trend in the year 2022, indicating a progressive increase in the utilization of this service over time. This input has the potential to

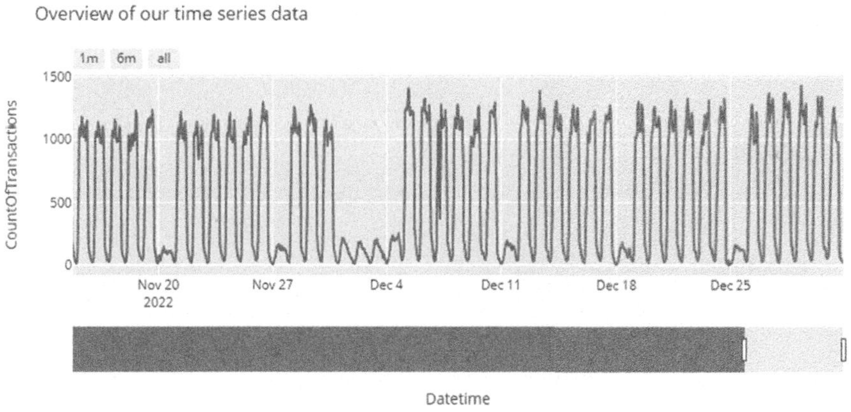

Fig. 2.2 A month timeseries overview

be valuable for capacity planning in the upcoming year. We saw a slight decline in the number of transactions throughout the months of May, July, and December. This decrease will be further analyzed when we develop our model (Fig. 2.2).

To determine the frequency of transaction count over time, it is beneficial to create a histogram graph. The histogram graph will provide valuable insights into the timeseries data. The histogram below illustrates that the values below 200 correspond to the most recent data point (Fig. 2.3).

SARIMAX Model

Our timeseries clearly exhibits seasonality, as we demonstrated throughout the data discovery process. For this reason, ARIMA is not the best model choice to reflect a fitted model; nevertheless, SARIMAX can be useful in these situations.

With our dataset, p, d, and q values of 1, 1 and 1, respectively, we will apply SARIMAX (p, d, q) × (P, D, Q). The values of P, D, and Q are 1, 1, and 0, correspondingly. Since our timeseries has a weekly seasonality, the final value is the s, or the seasonality value, which in our case is 7 (days).

The train dataset, which spans the years 2020–01–01 to 2022–10–01, and the test dataset, which spans the years 2022–10–01 to 2023–01–01, are the two subsets of our cleaned dataset.

The numerical results of our model are displayed in Fig. 2.4.

Additionally, Fig. 2.5 illustrates the contrast between the predicted vs observed sample.

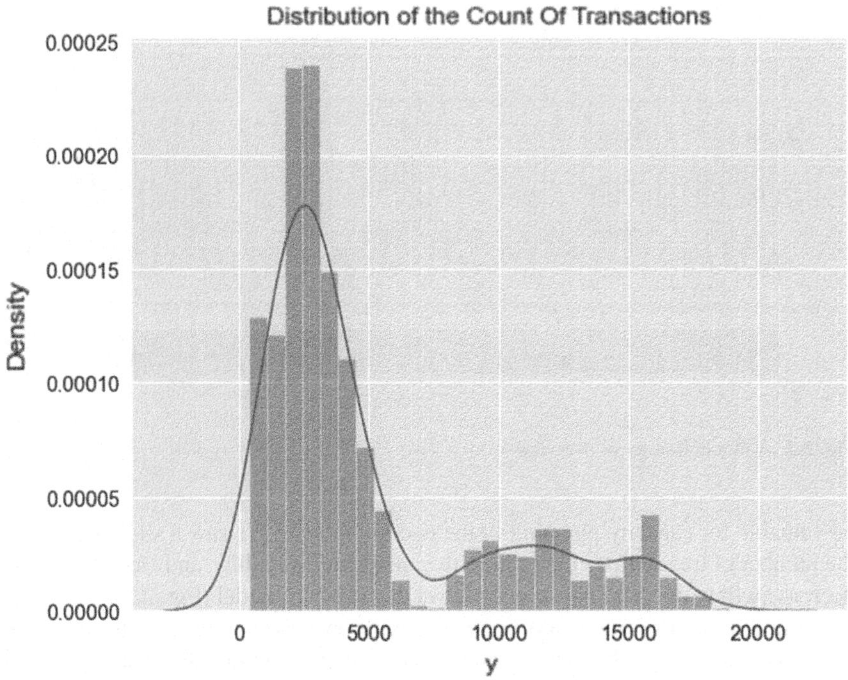

Fig. 2.3 Distribution of the count of transactions

```
                                    SARIMAX Results
==========================================================================================
Dep. Variable:                           value   No. Observations:                   1004
Model:             SARIMAX(1, 1, 1)x(1, 1, [], 7)   Log Likelihood                   -241.807
Date:                         Sun, 16 Apr 2023   AIC                              491.614
Time:                                 06:22:25   BIC                              511.229
Sample:                             01-01-2020   HQIC                             499.071
                                  - 09-30-2022
Covariance Type:                           opg
==============================================================================
                 coef    std err          z      P>|z|      [0.025      0.975]
------------------------------------------------------------------------------
ar.L1          0.6291      0.015     42.545      0.000       0.600       0.658
ma.L1         -1.0000      0.836     -1.196      0.232      -2.639       0.639
ar.S.L7       -0.5160      0.012    -43.741      0.000      -0.539      -0.493
sigma2         0.0943      0.079      1.192      0.233      -0.061       0.249
===================================================================================
Ljung-Box (L1) (Q):                   3.00   Jarque-Bera (JB):              4514.88
Prob(Q):                              0.08   Prob(JB):                         0.00
Heteroskedasticity (H):               2.37   Skew:                            -0.78
Prob(H) (two-sided):                  0.00   Kurtosis:                        13.31
===================================================================================
```

Fig. 2.4 SARIMAX model results

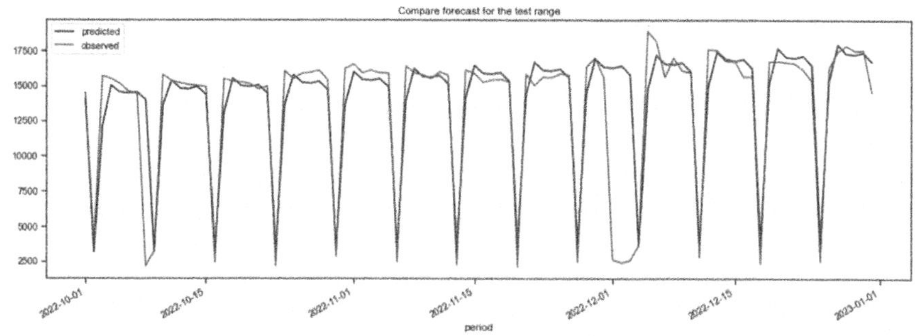

Fig. 2.5 Predicted versus observed sample

Facebook-Prophet Model

We began by configuring our model values as shown below Fitting the model is the next step (Fig. 2.6).

The model will be utilized to identify any anomalies, as shown below, where the anomalies are shown by red dots (Fig. 2.7).

	MAE_val	RMSE_val	Loss_val	RegLoss_val	epoch	MAE	RMSE	Loss	RegLoss
0	31611.388672	34557.273438	1.595960	0.0	0	23681.312500	28197.478516	0.944851	0.0
1	6726.849609	8573.799805	0.155113	0.0	1	12625.720703	15345.387695	0.376549	0.0
2	2766.475830	3667.225342	0.029396	0.0	2	4425.110352	5401.743164	0.051740	0.0
3	2250.795898	3781.328125	0.031270	0.0	3	2229.144531	2857.795898	0.014016	0.0
4	2225.123047	3696.863770	0.029881	0.0	4	1934.563354	2605.147217	0.012449	0.0
5	2324.956299	3779.929932	0.031238	0.0	5	1880.579712	2554.881592	0.012038	0.0
6	2209.526611	3736.277832	0.030520	0.0	6	1868.767944	2537.136963	0.011944	0.0
7	2507.412109	3785.434570	0.031321	0.0	7	1759.342529	2480.170654	0.011751	0.0
8	2437.001709	3717.328369	0.030205	0.0	8	1786.683594	2415.203857	0.011285	0.0
9	2415.369141	3719.059570	0.030233	0.0	9	1791.072266	2439.281494	0.011197	0.0

Fig. 2.6 Model fitting results

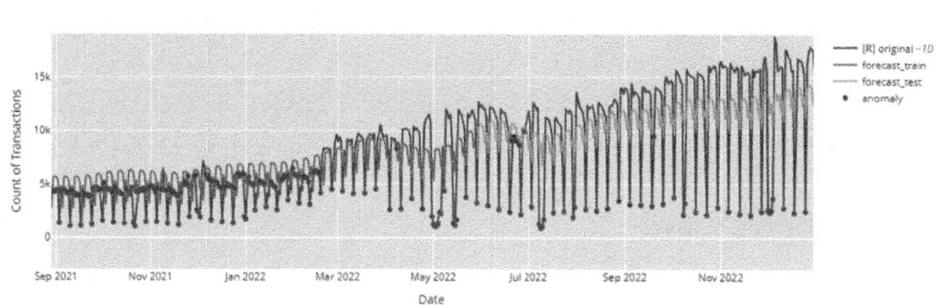

Fig. 2.7 Anomalies points representation

Support Vector Machines Model

We choose GridSearch CV with 5 k-folds since it is essential to determine the prober parameters for any machine learning model to maximize the accuracy of the anomaly detection capabilities. Conversely, the SVM general algorithm was utilized to adjust its parameters, such as the cost parameter and the decision border gamma curviness rate, gamma.

The GridSearchCV cross validation results are displayed in Fig. 2.8. With a score of 0.826, the best adjusted parameters were C at 15.85 and gamma at 0.032.

Afterwards, we plotted the observations from training, validation, and testing against the actual observed values. This shows that the testing and validation datasets have strong prediction, which will result in high performance accuracy (Fig. 2.9).

Providing insight into the validation set and the prediction outcomes in relation to the actual observed values, which show a strong degree of similarity (Fig. 2.10).

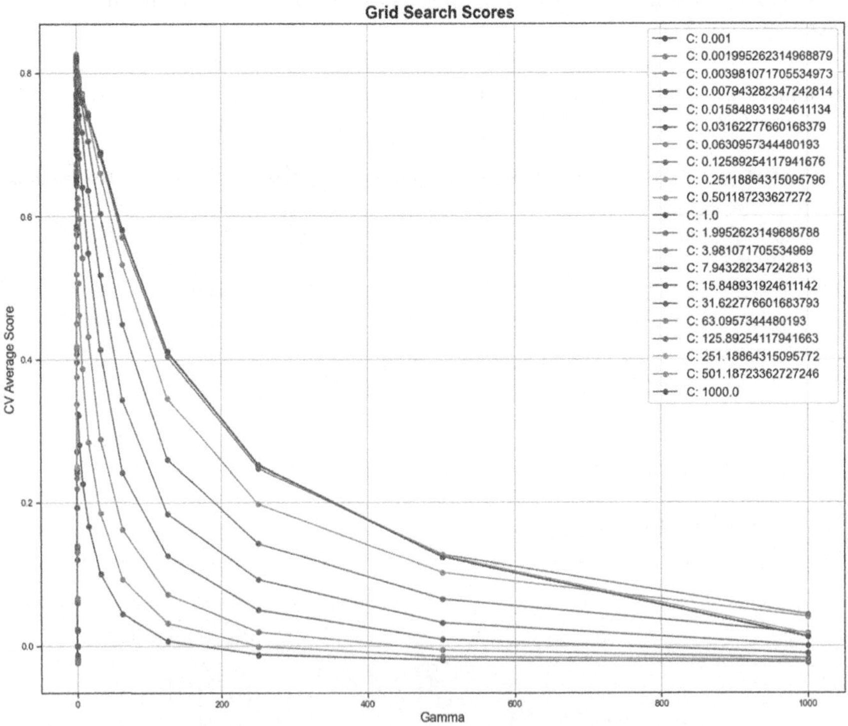

Fig. 2.8 GridSearchCV scores

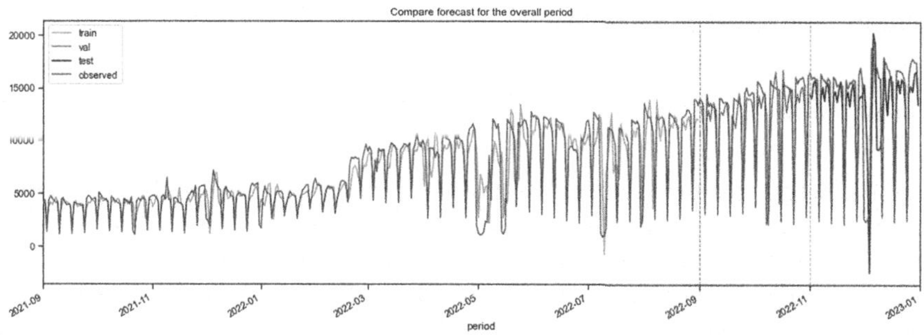

Fig. 2.9 Training, validation, and testing observations vs actual observed values

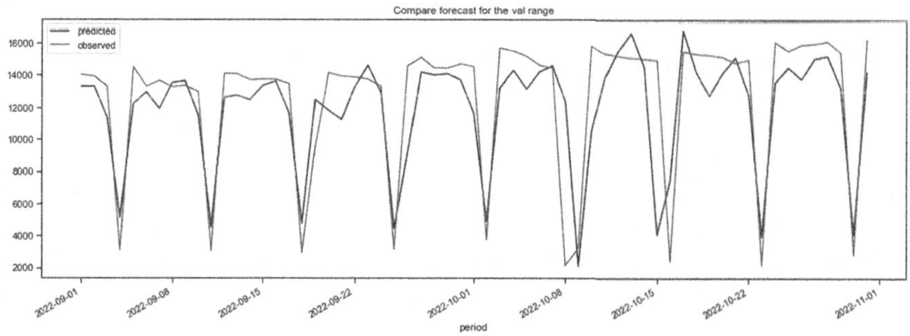

Fig. 2.10 Predicted results V.S actual observed points

Performance Evaluation

Root Mean Squared Error (RMSE), Mean Absolute Error (MAE), and Mean Absolute Percentage Error (MAPE), which are often used in the literature, are the three metrics employed in this study to assess performance. These are the main measures we use to evaluate the performance of the models given in the literature with the best-performing model we found throughout our investigation (Table 2.2).

In this work, we examined how machine learning algorithms can be used to identify anomalies in historical data while accounting for trend and seasonality. For our anomaly detection challenge, the Support Vector Machine model yielded the best results, with an accuracy rate of 80.2%.

Table 2.2 Forecast summary statistics of the time series models

Model	MAE	RMSE	R^2
SARIMA	1,365	2,980	67.6%
Facebook-prophet	3,128	3,755	NA
SVM	1,396	1,666	**80.2%**

Conclusion and Future Work

Conclusion

In this research, the time-series dataset of payment transactions is utilized to anticipate future anomalies. It is well established that traditional manual methods of projecting future transaction counts are not useful for forecasting future anomalies. We used the daily transaction count from January 1, 2020, to January 1, 2022, to train, validate, and test our machine-learning model. To achieve our key goal, which aims to identify anomalies in corporate systems proactively. In order to get some insights from the data, we began with Exploratory Data Analysis (EDA) using comprehensive descriptive analytics. We covered a wide range of techniques for feature engineering, data transformation, feature cleaning, and feature selection in the data preparation part. In the modeling part, we used the Python programming language to analyze three algorithms and then applied those models to forecast the anomaly points for the next year. In this research, we looked at using machine learning algorithms to look for anomalies in historical data while taking seasonality and trends into account. For our anomaly detection challenge, the Support Vector Machine model delivered the best results, with an accuracy rate of 80.2%. We highly recommend exposing these models on the internet via any cloud solution using big data real-time Spark tasks. Acquiring more historical data would improve the models' accuracy, but expanding the dataset size will increase the processing power required.

Future Work

Evolving technologies like big data capabilities are crucial to constructing a real-time anomaly detection system that can predict the next system outage in a scalable manner since the use of online services is growing with time, and the amount of data being collected will expand significantly.

References

1. Wang, Z., Tian, J., Fang, H., Chen, L., & Qin, J. (2022). LightLog: A lightweight temporal convolutional network for log anomaly detection on the edge. *Computer Networks, 203*, 108616.
2. Hisham, S., Makhtar, M., & Aziz, A. A. (2022). Combining multiple classifiers using ensemble method for anomaly detection in blockchain networks: A comprehensive review. *International Journal of Advanced Computer Science and Applications, 13*(8).
3. Nkongolo, M., van Deventer, J. P., & Kasongo, S. M. (2021). UGRANSOME1819: A novel dataset for anomaly detection and zero-day threats. *Information, 12*(10), 405.
4. Petrariu, I., Moscaliuc, A., Turcu, C. E., & Gherman, O. (2022). A comparative study of unsupervised anomaly detection algorithms used in a small and medium-sized enterprise. *International Journal of Advanced Computer Science and Applications, 13*(9).

5. Hajialian, H., & Toma, C. (2018). Network anomaly detection by means of machine learning: Random forest approach with apache spark. *Informatica Economica., 22*(4/2018), 89–98.
6. Zhao, X., & Keikhosrokiani, P. (2022). Sales prediction and product recommendation model through user behavior analytics. *Computers, Materials & Continua, 70*(2), 3855–3874.
7. Abdelrahman, O., & Keikhosrokiani, P. (2020). Assembly line anomaly detection and root cause analysis using machine learning. *IEEE Access, 8*, 189661–189672.
8. Pavlyshenko, B. (2019). Machine-learning models for sales time series forecasting. *Data, 4*(1), 15.
9. Kotios, D., Makridis, G., Fatouros, G., & Kyriazis, D. (2022). Deep learning enhancing banking services: A hybrid transaction classification and cash flow prediction approach. *Journal of Big Data, 9*(1).
10. Ibrahim, M., Alsheikh, A., Awaysheh, F., & Alshehri, M. (2022). Machine learning schemes for anomaly detection in solar power plants. *Energies, 15*(3), 1082.
11. Soro, F., Mellia, M., & Russo, N. (2020). Regular pattern and anomaly detection on corporate transaction time series. In *EDBT/ICDT workshops*.
12. Edholm, G. (2020). Anomaly detection and revenue loss estimation in accounting data.
13. Knutsen, E. E. (2020). *Detecting failures and locating faults in global scale online services using Bayesian networks* [PhD dissertation]. The George Washington University

Chapter 3
Impact of ChatGPT on Educational Strategies for Future-Proof Business Data Analyst: Machine Learning Code Generation in Teaching and Learning

Colin Fu and Joseph Damonte

Abstract This paper examines the role of ChatGPT in educational pedagogy, particularly its use in creating machine learning code for business applications. We assess ChatGPT Plus (version 4) against Google Bard's July-Aug version, focusing on its ability to develop code solutions for various business problems, i.e., Customer Churn Prediction for a Telecom Company; Purchase Prediction Based on Shopper Interactions; Yield Prediction of Blueberries and Sentiment Analysis of App Reviews. The study primarily explores ChatGPT's effectiveness in enhancing the teaching and understanding of complex machine learning concepts in higher education. The research evaluates ChatGPT's ability to create accurate and relevant code for four different business scenarios, highlighting its potential to transform educational methods and learning experiences. ChatGPT effectively generated functional code in three scenarios, showcasing its value in practical, business-focused education. However, its limitations in sentiment analysis tasks emphasize the ongoing need for human expertise in areas where AI is less effective. The study shows that ChatGPT Plus (version 4) provides more detailed feedback and practical code improvement suggestions than Google Bard. However, the need for educators to refine AI-generated outputs, especially due to outdated references and suboptimal code structures, is a key aspect to consider. ChatGPT can be concluded to show great promise in improving machine learning education, but it still requires human expertise to ensure content quality and relevance. Additionally, our findings indicate that ChatGPT could play a crucial role in developing advanced analytical solutions for complex business problems, highlighting its potential impact on AI-driven business decision-making.

Keywords Generative AI · Chat GPT · Machine learning

C. Fu · J. Damonte (✉)
University of Surrey, Guildford GU2 7XH, Surrey, UK
e-mail: j.damonte@icloud.com

© The Author(s), under exclusive license to Springer Nature Switzerland AG 2024
A. Emrouznejad et al. (eds.), *Business Analytics and Decision Making in Practice*,
Lecture Notes in Operations Research, https://doi.org/10.1007/978-3-031-61589-4_3

Introduction

Context

The recent emergence of generative artificial intelligence (AI) tools like ChatGPT has transformed possibilities for automating human-like content creation [1]. ChatGPT and similar generative models can produce natural language text, code, and other outputs after training on vast datasets [2]. With capabilities spanning content creation, reasoning, and problem-solving, the implications across education, business, and society are far-reaching [3, 4], which has it banned in 15 countries [5].

In particular, the ability of AI systems like ChatGPT to generate computer code for analytics and machine learning applications holds much promise. Machine learning has revolutionized prediction and pattern recognition, but developing models involves complex coding skills [6]. By automating parts of this process, generative AI could expand access and efficiency. However, given limitations like bias and inaccuracies, the actual viability for code generation remains uncertain. There has been little rigorous evaluation regarding if and how tools like ChatGPT can be effectively utilized for writing usable, novel machine learning code [7].

This study aims to provide robust assessment on ChatGPT's current capabilities for generating machine learning code tailored to solve authentic business problems. Using the August 2022 version of ChatGPT, it tests and compares code produced for four distinct problems against outputs from Google's Bard, the closest commercial competitor. Through quantitative analysis of code performance, opportunities to enhance teaching and learning of coding applications are also examined—both within higher education curricula and workplace settings involved in data science or business analytics.

Objectives

The key research objectives addressed are:

RO1: Evaluate effectiveness of ChatGPT-generated machine learning code across business problems of varying data complexity versus other AI.

RO2: Analyse inherent limitations in ChatGPT's autonomous analytical workflow orchestration through iterative modelling attempts.

RO3: Assess ChatGPT's potential for enhancing productivity and analytical skills of novice/intermediate machine learning practitioners compared to traditional coding.

By evaluating ChatGPT's effectiveness as well as fundamental gaps, findings will inform appropriate integration of generative AI to maximize productivity and innovation gains. This can guide adoption strategies by educational institutions and companies exploring leveraging technologies like ChatGPT for machine learning applications, while benefiting learners, employees and end users.

Literature Review

Year 2022

The recent introduction of ChatGPT by OpenAI in late 2022 has accelerated public and research attention on the potentials of generative artificial intelligence (AI) for automating a variety of complex text and coding tasks [2]. As neural network powered language models trained on vast datasets, tools like ChatGPT exhibit capabilities for conversational and customized content generation spanning text summarization to programming [7]. However, balancing productivity promises with ethical risks around accuracy, bias and critical skill development poses important challenges regarding appropriate integration [8–11].

ChatGPT Capabilities and Limitations

ChatGPT, with its advancements over its predecessor GPT-3, demonstrates improved capabilities in generating human-like responses, particularly in explanatory content. However, its effectiveness is not without limitations. Studies have pointed out significant error rates, with inaccuracies varying between 50 and 85% in complex reasoning or novel thinking scenarios [12, 13]. While [14] highlight ChatGPT's strengths in coding and debugging, this is contrasted by findings from [15], who reported a 52% error rate in ChatGPT-generated code.

This dichotomy in ChatGPT's performance is further elucidated when considering its user interaction capabilities. Research by [16] emphasizes ChatGPT's unique interactive features, setting it apart from other AI tools. Unlike static content generators, ChatGPT allows users to clarify objectives, evaluate preliminary outputs, and request specific refinements. This conversational engagement potentially enhances the reliability of its outputs. However, despite these advanced interactive abilities, the need for external validation remains critical to mitigate the risk of inaccuracies. This underscores the importance of a vigilant approach in utilizing ChatGPT, leveraging its interactivity while being cautious of its inherent limitations.

Implications for Education

In education contexts, AI tools like ChatGPT show promise for aiding learning through generating automatic explanations and tailored summaries, demonstrating their potential to enhance independent study [17]. Such capabilities can expand self-directed learning opportunities. However, an overdependence on these tools may hinder the development of crucial analytical abilities and the capacity for critical information evaluation [18]. Most researchers recommend careful application and additional training with these AI tools to ensure a balance between enhancing learning efficiency and maintaining educational depth.

Research exploring ChatGPT's coding capabilities in educational contexts is relatively new. Among first investigations, Chen [19] revealed that while ChatGPT's accuracy is promising and often exceeds previous standards, it remains dependent on specific contexts. The tool is effective for routine coding tasks but requires human input for more complex logical challenges. This finding highlights the importance of establishing frameworks that support learning and guide the integration of generative coding tools like ChatGPT in educational settings.

Business Analytics Productivity

The application of AI and machine learning has become crucial for enhancing decision-making. Despite this, challenges in coding, particularly in data processing and model development, persist [7]. Although AutoML technologies have advanced, they still require human oversight, particularly in interpreting objectives. Tools like ChatGPT, with their natural language capabilities, offer potential to make analytics more accessible and efficient, promoting wider usage in various business contexts [20].

Researchers, however, highlight the risk of a disconnect between ChatGPT's coding suggestions and specific organizational constraints, emphasizing the need for human guidance. Deng and Lin [21] note that while ChatGPT shows promise in assisting with coding tasks, its lack of embedded business insight hinders its ability to fully automate machine learning processes without human oversight. This underscores the necessity for detailed evaluations in specific application contexts to ensure effective and relevant use of such AI tools in business environments, such as improving efficiency and satisfaction of customers [22].

Combining the computational speed of AI like ChatGPT with human contextual understanding and judgment is a promising way forward. This approach, however, necessitates the development of proper expectations, training programs, and governance models. Hughes [23] emphasizes that these hybrid partnerships are essential for leveraging the full potential of AI tools while maintaining the crucial element of human oversight in various applications.

Trust and Ethics

Hughes [23] outlines five key areas–transparency, privacy, security, fairness, and user agency–as critical foundations for building trust in ecosystems using technologies like ChatGPT. These dimensions are crucial for ethical AI application and require the generation of solid evidence and the development of evolving standards to meet ethical imperatives, especially as technological innovation progresses rapidly. This comprehensive approach is essential to ensure responsible use and public trust in advanced AI systems.

Pataranutaporn et al. [24] focus on behavioural studies to understand how user perceptions, expectations, and decision-making are influenced by advancements in AI, aiming to enhance user-centric experiences. They propose moving beyond just accuracy metrics to more holistically assess the effectiveness of AI solutions throughout their lifecycle. Shen et al. [25] also emphasize the importance of evaluating AI solutions in a broader context, considering their overall efficacy and suitability for integration into various applications, thereby providing a more comprehensive understanding of their impact and effectiveness.

Generative AI, such as ChatGPT, marks significant advancement in automation capabilities. However, achieving an optimal balance between its vast opportunities and inherent responsibilities requires collaborative efforts across education, business, and societal domains. Despite challenges, the potential benefits and advancements offered by generative AI strongly advocate for continued exploration and development in this field. This exploration is essential to harness the full potential of AI while addressing ethical and practical concerns [26].

Methodology

Details on ChatGPT Version Used and Comparison

This study utilized ChatGPT Plus Version 4, released for public access in August 2022, for generating machine learning code tailored to the selected business problems. Version 4 represented the latest broadly available release at the project's commencement, superseding the initial ChatGPT 3 debut in June 2022. Progressive enhancements in conversational responsiveness, reasoning, and code capabilities underpinned its selection as an optimal balance of leading-edge sophistication with sufficient maturity for rigorous evaluation [27]. Commercial alternative Google Bard, first previewed in February 2023, served as the principal system for comparison given its competitive position as a code-generative service from a major technology provider.

Explanation of Business Problems Tested

Four distinct real-world business issues were defined to span a range of machine learning complexity and data types:

- Customer Churn Prediction for a Telecoms Firm [28]–Using customer activity records and attributes, develop a classification model forecasting churn likelihood. Inputs encompass call details, account statuses, and demographics like location and age. The customer churn records originated from an Iranian firm's subscriber base of 5000, with dozens of account usage attributes submitted to a 2020 machine learning competition. Labels indicated known churn outcomes. Realism came from operational origin, though localization merits caution generalizing cultural insights.
- Online Shopper Purchase Prediction [29]–Predict whether website visitors will complete an online purchase based on browse behaviours and interactions. Mixes categorical and continuously-valued inputs describing clickstreams. The online shopper dataset captured 12 million sessions from an e-commerce site transaction log in 2018. Labels reflect whether sessions finished in completed sales or abandoned carts. High volume aids robust modelling.
- Blueberry Crop Yield Prediction [30]–Regression task estimating harvest yields for wild blueberry growers utilizing weather, soil chemistry, and horticultural management details. Strictly quantitative inputs. The blueberry yield training series had 7000+ samples tracking 34 soil, climate and cultivation parameters from data partners of the USDA's Blueberry Council. Labels were yield rates.
- Mobile App Review Sentiment Analysis [31]–Detect sentiment polarity in textual app store user reviews. Combination of natural language review content and accompanying 1-5 star ratings. The mobile app review bank contained 1500 user comments in English drawing labels from accompanying 1-5 star ratings. It had no additional metadata. But freeform feedback is representative.

Problems were strategically chosen to compel creative, tailored solutions using datasets with varying statistical properties. They necessitate techniques spanning classification, regression, and text mining–all areas where claims of ChatGPT code competence warranted scrutiny.

Performance Metrics for Evaluation

Quantitative quality indicators for assessing and comparing code performance focused on standard measures of predictive accuracy:

- Accuracy–Percentage of correct model outcome predictions.
- Sensitivity–True positive rate identifying target class instances.
- Specificity–True negative rate identifying negative class instances.
- RMSE–Root mean squared error magnitude for numerical predictions.

Additionally, qualitative dimensions of interpretability, transparency, and responsiveness to prompt tuning were logged throughout interactive modelling processes with ChatGPT and Bard alternatives.

Data Sources and Analysis Process

All modelling leveraged established datasets from prior academic studies or open access repositories like Kaggle ensuring credible inputs representative of business contexts. After importing into R environments, a systematic workflow was undertaken:

1. Initial discussion in conversational prompts about optimal algorithm selections and pre-processing needs
2. Generation of machine learning code from ChatGPT and Google Bard alternatives
3. Execution of code with baseline scoring evaluations on accuracy, sensitivity etc.
4. Iterative refinement requests and code adjustments from generators
5. Comparison of metrics from AI-produced code against expert human coders given the same data

The above steps ensure human-in-the-loop approach integrating the generative code within an authentic effort to solve each business problem, while simultaneously evaluating objective performance metrics and subjectively assessing the strengths and weaknesses of the emerging processes.

Results

Summary of ChatGPT's Performance

Across the four business problems, ChatGPT demonstrated competent machine learning code generation capabilities for some tasks but also limitations in more complex contexts.

Customer Churn Prediction

The customer churn models displayed significant sensitivity improvement from the ChatGPT baseline of 70% accuracy, 49% sensitivity, and 95% specificity. After balancing the skewed dataset with SMOTE oversampling, sensitivity rose to 81% with 93% specificity maintained, more aligned with industry benchmarks above 80% [32]. Further regularization and under-sampling lifted sensitivity greater to 92% though with trade-offs on specificity decline. Overall gains over the baseline highlight adaptability tuning binary classification performance.

Online Shopper Purchase Prediction

For predicting shopper conversions, the baseline random forest model scored 70% accuracy. Conducting optimal parameter tuning via grid search with cross validation lifted this to an improved 87%, elevating sensitivity to 78% at 89% specificity. Industry analysis notes 80–87% accuracy rates for similar models [33], so ChatGPT's tuning adeptness proved impactful towards that parity. However, degradation remained for low sensitivity segments.

Mobile App Review Sentiment Analysis

All iterations of logistic regression and random forest models for the app review analysis failed on ChatGPT coded attempts. Without fundamental ability to parse text and extract features, any configurative tuning proved meaningless. Sensitivity and specificity remained pegged at 57% and accuracy below 70% compared to published benchmarks of 80% + [34]. Definitive weaknesses persist handling any unstructured data predictive tasks.

Blueberry Crop Yield Projection

ChatGPT regression codes for forecasting blueberry yields reduced RMSE error from 314 initially to 148 primarily via cross-validation based Support Vector Machines. But human tuned approaches lowering training-test data splits further reduced RMSE to 143 by systematically applying algorithms like XGBoost and optimizing parameters, which is similar to the metrics as stipulated in Jeong et al. [35]. While the initial attempt was exceptionally high, ChatGPT displayed adaptability matching more average agricultural yield prediction errors after systematic tuning and further refinement with human guidance achieved competitive accuracy.

Out of all the tests above, however, major weaknesses emerged in generating text analytics code. Across multiple attempts with two reviews datasets, ChatGPT failed at basic preprocessing like removing whitespace in strings. All logistic regression and random forest variants failed to run. This signals limitations in its training for more complex unstructured data.

Effectiveness Compared to Google Bard

When given identical prompts, Google Bard exhibited poorer performance than even ChatGPT's first baseline models. Its initial code for the churn prediction problem achieved only 65% accuracy with highly skewed sensitivity. Purchase prediction code failed to compile entirely. Unlike ChatGPT, Bard did not interact further to refine its outputs.

In contrast, ChatGPT's conversational nature [36] enabled it to clarify objectives, suggest improvements, execute them in code, and evaluate outcomes through user feedback. Quantitative metrics and qualitative responsiveness favours ChatGPT over current Bard offerings. However, Google's continued active development in this area could pose stiff competition soon.

Recurring Errors and Need for Interventions

Despite strengths, ChatGPT cannot rival human judgment for machine learning workflows overall. Persistent oversights like failing to check data balance or splitting appropriately required user oversight. Code often broke from outdated package references or mismatches applying generalized logic to novel datasets. This observation contradicts with what was reported by [37] in The Guardian.

Simple terminology ambiguities frequently confused ChatGPT. It struggled to transfer feedback on one model iteration into improved subsequent versions autonomously. Ultimately it generated reasonable baseline code fast but achieving robust refinements at par with an expert data scientist was unattainable without consistent human guidance [38].

Nuanced thinking linking code to context also lagged–while ChatGPT outlined sensible generic steps for model selection and improvement, it faltered tailoring these appropriately to business needs and priorities without assistance. Bridging this gap is essential for realizing performance potential.

For straightforward cases, ChatGPT speeds up initial coding. But for rigorous, optimized analytics aligned to organizational objectives, human-AI collaboration outshines either working in isolation currently. Ongoing advances and further training focused on this intersection could multiply the combined impact greatly.

Discussion

Interpreting Where ChatGPT Shows Promise

Considering the study findings holistically, ChatGPT flaunts legitimate potential to improve accessibility and efficiency of core machine learning coding–but falls short as a fully automated solution.

Across structured input problems of churn prediction and purchase likelihood modelling, it delivered accurate, well-generalized algorithms rivalling human Data Scientists upon first attempt with no outside help. Enhancements through rigorous tuning processes further solidified its capacity for simple binary classification tasks now commonplace for business forecasting.

Demonstrated adeptness also emerged applying regression, dimensionality reduction, and ensemble techniques to estimate blueberry harvests. This highlights an underlying aptitude, not merely mimicking previous patterns, but adaptively combining ML fundamentals with reasonable predictive outcomes.

In domains with well-defined features and established data, ChatGPT significantly enhances the speed of developing initial models. This acceleration allows for quicker iterations and more experimentation, leading to improved optimization of models.

Where Critical Gaps Around Complexity Persist

Despite ChatGPT's acclaim for coding, its struggle with text analytics mirrors limitations seen in its predecessor GPT-3, particularly in processing unstructured content and language [12, 13]. These challenges in handling complex, unstructured data highlight inherent architectural limitations and raise concerns about the model's ability to replicate nuanced human judgment in such areas. As a result, while ChatGPT excels in structured tasks, it falls short in more complex, language-based processing, necessitating further development and understanding.

The current challenges in ChatGPT's text analytics are rooted in its difficulty with complex, multifaceted contexts, a limitation that stems from its architectural design. This raises questions about the AI's capacity to fully emulate human judgment, particularly in nuanced, domain-specific tasks. While enhancements through extensive training might improve feature normalization and textual input vetting, it's uncertain if these adjustments can bridge the gap between AI-generated responses and the intricate, sophisticated understanding that human experts possess. These concerns highlight the ongoing debate about the extent to which AI can replicate human-like interpretive skills.

While ChatGPT significantly democratizes coding processes, entrusting it entirely for machine learning solutions could be risky without careful oversight. This is particularly true for individuals with limited data science expertise, as they may overrely on its outputs, potentially leading to misguided confidence. This stems from a lack of transparency and understanding of ChatGPT's limitations, underscoring the need for human judgment and verification in the use of AI-generated solutions.

Implications for Education and Business Integration

The integration of ChatGPT in educational and business settings carries significant implications. For users such as students and workers utilizing its coding capabilities, it's crucial to manage expectations and maintain a level of scepticism. This cautious approach is necessary until ChatGPT's limitations are fully addressed and understood. Emphasizing critical thinking and understanding its capabilities and boundaries will ensure more effective and responsible use of this AI tool in both educational and business contexts.

Enhancing curricula and training programs with a strong foundation in ethics, critical thinking, and statistics is crucial for balancing productivity gains with responsible AI use. Educators should focus on guiding learners to intelligently interpret when

AI solutions like ChatGPT are effective, and caution against the risks of blindly applying these technologies. This approach is key to ensuring learners are well-equipped to navigate the complexities and ethical considerations of using AI in practical scenarios.

Businesses aiming to enhance development speed with AI coding assistants like ChatGPT must be cautious of potential risks arising from inadequate governance. Establishing clear guidelines and quality checks for machine learning code produced by ChatGPT can help businesses leverage its advantages while minimizing potential drawbacks. By proactively creating criteria to assess the quality and origin of AI-generated code, businesses can better harness the benefits of AI tools and protect against possible negative consequences.

Future Research Directions

These findings highlight important areas for further research, emphasizing the need to move beyond just assessing the accuracy and speed of coding by generative AI like ChatGPT. It's crucial to evaluate how these AI systems perform throughout the entire lifecycle of business analytics solutions. This comprehensive approach will help understand the nuanced impacts of generative AI in real-world scenarios, contributing to more effective and informed applications of these technologies in various fields.

Enhancing integration between users and AI, particularly in hybrid human-chatbot teams, is crucial. It was suggested that the following Cognitive-AI Framework (Fig. 3.1) which was tested extensively at University of Surrey to be implemented more widely.

Moving away from simplistic interactions to more complex, integrated use cases can better serve user needs. Additionally, complementing technical assessments of AI with behavioural research—focused on understanding user perceptions, trust, and acceptance—is vital. This holistic approach not only maximizes the potential of future AI breakthroughs but also ensures rigorous adherence to ethical standards and practices.

In the dynamic space where generative AI's capabilities meet specific user needs, numerous opportunities arise. ChatGPT's coding skills, currently showing promise, necessitate a strategic partnership for research to effectively manage its limitations. The potential advantages of embracing this nuanced approach to AI technology, despite its present imperfections, significantly surpass the risks, advocating for continued exploration and advancement in the field.

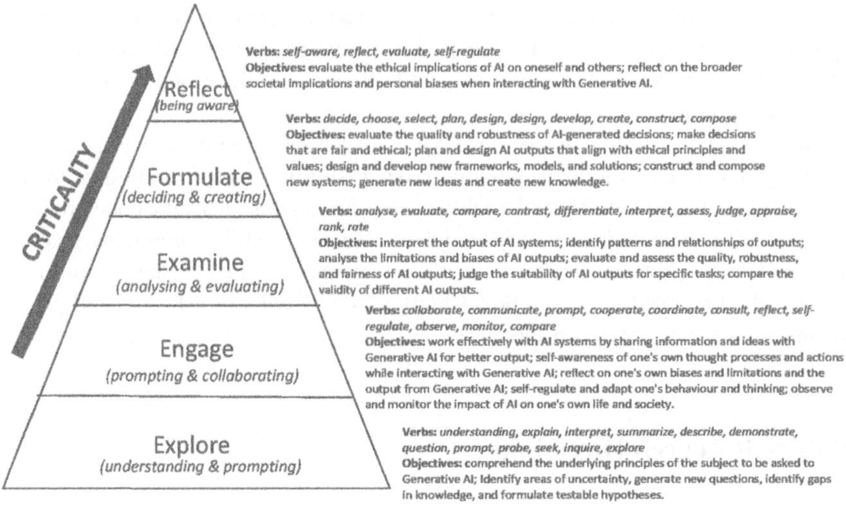

Fig. 3.1 Cognitive AI-framework [39] evolved from [40–43]

Conclusion

This study assessing ChatGPT's capability to generate machine learning code in business settings reveals both promising aspects and notable shortcomings that call for further enhancements. The research uncovers substantial potential in the application of ChatGPT in these contexts, yet highlights areas needing improvement for more effective utilization [44].

For customer churn and purchase prediction classification tasks, ChatGPT produced functional random forest models exceeding 72% accuracy upon first attempt, which increased to 83% or greater through techniques it suggested like grid search tuning [15, 45]. However, when processing mobile app reviews text, it failed in executing basic preprocessing steps or modelling code [46], evidencing weaknesses handling unstructured data.

For educational institutions and businesses considering the adoption of ChatGPT, its role as a coding assistant in defined analytical tasks is promising. However, relying solely on it without proper oversight can lead to unfavourable outcomes [47]. Therefore, creating educational programs that enable individuals to effectively use ChatGPT's strengths, particularly on well-established datasets, while also teaching them to address its limitations in more complex or uncertain scenarios, is essential for successful integration [48].

As rapid generative AI advances continue, future research should focus on evaluating holistic solutions rather than just measuring coding accuracy [25]. This approach will provide deeper insights into how best to combine human expertise with ChatGPT's capabilities, thereby unlocking their collective potential for more effective solutions.

The conclusion emphasizes the significant advancements made by generative technologies like ChatGPT in coding. However, it underlines the need for setting realistic expectations and fostering responsible collaboration to maximize their benefits for both business and society. Despite the challenges, when used judiciously, the acceleration and impact of technologies like ChatGPT remain substantial and influential, especially in this case, when there was report about shortage of programmer [49].

References

1. Gewirtz, D. (2023). How does ChatGPT actually work? ZDNET. Retrieved October 15, 2023, from https://www.zdnet.com/article/how-does-chatgpt-work/
2. Cao, Y., Li, S., Liu, Y., Yan, Z., Dai, Y., Yu, P., & Sun, L. (2023). A comprehensive survey of AI-Generated Content (AIGC): A history of generative AI from GAN to ChatGPT. Retrieved September 13, 2023, from https://www.researchgate.net/publication/369090930_A_Comprehensive_Survey_of_AI-Generated_Content_AIGC_A_History_of_Generative_AI_from_GAN_to_ChatGPT
3. Kahneman, D. (2021). Interview on 'clearly ai is going to win. how people are going to adjust is a fascinating problem'. Interviewed by Tim Adams. The Guardian. Retrieved May 16, 2021, from https://www.theguardian.com/books/2021/may/16/daniel-kahneman-clearly-ai-is-going-to-win-how-people-are-going-to-adjust-is-a-fascinating-problem-thinking-fast-and-slow
4. Yenduri, G., Srivastava, G., Maddikunta, P. K. R., Jhaveri, R. H., Wang, W., Vasilakos, A. V., & Gadekallu, T. R. (2023). Generative pre-trained transformer: A comprehensive review on enabling technologies, potential applications, emerging challenges, and future directions. Retrieved September 21, 2023, from https://www.researchgate.net/publication/370869544_Generative_Pre-trained_Transformer_A_Comprehensive_Review_on_Enabling_Technologies_Potential_Applications_Emerging_Challenges_and_Future_Directions
5. Ijaz, S. (2023). 15 countries that banned CHATGPT. Yahoo! Finance. Retrieved September 11, 2023, from https://finance.yahoo.com/news/15-countries-banned-chatgpt-204342617.html?guccounter=1
6. Scott, A. (2023). How can data scientists use CHATGPT for developing machine learning models? In Data Science Central. Retrieved September 11, 2023, from https://www.datasciencecentral.com/how-can-data-scientists-use-chatgpt-for-developing-machine-learning-models/
7. Brynjolfsson, E., Li, D., & Raymond, L. R. (2023). Generative AI at work. National Bureau of Economic Research. Retrieved September 11, 2023, from https://www.nber.org/papers/w31161
8. Borji, A. (2023). A categorical archive of ChatGPT failures. Retrieved September 10, 2023, from https://www.researchgate.net/publication/368333200_A_Categorical_Archive_of_ChatGPT_Failures
9. Chomsky, N., Roberts, I., & Watumull, J. (2023). Noam Chomsky: The false promise of ChatGPT. The New York Times, 8th March 2023. Retrieved September 13, 2023, from https://www.nytimes.com/2023/03/08/opinion/noam-chomsky-chatgpt-ai.html
10. Stokel-Walker, C., & Van Noorden, R. (2023). What ChatGPT and generative AI mean for science. *Nature*. Retrieved October 19, 2023, from https://www.nature.com/articles/d41586-023-00340-6
11. Thornhill, J. (2023). The promise—and peril—of generative AI. Financial Times, 28 Sep. Retrieved September 23, 2023, from https://www.ft.com/content/e6a391c7-bfd2-4eb1-82e5-6bc4eac9b131

12. Bang, Y., Cahyawijaya, S., Lee, N., Dai, W., Su, D., Wilie, B., Lovenia, H., Ji, Z., Yu, T., Chung, W., & Do, Q. V. (2023). A multitask, multilingual, multimodal evaluation of ChatGPT on reasoning, hallucination, and interactivity. Retrieved September 11, 2023, from https://arxiv.org/pdf/2302.04023
13. Gozalo-Brizuela, R., & Garrido-Merchan, E. C. (2023). ChatGPT is not all you need. A state of the art review of large generative AI models. Retrieved October 15, 2023, from https://www.researchgate.net/publication/367088266_ChatGPT_is_not_all_you_need_A_State_of_the_Art_Review_of_large_Generative_AI_models
14. Cribben, I., & Zeinali, Y. (2023). The benefits and limitations of ChatGPT in business education and research: A focus on management science, operations management and data analytics. *Operations Management and Data Analytics.*
15. Kabir, S., Udo-Imeh, D., Kou, B., & Zhang, T. (2023). Who answers it better? An in-depth analysis of ChatGPT and stack overflow answers to software engineering questions. Retrieved September 11, 2023, from https://www.researchgate.net/publication/372951026_Who_Answers_It_Better_An_In-Depth_Analysis_of_ChatGPT_and_Stack_Overflow_Answers_to_Software_Engineering_Questions
16. Weisz, J. D., Muller, M., Houde, S., Richards, J., Ross, S. I., Martinez, F., Agarwal, M., & Talamadupula, K. (2021). Perfection not required? Human-AI partnerships in code translation. In *26th International conference on intelligent user interfaces, Texas, United States* (pp. 402–412). New York, United States
17. Savelka, J., Agarwal, A., Bogart, C., Song, Y., & Sakr, M. (2023). Can generative pre-trained transformers (GPT) pass assessments in higher education programming courses? In *ITiCSE 2023: Proceedings of the 2023 conference on innovation and technology in computer science education V1* (pp 117–123). Retrieved October 8, 2023, from https://doi.org/10.1145/3587102.3588792
18. Pavlik, J. V. (2023). Collaborating with ChatGPT: Considering the implications of generative artificial intelligence for journalism and media education. *Journalism & Mass Communication Educator, 78*(1), 84–93.
19. Chen, T. J. (2023). ChatGPT and other artificial intelligence applications speed up scientific writing. *Journal of the Chinese Medical Association, 86*(4), 351–353.
20. George, A. S., & George, A. H. (2023). A review of ChatGPT AI's impact on several business sectors. *Partners Universal International Innovation Journal, 1*(1), 9–23.
21. Deng, J., & Lin, Y. (2022). The benefits and challenges of ChatGPT: An overview. *Frontiers in Computing and Intelligent Systems, 2*(2), 81–83.
22. Maya, V. (2023). ChatGPT use cases for customer service: Improving efficiency and satisfaction. Retrieved September 11, 2023, from https://customersfirstacademy.com/chatgpt-use-cases-for-customer-service/
23. Hughes, A. (2023). CHATGPT: Everything you need to know about OpenAI's GPT-4 tool. BBC Science Focus. Retrieved September 9, 2023, from https://www.sciencefocus.com/future-technology/gpt-3
24. Pataranutaporn, P., Liu, R., Finn, E., et al. (2023). Influencing human–AI interaction by priming beliefs about AI can increase perceived trustworthiness, empathy and effectiveness. *Nature Machine Intelligence, 5*, 1076–1086. https://doi.org/10.1038/s42256-023-00720-7
25. Shen, Y., Heacock, L., Elias, J., Hentel, K. D., Reig, B., Shih, G., & Moy, L. (2023). ChatGPT and other large language models are double-edged swords. *Radiology, 307*(2). Retrieved September 21, 2023, from https://doi.org/10.1148/radiol.230163
26. Maciel, L. (2023). ChatGPT and the ethical aspects of artificial intelligence. *Revista de Gestão, 30*(2), 110–112.
27. OpenAI. (2022). Introducing ChatGPT. Retrieved September 11, 2023, from https://openai.com/blog/chatgpt
28. Iranian churn dataset: UCI machine learning repository. Retrieved September 11, 2023, from https://archive.ics.uci.edu/dataset/563/iranian+churn+dataset
29. Sakar, C., & Yomi, K. (2018). Online shoppers purchasing intention dataset. UCI Machine Learning Repository. Retrieved September 11, 2023, from https://archive.ics.uci.edu/dataset/468/online+shoppers+purchasing+intention+dataset

30. Kaggle. (2023). Prediction of wild blueberry yield. Retrieved September 11, 2023, from https://www.kaggle.com/competitions/playground-series-s3e14/data
31. Jhalani, S. (2023). Threads, an Instagram app reviews. Kaggle. Retrieved September 11, 2023, from https://www.kaggle.com/datasets/saloni1712/threads-an-instagram-app-reviews
32. Dal Pozzolo, A., Caelen, O., Johnson, R. A., & Bontempi, G.: Calibrating probability with undersampling for unbalanced classification. In *2015 IEEE symposium series on computational intelligence (SSCI). Cape Town, South Africa* (pp. 159–166).
33. Agustyaningrum, C. I., Haris, M., Aryanti, R., & Misriati, T. (2021). Online shopper intention analysis using conventional machine learning and deep neural network classification algorithm. *Jurnal Penelitian Pos dan Informatika., 11*(1), 89–100.
34. Conneau, A., Kiela, D., Schwenk, H., Barrault, L., & Bordes, A. (2017). Supervised learning of universal sentence representations from natural language inference data. In M. Palmer, R. Hwa, & S. Riedel (Eds.), *Proceedings of the 2017 conference on empirical methods in natural language processing. conference on empirical methods in natural language processing, 07–11 Sep 2017, Copenhagen, Denmark* (pp. 670–680). Association for Computational Linguistics
35. Jeong, J. H., Resop, J. P., Mueller, N. D., Fleisher, D. H., Yun, K., Butler, E. E., Timlin, D. J., Shim, K. M., Gerber, J. S., Reddy, V. R., Kim, S. H. (2016). Random forests for global and regional crop yield predictions. *PLoS One, 11*(6).
36. Gottlieb, M., Kline, J. A., Schneider, A. J., & Coates, W. C. (2023). ChatGPT and conversational artificial intelligence: Friend, foe, or future of research? *The American Journal of Emergency Medicine, 70*, 81–83.
37. Naughton, J. (2023, April 1). Programmers, beware: ChatGPT has ruined your magic trick. In J. Naughton (ed.), The Guardian. Retrieved September 9, 2023, from https://www.theguardian.com/commentisfree/2023/apr/01/chatgpt-write-code-computer-programmer-software
38. McManus, S. (2023, March 31). Friend or foe: Can computer coders trust ChatgGPT? BBC News. Retrieved September 11, 2023, from https://www.bbc.co.uk/news/business-65086798
39. Fu, C. (2023d). Revolutionize the education paradigm in the generative AI Era. In QS Higher Education Summit Aisa Pacific 2023. Kuala Lumpur, Malaysia, 7 - 9 November, 2023. QS.
40. Fu, C. (2023b) Effective assessment strategies in the Era of artificial intelligence. In *The chartered ABS's annual LTSE conference, 2023. Wales, United Kingdom, 22–23 May 2023.* Chartered Association of Business Schools
41. Fu, C. (2023c). Redefine the theoretical framework of bloom's taxonomy in the era of artificial intelligence for positive learning outcome. In *The chartered ABS's annual LTSE conference, 2023. Wales, United Kingdom, 22–23 May 2023.* Chartered Association of Business Schools.
42. Fu, C. (2023a). Applications of AI on business in the digital era. In N. Amirah, H. Malik, & A. Afthanorhan (Eds.), *A new paradigm for the sustainability management in digital age: Advances in data analytics.* Springer Nature.
43. Fu, C., Vaiksnoras, V., Kern, C. (2023). Make your assignments work. In: Festival of learning and teaching 2023, effective assessment for learner success. University of West London.
44. Dwivedi, Y. K., Kshetri, N., Hughes, L., et al. (70 more authors). (2023). So what if ChatGPT wrote it?: Multidisciplinary perspectives on opportunities, challenges and implications of generative conversational AI for research, practice and policy. *International Journal of Information Management, 71*, 102642.
45. Laskar, M., Bari, M., Rahman, M., Bhuiyan, M., et al. (2023). A systematic study and comprehensive evaluation of ChatGPT on benchmark datasets. Retrieved September 11, 2023, from https://arxiv.org/abs/2305.18486
46. Geron, A. (2019). *Hands-on machine learning with Scikit-Learn, Keras, and TensorFlow: Concepts, tools, and techniques to build intelligent systems* (2nd ed.). Oreilly.
47. Hacker, P., Engel, A., & Mauer, M. (2023). Regulating ChatGPT and other large generative AI models. In *Proceedings of the 2023 ACM conference on fairness, accountability, and transparency* (pp. 1112–1123).
48. Baidoo-Anu, D., & Ansah, L. (2023). Education in the Era of generative artificial intelligence (AI): Understanding the potential benefits of ChatGPT in promoting teaching and learning.

Retrieved September 11, 2023, from https://www.researchgate.net/publication/369385210_Education_in_the_Era_of_Generative_Artificial_Intelligence_AI_Understanding_the_Potential_Benefits_of_ChatGPT_in_Promoting_Teaching_and_Learning
49. Phillips, T. (2023). Is there a shortage of developers? Developer shortage statistics in 2022. In: CodeSubmit Blog. Retrieved September 11, 2023, from https://codesubmit.io/blog/shortage-of-developers/

Chapter 4
NER-IPL: Indian Legal Prediction Dataset for Named Entity Recognition

Sarika Jain and Pooja Harde

Abstract Identifying Named Entities from unstructured text is difficult, especially for domain-specific data. Legal documents are usually very lengthy and highly unstructured. We can use two approaches for extracting the named legal entities from the legal documents: the Rule-based approach and the Machine-learning (ML) approach. This paper introduces NER-IPL, an Indian Legal Prediction Dataset for Named Entity Recognition. The dataset consists of 213481 sentences with 123193 annotated entities and 6198700 tokens. Different ML models take different encoding schemes to process the dataset; therefore, we use three different encoding schemes (BILOU, BOI, IOEBS) on the named entities for tagging, allowing the corpus to be trained on any machine learning model for automatic extraction of named entities. To validate our dataset, we have created a battery of baseline models to test the suitability of NER tasks using language models. All the different experiments with scores, detailed analysis, and the scope of improvements are elaborated in detail. Amongst the experimented baseline models, the InLegalBERT model gives the best F1 score of 0.67 on our dataset.

Keywords Indian legal dataset · Corpus creation · NER · Legal domain · Semantic web · NLP · Entity extraction

Introduction

India has a lot of pending cases on every level of the judicial system, whether it is the Supreme Court of India (SCI), High Court, or a District Court (approximately 40 million pending cases[1]). Legal stakeholders and automated tools like document

[1] https://njdg.ecourts.gov.in/njdgnew/?p=main/pend_dashboard.

S. Jain · P. Harde (✉)
National Institute of Technology Kurukshetra, Kurukshetra, India
e-mail: pooja_jrf@nitkkr.ac.in

S. Jain
e-mail: jasarika@nitkkr.ac.in

© The Author(s), under exclusive license to Springer Nature Switzerland AG 2024

A. Emrouznejad et al. (eds.), *Business Analytics and Decision Making in Practice*, Lecture Notes in Operations Research, https://doi.org/10.1007/978-3-031-61589-4_4

classification, searching for relevant documents, and identifying important clauses face various challenges while deciphering legal documents, as they are usually lengthy and highly unstructured. A legal information extraction system is required to make this process of finding a piece of information an easy task. Information extraction is a technique that automatically extracts the relevant data from structured, semi-structured, and unstructured text. This can help develop an advanced legal information retrieval system to perform various tasks for legal stakeholders. One important sub-task of information extraction is named entity recognition (NER), which involves identifying and categorizing entities within text into predefined categories such as names of persons, organizations, locations, dates, and many more. NER for legal texts is more cumbersome than for open-domain texts as it requires a deep understanding of the legal domain.

Although open data initiatives like the National Judicial Data Grid (NJDG) provide some metadata on Indian court judgments, they are not exhaustive. NER from court judgments will provide exhaustive metadata, and the dataset thus created will be useful for many downstream tasks like knowledge graph construction and predictions. Once we get the named legal entities, it can help automatically process legal documents. This automatic processing will make information access easier and incorporate in developing different legal applications such as document classification, legal document recommendation systems, and so on, which can help reduce the workload of judges and lawyers, leading to more efficient proceedings and quicker verdicts.

The process of NER can be rule-based [6, 8] and machine learning-based [7]. Rule-based NER depends on using a set of linguistic rules, vocabulary, regular expressions, prefixes, etc., that help identify a pattern in the textual data. As shown in [6, 8], these rules can also be used on the legal documents to extract the named legal entities that follow a certain pattern of text surrounding them. However, due to the legal documents' different legal data representation, interpretation, complexity, and highly unstructured nature, the rule-based approach becomes a challenging task for extracting and identifying all the named legal entities from the documents.

With the buildup of the Digital India Mission, there has been a lot of advancement in natural language processing and computer vision research. However, with the rise in the campaigning of AI for social good, machine learning techniques have been applied increasingly in the fields of social science as well as administration departments. If we want more accuracy and recognition of the complex named entities in the legal domain, machine learning can provide better learning of the named legal entities.

In this paper, we are proposing a dataset for the Indian Named Legal Entity Recognition task. Though many Legal NER datasets are available, the distinctness among different judicial systems worldwide induces the requirement to create a different legal NER dataset specific to Indian Legal Documents. This paper extends the work of [8], where previously, an Indian Legal NER dataset on 50 documents was released and tested using the Rule-based approach by writing simple Python Regular Expressions. We have downloaded the Supreme Court Judgments from the

IndianKanoon[2] website. Extracted documents were given as input to the pipeline architecture, as shown in Fig. 4.1. Our research aims to release different datasets for different legal application tasks. Likewise, this paper introduces one such dataset for Named Legal Entity Recognition for Indian Legal documents to be used for various predictions, named NER-IPL.

There are three major contributions made in this paper:

- We have developed an Indian Legal NER (NER-IPL) Dataset out of 18412 Supreme Court of India judgments constituting 126919 annotated entities categorized into seven entities.
- We propose a battery of baseline models to validate the created dataset.
- Expert validation is carried out for NER-IPL.

To our knowledge, the dataset described in this paper is one of the important initial attempts at named entity recognition that covers the Indian Legal Domain. On this account, section "Related Work" describes some of the previous works in the legal NER domain. Section "NER-IPL Construction" elaborates on the methodology followed for NER-IPL construction, with NER-IPL Validation discussed in section "NER-IPL Validation". Finally, section "Conclusion and Future Scope" concludes the work and presents future scope.

Related Work

A high-quality dataset is the key to a successful NLP task. A diverse set of NER datasets are available online and can be procured based on the application. Domain-specific datasets are very limited in number, and the same is the case with the legal NER. The following section summarizes some of the existing legal NER datasets. Two of them are for the Indian judicial system.

GermEval [2] The German NER dataset comprises 31000 German sentences and 590,000 tokens collected from German Wikipedia and news corpora. It was initially released in 2014, and many revisions have been made. Tagged entity spans are in the BIO encoding scheme.

Han Huang [4] introduced the Chinese legislative domain dataset, in which they obtained annotated 260,000 word manually of judicial documents from the Chinese Justice Document Network, of which 40737 total named legal entities. These documents fall into different categories: people, organizations, crimes, laws and regulations, and penalties.

LeNER-br [10] The Brazilian Legal NER dataset was built using 70 legal documents from different Brazilian Courts. Fifty random documents were used for the training sample, and ten were used for testing and development each. BOI encoding was used for the tagging scheme.

[2] https://indiankanoon.org/.

BSNLP-2019 [14] Balto-Slavic Natural Language Processing-2019 is a multilingual dataset created in 4 Slavic languages, namely Bulgarian, Czech, Polish, and Russian for Named Entity Recognition, Lemmatization, and Entity Linking. This work is an extended version of their work presented in [15].

SemEval 2022 Task 11 [11] SemEval 2022 Task 11, also referred to as Multi-CoNER consists of 6 very standard entity classes PER, LOC, CORP, GRP, PROD, CW varying over 11 different languages - Bangla (BN), Chinese (ZH), Dutch (NL), English (EN), Farsi (FA), German (DE), Hindi (HI), Korean (KO), Russian (RU), Spanish (ES), and Turkish (TR). The dataset has been divided into 13 subsets—Monolingual (1 subset for each language), Multilingual (Mixed random sample), and Code-mixed (text and entity of different languages). All subsets have varying numbers of entities based on language, crossing the number of instances to around 2.3 million.

MSRA-2006 [9] MSRA-2006 is a Chinese NLP Dataset compiled from 5 different sources. The dataset consists of only three entity classes in the BIO Scheme: PER, ORG, and LOC. The scope of the dataset is for NER and Word Segmentation (WS). Evaluation Dataset is divided into 3 and 2 Corpora, Simplified Characters, and Traditional Characters, respectively.

ICDAC [8] provides the very first Indian Legal NER dataset developed on only 50 documents of the Supreme Court of India with three encoding schemes (i.e., BIO, IOBES and BILOU). This is a gold standard dataset as it is manually annotated with two semantic classes, namely, coarse-grained and fine-grained classes, consisting of 10 and 30 entities respectively.

Kalamkar et al. [7] released another Legal NER dataset for Indian Court Judgments. The authors took the judgments from the entire judicial hierarchy, such as the Supreme Court of India, High Court, and District Court judgments. They provided a dataset of 11970 judgments annotated with 14 legal entities and 13365 annotated entities.

After analyzing the existing Indian Legal NER datasets, we discovered no Indian Legal NER dataset specific to the Supreme Court of India judgments. Such a dataset will improve the efficacy of identifying the named legal entities. Though the results published by [7] gives the accuracy of 91.1% on their proposed dataset, after doing the ablation study, we found that, if a new case apart from the dataset is given as an input to their model, it correctly identifies only a few of entities like statutes, provisions, few occurrences of judges, geographic location, and organization. We believe that the documents being considered for the dataset creation play an important role in black-box machine learning systems. Therefore, we are releasing a new Indian Legal NER dataset comprising annotated named legal entities for only the Supreme Court of India judgments.

NER-IPL Construction

This section briefly describes the methodology for creating the NER-IPL dataset and the annotations and statistics. The dataset is constructed using a semi-automatic procedure for identifying the named legal entities.

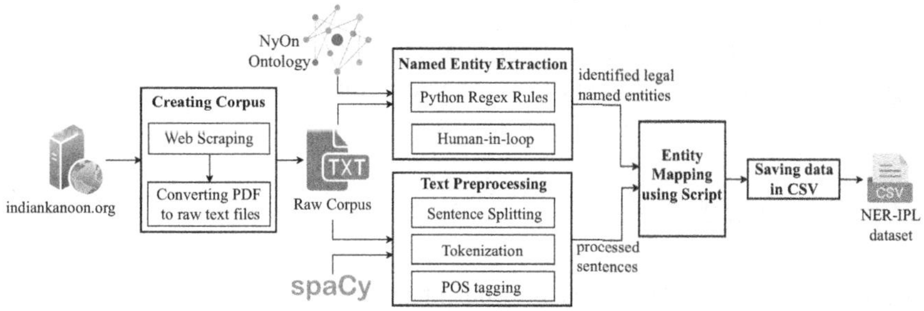

Fig. 4.1 NER-IPL construction pipeline

Materials and Methods

Figure 4.1 illustrates the diagrammatic representation of the procedure followed for NER-IPL construction. We collected the legal documents for creating the corpus from the Indian Kanoon[3] website, which is an online repository and search engine for Indian legal documents. The documents collected originated from the Supreme Court of India (SCI). Collected documents were in PDF format and had unique document numbers associated with them. We converted each PDF judgment into raw text files.

The different legal terms were borrowed from the Nyaya Ontology (NyOn) [5], a modular ontology to describe court judgments. NyOn has been published on the web adhering to the Semantic Web best practices and FAIR principles. NyOn contains 14 main classes, seven of which are incorporated in NER-IPL. Thus, each raw text file is passed through the SPACY tool[4] for text processing steps such as sentence splitting, tokenization, and POS tagging. Similarly, the same raw text file is processed through a Python Script we developed to extract the simple legal entities. This Python Script contains different regular expressions designed to extract legal entities that can be extracted by identifying a textual pattern around the entity present in the text. The human-in-loop is also used for verifying the rule-based extracted legal entities. We have analyzed that the regular expressions we designed sometimes could not extract some entities due to noise and different patterns available in the text data. Later, after the text processing step and extraction of legal entities, the output data from these levels are combined with each token per sentence encoded with the respective encoding scheme. Lastly, this mapped data is saved into the CSV file. The three encoding schemes used in NER-IPL are BILOU (B-Beginning, I-Internal, L-Last, O-Outside, U-Unit), BOI (B-Beginning, O-Outside, I-Inside), and IOEBS (I-Inside, O-Outside, B-Beginning, E-End, S-Singleton).

[3] https://indiankanoon.org.

[4] https://spacy.io/.

The named entities are annotated only using the preamble part of the judgment rather than considering the whole judgment. The generated dataset follows CONLL 2003 format widely used for datasets for named entity recognition task.

NER-IPL Statistics

The dataset comprises 18412 legal documents with 213481 sentences and 6198700 tokens excluding the <sof>, <eof>, and header of the dataset. Here, <sof> and <eof> tags refer to the starting and end of the individual documents, respectively, for identifying the start and end of the document text. The dataset contains five feature columns: Token, POS, BILOU, IOB, and IOEBS. Table 4.1 provides the statistics of the legal tags and there occurrences present in the dataset.

NER-IPL Validation

Expert Validation

Experts in the legal domain have validated the developed dataset to verify the annotated legal entities' efficacy and correctness. A team of 5 students and one faculty from the National Forensic Sciences University, India, did the expert validation. As per the legal expert examination, the identified and annotated entities are correct with 90% accuracy. As stated by the expert, the reason behind 10% less accuracy is that in some cases, though the plaintiff and appellant were mentioned, we have annotated them as petitioners. According to legal experts, there are differences in the definitions of these entities; thus, they should be treated accordingly. The experts also stated that we could add more legal entities into the dataset, such as evidence, weapons mentioned in case of criminal cases, statutes, provisions, citations, and so on.

Table 4.1 Named legal entity statistics in NER-IPL

S. No.	Tag name	Tag	Tag count
1	Bench	BCH	29191
2	CaseType	CTYP	15436
3	Court	CRT	29975
4	Jurisdiction	JURD	7985
5	Location	LOC	5335
6	Petitioner	PETR	18525
7	Respondent	RES	16746
Total no. of tags			123193

Experimental Setup

All the models are trained on NVIDIA Tesla V100 GPU with an average of 2.5 h for each model to complete the training. The parameters used are described in Table 4.3 for all the models.

The dataset is split into three parts: train, test, and dev, as mentioned in Table 4.2. To evaluate the Named Entity Recognition models on the NER-IPL dataset, we used the standard metrics of Precision, Recall, and F1 score. These are the common metrics used for the evaluation of the classification models. Precision calculates the reliability of the model for correctly identifying entity tags out of the total identified entity tags (i.e., True Positives); Recall is calculated for checking the reliability of the model for correctly identifying entity-tag out of actual entity tags, which helps in minimizing the False Negatives. F1 score, also known as the harmonic mean, is used to find a good balance between the Precision and Recall.

The already manually tokenized dataset is grouped based on sentences and passed through the respective ModelTokenizer to get word_ids, which aligns labels to the further tokenized words. After padding the input IDs, tags, and attention masks, they were all split into training and validation sets.

All transformer models were executed on the same configuration to compare the results obtained after 10 epochs with a learning rate of 3e-5 with Adam Optimizer to optimize the gradient values to correct the model.

DistilBERT [1] (Distilled Version of BERT) is a lighter version of BERT trained on the distilbert-base. It is smaller, takes less computational power, and runs faster than the model trained on the bert-base-uncased, as it takes fewer parameters. They

Table 4.2 Train and test statistics

	Sentences	Entities
Train	149437	87601
Test	42696	24142
Dev	21348	11450
Total	213481	123193

Table 4.3 Model training parameters

Parameter	Value
Transformer model	DistilBERT
	DeBERTa
	RoBERTa
	InLegalBERT
	InCaseLawBERT
Batch size	32
Epochs	10
Learning rate	3e-5
Optimizer	Adam optimizer

have used i2b2, a Biomedical Informatics dataset, to do domain-specific NER. We have used the distilbert-base-uncased model for our experiments.

DeBERTa [12] (Decoding-enhanced BERT with disentangled attention) aims to improve BERT and RoBERTa models both through techniques such as disentangled attention and mask decoder.

RoBERTa [16] (Robustly Optimized BERT Pretraining Approach) is a BERT variant that is trained on the extensive dataset in large batches and has a dynamic masking pattern and removal of NSP (next sentence prediction) objective. RoBERTa is one of the most popular versions of BERT variants, and in turn, RoBERTa also has many pre-trained models. We have used the standard roberta-base.

InLegalBERT [13] and **InCaseLawBERT** [13] are the Indian variants of the models LegalBERT [3] and CaseLawBERT[17] trained on Indian Legal corpus. The authors collected around 5.4 million documents, comprising of around 27GB raw text and trained the models from the scratch with a vocabulary based on Indian legal text.

These models, as mentioned above are considered the baseline models for the NER-IPL dataset, and the models are trained on the dataset as per the statistics mentioned in Table 4.2.

Results

Table 4.5 briefs about the scores obtained by the models while training the dataset. Amongst the four models, InLegalBERT has achieved the highest F1 score of 0.61 on the NER-IPL dataset for identifying Named Legal Entities for the Indian Judgments.

Table 4.4 provides the performance of models for individual entities present in the dataset and the scores obtained by each model concerning Precision, Recall, and F1 scores. As the Indian judgments are very lengthy and unstructured and contains unique words, InLegalBERT already being stated in [13] being more adaptable to

Table 4.4 Model performance for individual entities

	DistilBERT			DeBERTa			RoBERTa			InLegalBERT			InCaseLawBERT		
Tags	Prec	Rec	F1	Prec	Rec	F1	Prec	Rec	F1	Prec	Rec	F1	Prec	Rec	F1
BCH	0.85	0.66	0.74	0.51	0.35	0.41	0.55	0.34	0.42	0.86	0.70	0.77	0.83	0.70	0.76
CTYP	0.70	0.61	0.65	0.85	0.14	0.24	0.71	0.20	0.31	0.70	0.70	0.70	0.70	0.70	0.70
LOC	0.58	0.20	0.30	0.64	0.09	0.16	0.49	0.34	0.40	0.63	0.40	0.49	0.53	0.29	0.38
CRT	0.87	0.90	0.89	0.74	0.61	0.67	0.66	0.32	0.43	0.87	0.91	0.89	0.87	0.92	0.90
JURD	0.72	0.77	0.75	0.54	0.61	0.57	0.51	0.14	0.22	0.71	0.75	0.73	0.76	0.80	0.78
PETR	0.60	0.52	0.55	0.21	0.06	0.10	0.31	0.12	0.17	0.66	0.60	0.63	0.66	0.59	0.62
RES	0.43	0.38	0.41	0.06	0.01	0.02	0.24	0.11	0.15	0.52	0.47	0.49	0.49	0.45	0.47

Table 4.5 Model performance on test dataset

Domain	Baseline models	Precision	Recall	F1
Pretrained on open domain	DistilBERT	0.68	0.58	0.61
	DeBERTa	0.51	0.27	0.31
	RoBERTa	0.49	0.22	0.30
Pretrained on indian legal domain	**InLegalBERT**	**0.71**	**0.65**	**0.67**
	InCaseLawBERT	0.69	0.64	0.66

Indian data, also proved to be performing better with NER-IPL data for identifying named legal entities.

Conclusion and Future Scope

We outline a dataset that contains legal documents from India. We defined various distinctive semantic categories (such as Court, CourtOfficial, CaseType, Judgement, etc.) pertinent to court decisions and appropriate annotation rules. The models experimented on are only the transformer models, and it turns out that InLegalBERT outperforms other models for the NER-IPL dataset with the F1 score of 0.67.

We plan to extend this dataset for future work with more legal entities such as evidence, witnesses, statutes provisions involved, etc. Along with the number of entities to be extended, we plan to work for more NLP models such as LSTM, BiLSTM, and Hierarchical models to achieve more accuracy on the dataset so that the model can be deployed on the real-time system to be utilized as one of the applications of Named Legal Entity Recognition for the Legal Research Systems.

Acknowledgements We acknowledge the support of the IHUB-ANUBHUTI-IIITD FOUNDATION set up under the NM-ICPS scheme of the Department of Science and Technology, India.

References

1. Abadeer, M. (2020). Assessment of DistilBERT performance on named entity recognition task for the detection of protected health information and medical concepts. In *Proceedings of the 3rd clinical natural language processing workshop* (pp. 158–167). Association for Computational Linguistics. https://doi.org/10.18653/v1/2020.clinicalnlp-1.18
2. Benikova, D., Biemann, C., Kisselew, M., & Padó, S. (2014). Germeval 2014 named entity recognition shared task. In *Proceedings of the KONVENS GermEval workshop* (pp. 104–112). Hildesheim, Germany
3. Chalkidis, I., Fergadiotis, M., Malakasiotis, P., Aletras, N., & Androutsopoulos, I. (2020). LEGAL-BERT: The muppets straight out of law school. Preprint at arXiv:2010.02559

4. Huang, W., Hu, D., Deng, Z., & Nie, J. (2020). Named entity recognition for Chinese judgment documents based on BiLSTM and CRF. *EURASIP Journal on Image and Video Processing, 2020*(1), 1–14.
5. Jain, S., Harde, P., & Mihindukulasooriya, N. (2023). Nyon: A multilingual modular legal ontology for representing court judgements. In *Semantic intelligence*. Select Proceedings of ISIC 2022 (pp. 175–183). Springer.
6. Jain, S., Harde, P., Mihindukulasooriya, N., Ghosh, S., Dubey, A., & Bisht, A. (2022). Constructing a knowledge graph from Indian legal domain corpus. In *International workshop on knowledge graph generation from text (Text2kg) co-located with the extended semantic web conference (ESWC 2022)*, CEUR Workshop Proceedings (Vol. 3184, pp. 80–93).
7. Kalamkar, P., Agarwal, A., Tiwari, A., Gupta, S., Karn, S., & Raghavan, V. (2022). Named entity recognition in Indian court judgments. Preprint at arXiv:2211.03442
8. Kamat, P., Kalson, S., Suraj, S., Harde, P., Mihindukulasooriya, N., & Jain, S. (2022). An indian court decision annotated corpus and knowledge graph. In *International workshop on artificial intelligence technologies for legal documents and the 1st international workshop on knowledge graph summarization*
9. Levow, G. A. (2006). The third international Chinese language processing bakeoff: Word segmentation and named entity recognition. In *Proceedings of the fifth SIGHAN workshop on chinese language processing* (pp. 108–117). Association for Computational Linguistics, Sydney, Australia. https://aclanthology.org/W06-0115
10. Luz de Araujo, P. H., de Campos, T. E., de Oliveira, R. R. R., Stauffer, M., Couto, S., & Bermejo, P. (2018). LeNER-Br: A dataset for named entity recognition in Brazilian legal text. In *International conference on the computational processing of Portuguese (PROPOR)* (pp. 313–323). Lecture Notes on Computer Science (LNCS), Springer, Canela, RS, Brazil, 24–26 September 2018. https://doi.org/10.1007/978-3-319-99722-3_32, https://teodecampos.github.io/LeNER-Br/
11. Malmasi, S., Fang, A., Fetahu, B., Kar, S., & Rokhlenko, O. (2022). SemEval-2022 task 11: Multilingual complex named entity recognition (MultiCoNER). In *Proceedings of the 16th international workshop on semantic evaluation (SemEval-2022)* (pp. 1412–1437). Association for Computational Linguistics, Seattle, United States. https://doi.org/10.18653/v1/2022.semeval-1.196
12. Martin, C., Yang, H., & Hsu, W. (2022). KDDIE at SemEval-2022 task 11: Using DeBERTa for named entity recognition. In *Proceedings of the 16th international workshop on semantic evaluation (SemEval-2022)* (pp. 1531–1535). Association for Computational Linguistics, Seattle, United States. https://doi.org/10.18653/v1/2022.semeval-1.210
13. Paul, S., Mandal, A., Goyal, P., & Ghosh, S. (2023). Pre-trained language models for the legal domain: a case study on Indian law. In *Proceedings of the nineteenth international conference on artificial intelligence and law* (pp. 187–196)
14. Piskorski, J., Laskova, L., Marcińczuk, M., Pivovarova, L., Přibáň, P., Steinberger, J., & Yangarber, R. (2019). The second cross-lingual challenge on recognition, normalization, classification, and linking of named entities across Slavic languages. In *Proceedings of the 7th workshop on balto-slavic natural language processing* (pp. 63–74). Association for Computational Linguistics, Florence, Italy. https://www.aclweb.org/anthology/W19-3709
15. Piskorski, J., Pivovarova, L., Šnajder, J., Steinberger, J., & Yangarber, R. (2017). The first cross-lingual challenge on recognition, normalization, and matching of named entities in Slavic languages. In *Proceedings of the 6th workshop on balto-slavic natural language processing* (pp. 76–85). Association for Computational Linguistics, Valencia, Spain. https://doi.org/10.18653/v1/W17-1412, https://aclanthology.org/W17-1412
16. Wang, Y., Sun, Y., Ma, Z., Gao, L., Xu, Y., & Sun, T. (2020). Application of pre-training models in named entity recognition. Preprint at arXiv:2002.08902
17. Zheng, L., Guha, N., Anderson, B. R., Henderson, P., & Ho, D. E. (2021). When does pretraining help? Assessing self-supervised learning for law and the casehold dataset of 53,000+ legal holdings. In *Proceedings of the eighteenth international conference on artificial intelligence and law* (pp. 159–168).

Chapter 5
Benchmarking the Social Media Performance of E-Commerce Websites: A Case Study of the UAE E-Commerce Market

Roula AlBaroudi, Mohamad Badran, Ousha Awad AlNeyadi, and Gurdal Ertek

Abstract This study presents a methodology for benchmarking e-commerce websites in terms of social media performance. The methodology is applied in a case study of the United Arab Emirates e-commerce market with data from the top 10,000 e-commerce websites in the UAE. The extensive dataset acquired for this research allowed the derivation of a multitude of actionable insights, such as considerable gaps with respect to social media presence. Furthermore, a visual analytics-based decision support system (DSS) that enables e-commerce companies to assess their performance in comparison to peers was developed. The DSS, using the "followers per post" metric, suggests how many posts would be needed on average to reach the targeted percentiles in the market. While the developed DSS can be used by existing and pre-market e-commerce firms to enable benchmarking before launching e-commerce platforms and during operations in the UAE, the presented methodology documents how such a system can be created for any country or region of operations.

Keywords E-commerce · Social media · Benchmarking · Digital marketing · Data analytics · Decision support system (DSS)

Supplementary Information The online version contains supplementary material available at (https://doi.org/10.1007/978-3-031-61589-4_5).

R. AlBaroudi · M. Badran · O. A. AlNeyadi · G. Ertek (✉)
College of Business and Economics, United Arab Emirates University, Al Ain, UAE
e-mail: gurdal@uaeu.ac.ae

R. AlBaroudi
e-mail: 700040163@uaeu.ac.ae

M. Badran
e-mail: 201450124@uaeu.ac.ae

O. A. AlNeyadi
e-mail: 201211995@uaeu.ac.ae

© The Author(s), under exclusive license to Springer Nature Switzerland AG 2024
A. Emrouznejad et al. (eds.), *Business Analytics and Decision Making in Practice*, Lecture Notes in Operations Research, https://doi.org/10.1007/978-3-031-61589-4_5

Introduction

This study examines social media platform presence and performance in e-commerce, offering a thorough methodology specifically designed for performance benchmarking across social media sites. The methodology and the development of a Decision Support System (DSS) for benchmarking are demonstrated through a case study focusing on the United Arab Emirates' (UAE's) dynamic e-commerce market.

E-commerce refers to the online exchange of goods and services [1]. The global e-commerce market is expected to reach more than $8.1 trillion by 2028 [2]. Continuing growth brings new opportunities and challenges for companies seeking to boost sales and build strong online presence.

The present study was conducted in and by using e-commerce market data for UAE. According to the sectoral and research data, the e-commerce market in the UAE has experienced significant growth in recent years. Notably, the number of individuals participating in online shopping has doubled over the past five years [3].

The topic of this study is the analysis of social media metrics, and the objectives are to obtain actionable insights and a DSS that can be used by decision makers. Social media followers, who actively engage with and follow e-commerce brands, play a significant role in shaping the perceived brand value and success of businesses [4]. Engaging in marketing activities can result in increasing brand visibility, helping reach the marketing strategy plan effectively, and having higher follower engagement [5]. Furthermore, these platforms function as significant channels for obtaining feedback, enabling businesses to gain a deeper understanding of customer preferences and dissatisfaction. This is where internal self-assessment and external benchmarking are relevant. E-commerce businesses typically provide transparent information pertaining to their products, pricing, shipping, and return policies, while also implementing warranties and practicing fair trade. Such practices cultivate strong customer relationships and contribute to the growth of the digital and real economy in the United Arab Emirates [6], as in other countries. The open availability of data on e-commerce platforms can also enable external benchmarking with other e-commerce companies.

The DSS developed in this study was expected to provide the capability for simple yet effective benchmarking analysis. Definitions of benchmarking include themes of measurement, comparison, identification of best practices, implementation, and improvement [2]. Frequently employed to identify bottlenecks, strengths and weaknesses, optimal methodologies, and to enhance performance, and make well-informed judgments, benchmarking enables businesses to assess their performance and practices, including social media presence and performance, in comparison with industry competitors and performance benchmarks. The benchmarking process enables organizations of all types (private, public, and non-profit) to assess their strengths and weaknesses, identify areas in need of improvement, and make informed decisions based on data to enhance the effectiveness of their strategies.

This study employs data from the top 10,000 e-commerce websites in the UAE ranked based on their estimated sales revenue. The DSS effectively evaluates the social media performance of e-commerce websites by harnessing data on the number of social media followers and posts and by a key derived metric, namely, "followers per post." Additionally, the DSS provides valuable insights into the number of posts necessary to attain specific percentiles within the market.

Literature

Social media have a significant impact on e-commerce, shaping consumer preferences, behavior, and transactions. The increasing use of social media platforms such as Instagram and TikTok has allowed consumers to interact more actively and make more informed purchasing decisions through product comments and reviews.

Businesses have recognized the importance of social media in advertising and have utilized it to manage brand performance and engage with consumers. Social media tools significantly impact the performance of online retail businesses in the consumer electronics industry. Perceived ease of use, social media advertising, and electronic word-of-mouth (eWOM) positively affect business performance [7]. These findings can benefit aspiring online business owners, consumers, and future researchers by maximizing social media use. For businesses that operate both dual channels of both brick and mortar and online, referred to as "click and mortar" operations, decisions regarding one channel are typically interdependent on decisions regarding the other. This requires such businesses to make decisions for multiple channels simultaneously [8].

According to [9], there are multiple factors that influence consumers' adoption of social commerce, including familiarity, satisfaction, social presence, and uncertainty. [10, 11] reveal that "intention" is a significant mediating factor for actual usage. Intention was significantly and positively related to perceived trust, enjoyment, social pressure, satisfaction, and awareness. According to [11], surface credibility helps develop consumer trust, which in turn positively influences consumer satisfaction with food and beverage (F&B) product offerings. Surface credibility refers to the degree of credibility and trustworthiness people or things exhibit in certain situations.

Social media followers are considered highly valuable by businesses, and these platforms enhance the movement of data, encourage customer engagement, and provide opportunities to collect customer feedback. Furthermore, social media influencers can develop strong relationships with their followers. Social media platforms and influencers offer significant potential for marketing and advertisement [12]. Social media followers can play a crucial role in enhancing brand visibility and providing businesses with a multitude of opportunities to engage potential customers and generate value [13]. While the success of pure social media businesses is almost completely based on driving the support and engagement of their followers [14], e-commerce companies must also plan for social media presence.

Social media analytics is the assessment of social media data to learn more about customer behavior, the effectiveness of campaigns, and their overall influence. Metrics such as sentiment, reach, and engagement are tracked, helping organizations improve their online strategies, make well-informed decisions, and maximize content [6, 15]. Product review under the e-commerce websites can also be considered as social platforms because users can not only post their own reviews but can also address issues or provide feedback for other users' reviews. This makes sentiment analysis of product reviews on e-commerce websites an important research stream [16].

In this study, in addition to insights into the e-commerce market in the UAE, a basic yet potentially beneficial DSS has been developed for social media performance benchmarking. DSS for digital marketing can help marketers in the ever-changing digital marketplace by providing data integration, analytics, and trend prediction to support online decision-making [17, 18]. DSS can help develop customized marketing strategies, thereby simplifying analytics and integrating a variety of data. Through CRM connections, DSS can help improve customer insights, simplify campaign management, and give marketers the ability to make proactive, data-driven decisions that lead to overall campaign success [19, 20].

Utilizing social media platforms for communication can improve companies' cost bottomline, marketing operations, customer satisfaction, and performance [21, 22]. Furthermore, social media platforms provide businesses with valuable insights into consumer behavior and preferences, allowing them to tailor their marketing strategies accordingly. This data-driven approach can lead to higher conversion rates and increased sales of e-commerce platforms. Additionally, social media allows for real-time communication and feedback, enabling businesses to promptly address customer concerns and build stronger relationships with their target audience (Fig. 5.1).

Methodology

Data Collection and Preparation

The data used in this study were acquired on September 30, 2023, from My Lead Fox [23], an online data provider that supplies data related to e-commerce platforms worldwide. While the platform offers much more extensive data, for the present study, data for the UAE and for the top 10,000 e-commerce platforms in the country were filtered as source Dataset A. The research team was granted direct approval by the founder of MyLeadFox.com platform to use the data in the present research study and publish the data openly as public data along with the paper.

The full dataset acquired from My Lead Fox consisted of 28 columns that included textual, categorical, and numerical attributes for e-commerce platforms. Attributes include website URL (hyperlink text), product category (categorical) (e.g., food, apparel), location of the businesses (categorical) like their country and state, as well as data pertaining to social media platforms, particularly the number of followers and posts.

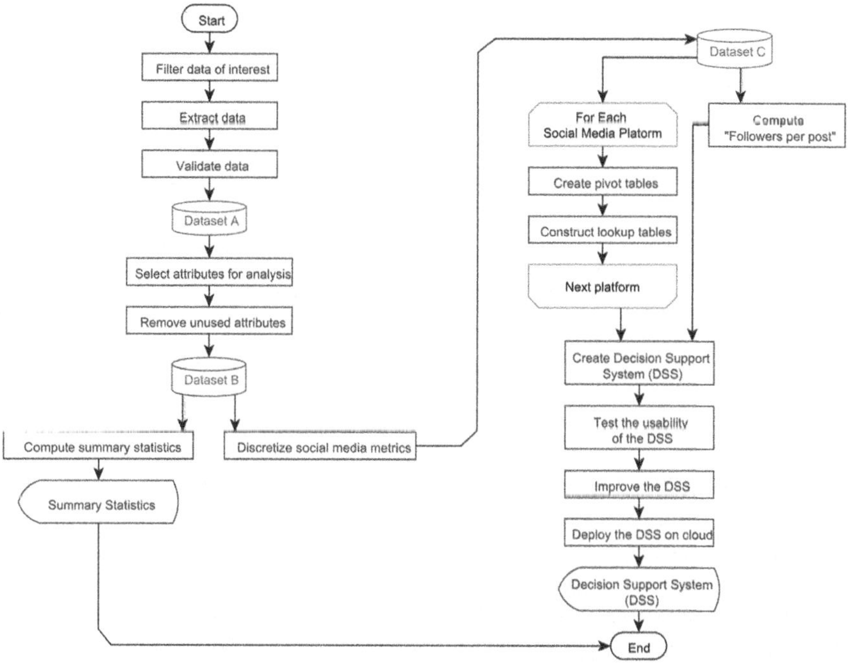

Fig. 5.1 The data analytics and DSS development methodology applied in the research study

In this study, the research term focused exclusively on selected quantitative attributes. Data regarding popular social media platforms such as Facebook, Instagram, and Twitter (which was rebranded as "X" in April 2023 [24]), YouTube, Pinterest, and TikTok were of special interest. These are the top social media platforms used in most countries when any business or company likes to interact and communicate with customers. These numerical attributes represent each social media platform's number of followers and posts, except for the number of Facebook and YouTube posts, which were not available in the source data. The attributes considered in this study were ID (primary key) and the numerical (integer) attributes facebook_followers, instagram_followers, instagram_posts, twitter_followers, twitter_posts, youtube_followers, pinterest_followers, pinterest_posts, and tiktok_followers, which together constituted the filtered Dataset B of interest.

Data Exploration

Once the data were filtered, cleaned, and prepared for analysis, they were loaded into the Orange data mining software [25, 26]. The Orange data exploration widgets were used to gain insights into the dataset, starting with the summary statistics shown in Table 5.1.

Table 5.1 Summary statistics for the social media metrics

Feature	Mean	Median	Missing
DataType	Numerical	Numerical	Numerical
Facebook_followers	113,716	**1,081**	**4421**
Instagram_followers	85,257	**3,481**	**5003**
Instagram_posts	780	321	5067
Twitter_followers	34,599	68	7423
Twitter_posts	2,167	192	7439
Youtube_followers	15,623	107	7813
Pinterest_followers	6,585	18	**8995**
Pinterest_posts	93,390	108	**8919**
Tiktok_followers	44,339	**627**	**9287**

The top platforms with respect to the median number of followers are Instagram (median of 3481 followers), Facebook (1081), and TikTok (627). However, it is important to consider that these statistics are calculated by ignoring missing data. Therefore, both the mean and median statistics are only for websites that have a presence on the respective social media platforms. To this end, one would also need to take into account the last column in Table 5.1, which gives the number of missing values for that attribute. For example, while e-commerce websites with TikTok accounts have the third highest median number of followers, only (10,000–9287 =) 713 out of the 10,000 websites actually have a TikTok account. Therefore, there is a significant gap with respect to the TikTok presence of UAE e-commerce businesses. There were also many missing values for Pinterest (8995 missing values), YouTube (7813 missing values), and Twitter (7423 missing values). Thus, TikTok, Pinterest, and Twitter had the most missing values. In other words, these three platforms are mostly unutilized by UAE e-Commerce businesses.

Even for platforms with the least missing data, namely Facebook (4421 missing values) and Instagram (5003 missing values), nearly half of the businesses do not have accounts. This signifies the lack of presence on all social media platforms, which could be the first consideration for UAE based e-commerce businesses. It is also important to note the overwhelming absence of UAE-based e-commerce businesses on YouTube, which is not only a social media platform but also the third most widely used search engine in the US [27] and among the top search engines globally. Therefore, there is a big opportunity to establish the presence of YouTube, which would already put a UAE e-commerce business ahead of ~78% of its peers.

Table 5.1 also provides direct metrics that businesses can use to see where they rank with respect to establishing their presence on social media. For example, similar to the discussion on YouTube, establishing an active Instagram account alone would put a UAE e-commerce business ahead of approximately half of the 10,000 peers. Similarly, establishing an active TikTok account would place business ahead of approximately 93% of peers.

Data Preparation for DSS

The "Discretize" widget in the Orange software [28] was used with "equal frequency" setting with 10 intervals (bins) to discretize the continuous attributes into categorical data to enable further analysis based on discrete percentile values. This setting splits the attribute into a given number of intervals (10 in our case), with each interval having approximately the same number of data points (instances). While one might discretize the numerical attributes into fuzzy ordinal values categories like "Low," "Medium," and "High," in our case, we discretized the data into ranges (bins) which included 10 equal frequency intervals for each social media platforms to be able to analyze and map each e-commerce platform under a certain range/basket out of the 10 ranges created by discretization.

The discretized data were analyzed to uncover patterns or insights related to social media followers and number of posts. Specifically, through MS Excel, a cross-tabular analysis was conducted for the number of followers of every social media platform available in the dataset, reporting the number of websites that were included under each of the 10 resulting ranges, as shown in Fig. 5.2. Then, a rank (Peer Group) was set for each of the ranges, reporting the first rank to the range with the largest numerical values and the lowest rank to the range with the smallest numerical values. This indicates that the more followers a business has, the better its rank will be.

The proportions of the ranges are as follows: percentiles, and peer groups (Fig. 5.2), and the calculation of followers per post (Table 5.2), a spreadsheet model was built in MS Excel for each metric using MS Excel function "VLookUp" to look up the positions of websites and where they stand currently in the market. The model also reports the number of followers a website is away from reaching a certain desired rank and how many posts are required to do so, assuming constant follower returns per post. This benchmarking and recommendation model can be used with input data about the number of followers a website has across different platforms, to see where they stand regarding rank and which range they fall into. Moreover, the user can enter the targeted rank, and the model can display the number of followers needed to reach that rank, counting on the results of the analysis, and recommend an approximate number of posts needed to reach that number.

Row Label	Count of instagram_followers
<131	502
≥49306	500
1047.5 - 1952	499
11134 - 1958	499
131 - 462.5	498
1952 - 3482	500
19588 - 4930	500
3482 - 6408.5	500
462.5 - 1047.	500
6408.5 - 1113	499
(blank)	
Grand Total	4997

instagram_followers_Range				
From	To		Percentile	PeerGroup
0	131	502	100	10
131	463	498	90	9
463	1048	500	80	8
1048	1952	499	70	7
1952	3482	500	60	6
3482	6409	500	50	5
6409	11134	499	40	4
11134	19588	499	30	3
19588	49306	500	20	2
49306		500	10	1

Fig. 5.2 The benchmarking decision support system (DSS) developed in the study

Table 5.2 Followers per post analysis results

	Instagram followers per post	Twitter followers per post	Pinterest followers per post
Median	10.5976	0.4454	0.2059
Average	170.7826	5.2331	6.9042

Furthermore, another analysis was conducted utilizing the number of followers and posts on a specific social media platform to calculate the number of posts a business would be required to post, gain a certain number of followers, or reach a certain peer group. The analysis included dividing the number of followers by the number of posts per business and then calculating the median for each platform, as presented in Table 5.2. The results indicated a median of 10.5976 followers per post on Instagram in the UAE, a median of 0.4454 followers per post on Twitter and a median of 0.2059 followers per post on Pinterest. The assumption for this analysis is that the return on posts, in terms of additional followers, is constant, indicating that each post will bring an equal number of additional followers.

The analysis enabled us to propose an approximate number of followers and posts that a business would need to achieve a desired rank. This was done by subtracting the number of followers that represent a certain rank from the total number of current followers and then dividing that number by the ratio and median acquired from our analysis. This allows for the determination of the number of posts required to reach the target.

Finally, all these models were integrated under a single decision support system (DSS), as displayed in Fig. 5.3.

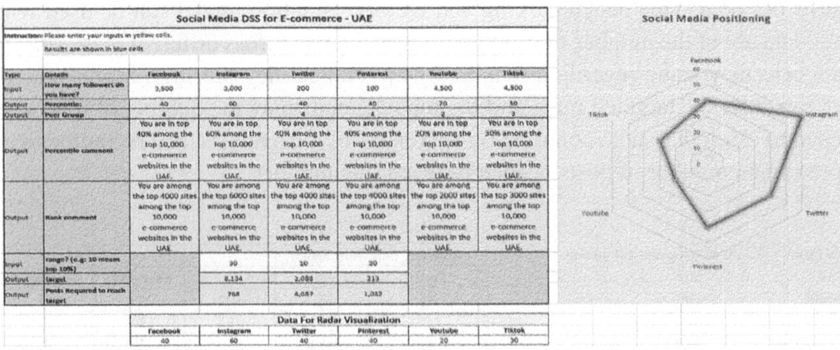

Fig. 5.3 The benchmarking decision support system (DSS) developed in the study [29]

Decision Support System (DSS)

The simple yet potentially highly useful DSS, based on a spreadsheet model, is shown in Fig. 5.3. The model is also shared as an online sheet using MS Office 365 [29] and in the supplement package as an MS Excel file [30]. Any e-commerce business can use this model to benchmark and see the current position of its social media presence across platforms. In addition, a radar chart visualizes the position of the business on each social media platform using the percentile for further clarity. The radar chart would show if a certain data point was positioned towards the outer corners, meaning that the specific platform scored lower than the rest. The scale increased from 10 to 100 when moving away from the center of the chart. The chart suggests a better rank position on a certain platform when it is closer to the center.

The created DSS was tested by the research team to validate whether all formulas were working correctly by providing sample ranges of inputs for confirmation. A few errors in the formulas identified during testing were corrected before deployment. Moreover, further enhancements were made to improve the model interface and design, including clearer instructions to direct the users while utilizing the DSS and having the model scheme sharper and concise, such as deleting unnecessary columns, resizing them, and providing succinct instructions.

When using the model, the percentile and rank values represent the current position on social media; hence, the lower the percentile or rank, the better the business is positioned, and the greater the exposure it has on social media. For example, if a business was ranked in peer group 1, it means that it was within the top 10% (in the top 1000 out of the top 10,000) e-commerce platforms in the UAE that are present on that social media platform, while the lower the rank, the smaller the exposure and presence on social media. Furthermore, in the radar chart, if a certain company is positioned towards the outer corners, the specific platform scored lower than the rest as the scale increased from 10 to 100 when moving away from the center of the chart. In contrast, being closer to the center would suggest a better rank/position.

The developed model, data analyzed, DSS, and results have been readily shared openly online [30] to benefit businesses by enabling them to benchmark their presence and performance across social media platforms and help them take action to reach the targets they aim.

Future Work

A step still needed to complete the study is to conduct a usability study with potential users of the DSS. These users would primarily be the founders and managers of e-commerce businesses in the UAE. Such a study is necessary to test the applicability, ease of use, and potential benefits of DSS. Practitioners suggest that testing should be a high priority [31] and that even a small user group of participants, such as five, can help identify a large majority of possible problems [32]. Because of project time constraints, as well as the need to complete a rigorous process to obtain ethics approval, the usability study was not carried out within the scope of the present study.

Conclusion

In summary, this study introduced a comprehensive approach for evaluating the social media effectiveness of e-commerce platforms, specifically within the context of the UAE market [33]. This study utilizes data from the top 10,000 e-commerce websites in the United Arab Emirates (UAE) to create an open online decision support system (DSS) that relies on visual analytics. This tool enables e-commerce companies to evaluate and benchmark their social media performance and to gain valuable insights into the number of posts necessary to attain specific market percentiles. Consequently, DSS can empower businesses to make well-informed decisions aimed at improving their digital marketing and growth strategies through benchmarking [34]. Given the growing influence of social media platforms, e-commerce businesses are equipped with practical resources to effectively navigate the dynamics of digital landscapes would be at advantage compared to peers.

Acknowledgements The authors thank Mr. Sumit Ghughorwol, founder of My Lead Fox, for granting the research team the rights to analyze the data obtained from My Lead Fox, publish results as a research paper, and publish the filtered data openly online as public data.

References

1. Semerádová, T., & Weinlich, P. (2022). The broad and narrow definition of E-commerce. In T. Semerádová, P. Weinlich (Eds.), *Achieving business competitiveness in a digital environment*. Contributions to management science. Springer. https://doi.org/10.1007/978-3-030-93131-5_1
2. Baluch, A., Main, K. (2023). *38 E-commerce statistics of 2023*. Retrieved December 26, 2023, from https://www.forbes.com/advisor/business/ecommerce-statistics/
3. Virto Commerce. (2023). *eCommerce in UAE & Dubai: Statistics, trends, platforms*. Retrieved November 4, 2023, from https://virtocommerce.com/blog/ecommerce-in-uae
4. Chauhan, K., & Abbasi, F. (2021). Social media and e-commerce: A theoretical study of factors affecting consumer behavior of social media buyers. *Global Journal of Management and Business Research,* 71–76. https://doi.org/10.34257/gjmbrevol21is3pg71
5. Putri, N. N., Sadewo, T. S. P., & Ananda, A. S. (2022). The impact of endorser on social media engagement and consumers' purchase intention: Case of Indonesian footwear brands. In *Proceedings of the international conference on industrial engineering and operations management*. https://ieomsociety.org/proceedings/2022istanbul/331.pdf
6. Brooker, P., Barnett, J., & Cribbin, T. (2016). Doing social media analytics. *Big Data & Society, 3*(2), 2053951716658060. https://doi.org/10.1177/2053951716658060
7. Pre, D., Tolentino, J., Varquez, K., & Etrata, A. (2022). The effects of social media tools on online retail businesses in the consumer electronics industry. *Millennium Journal of Humanities and Social Sciences,* 68–87. https://doi.org/10.47340/mjhss.v3i2.7.2022

8. Wei, C., Asian, S., Ertek, G., & Hu, Z. H. (2020). Location-based pricing and channel selection in a supply chain: A case study from the food retail industry. *Annals of Operations Re-search, 291*, 959–984.
9. Friedrich, T. (2016). On the factors influencing consumers' adoption of social commerce – a review of the empirical literature. *Pacific Asia Journal of the Association for Infor-mation Systems,* 1–32. https://doi.org/10.17705/1pais.08401
10. Akman, I., & Mishra, A. (2017). Factors influencing consumer intention in social commerce adoption. *Information Technology & People, 30*(2), 356–370. https://doi.org/10.1108/itp-01-2016-0006
11. Tseng, H., Shanmugam, M., Magalingam, P., Shahbazi, S., & Featherman, M. (2022). Managing enterprise social media to develop consumer trust. *British Food Journal, 124*(12), 4626–4643. https://doi.org/10.1108/bfj-11-2020-0995
12. Yuan, S., & Lou, C. (2020). How social media influencers foster relationships with followers: The roles of source credibility and fairness in parasocial relationship and product interest. *Journal of Interactive Advertising, 20*(2), 133–147. https://doi.org/10.1080/15252019.2020.1769514
13. Carlson, A., & Lee, C. C. (2015). Followership and social media marketing. *Academy of Marketing Studies Journal, 19*(1), 80–101.
14. Church, E., Zhao, X., & Iyer, L. (2021). Media-generating activities and follower growth within social networks. *Journal of Computer Information Systems, 61*(6), 551–560. https://doi.org/10.1080/08874417.2020.1824597
15. Fan, W., & Gordon, M. D. (2014). The power of social media analytics. *Communications of the ACM, 57*(6), 74–81. https://doi.org/10.1145/2602574
16. Abou-Kassem, T., Alazeezi, F. H. O., & Ertek, G. (2023). A data analytics methodology for benchmarking of sentiment scoring algorithms in the analysis of customer reviews. In X. S. Yang, R. S. Sherratt, N. Dey, & A. Joshi, A. (Eds.), *Proceedings of eighth international congress on information and communication technology. ICICT 2023. Lecture notes in networks and systems* (vol 693). Springer. https://doi.org/10.1007/978-981-99-3243-6_46
17. Noori, B., & Hossein Salimi, M. (2005). A decision-support system for business-to-business mar-keting. *Journal of Business & Industrial Marketing, 20*(4/5), 226–236. https://doi.org/10.1108/08858620510603909
18. Saura, J. R. (2021). Using data sciences in digital marketing: Framework, methods, and performance metrics. *Journal of Innovation & Knowledge, 6*(2), 92–102. https://doi.org/10.1016/j.jik.2020.08.001
19. Lei, N., & Moon, S. K. (2015). A decision support system for market-driven product positioning and design. *Decision Support Systems, 69*, 82–91. https://doi.org/10.1016/j.dss.2014.11.010
20. Lilien, G. L., Rangaswamy, A., Van Bruggen, G. H., & Starke, K. (2004). DSS effectiveness in marketing resource allocation decisions: Reality versus perception. *Information Systems Research, 15*(3), 216–235. https://doi.org/10.1287/isre.1040.0026
21. Parveen, F., Jaafar, N., & Sulaiman, A. (2015). Social media usage and organizational performance: Reflections of Malaysian social media managers. *Telematics and Informatics, 32*(1), 67–78. https://doi.org/10.1016/j.tele.2014.03.001
22. Paniagua, J., & Sapena, J. (2014). Business performance and social media: Love or hate? *Business Horizons, 57*(6), 719–728. https://doi.org/10.1016/j.bushor.2014.07.005
23. My Lead Fox. Retrieved December 25, 2023, from https://myleadfox.com
24. Britannica, X. (2023). Retrieved December 26, 2023, from https://www.britannica.com/topic/Twitter
25. Orange Data Mining. Retrieved December 26, 2023, from https://orangedatamining.com
26. Demsar, J., Curk, T., Erjavec, A., Gorup, C., Hocevar, T., Milutinovic, M., Mozina, M., Polajnar, M., Toplak, M., Staric, A., Stajdohar, M., Umek, L., Zagar, L., Zbontar, J., Zitnik, M., & Zupan, B. (2013). Orange: Data mining toolbox in python. *Journal of Machine Learning Research,* 2349−2353. Retrieved December 26, 2023, from http://jmlr.org/papers/volume14/demsar13a/demsar13a.pdf

27. Tubics. (2023). YouTube 2nd biggest search engine–The myth that just won't die. Retrieved December 26, 2023, from https://www.tubics.com/blog/youtube-2nd-biggest-search-engine
28. Orange Widget Catalog. Discretize. Retrieved December 26, 2023, from https://orangedatamining.com/widget-catalog/transform/discretize/
29. Social Media DSS for E-commerce–UAE.Retrieved December 26, 2023, from https://qr.link/PSzl67
30. AlBaroudi, R., Badran, M., Awad, & Ertek, G. (2024). Supplement to "Benchmarking of social media performance of E-commerce websites: A case study of the UAE e-commerce market. Retrieved December 26, 2023, from https://ertekprojects.com/ftp/supp/21.zip
31. Praxent. (2023). Why do usability testing? 4 reasons to make it a priority. Retrieved December 26, 2023, from https://praxent.com/blog/usability-testing-4-reasons-top-priority
32. Sauro, J. Why you only need to test with five users (explained). Retrieved December 26, 2023, from https://measuringu.com/five-users/
33. Supra, J. D. (2023, November 07). *UAE regulations for E-commerce businesses: A comprehensive overview*. Retrieved December 26, 2023, from https://www.jdsupra.com/legalnews/uae-regulations-for-e-commerce-6512875/
34. Anand, G., & Kodali, R. (2008). Benchmarking the benchmarking models. *Benchmarking: An International Journal, 15*(3), 257–291. https://doi.org/10.1108/14635770810876593

Chapter 6
A Graph Analytics Methodology for Analyzing Startup Ecosystems

Ali AlKatheeri, Saif Abdulmajeed, Abdulla Albedwawi, Abdulalrheem AlSheebany, Samir Safi, and Gurdal Ertek

Abstract Startup companies are engines of economic growth and diversity. A unicorn company is a privately held start-up company that has grown to a valuation of over $1 billion. As of January 2023, there were over 1,200 unicorn companies worldwide. In this study, we present a custom-developed graph analytics methodology for the systematic analysis of unicorn and startup ecosystems. We show the applicability of and insights obtained by the methodology through a case study using a public dataset on unicorn companies. Graph analytics through algorithmically generated visualizations can enable the derivation and understanding of meaningful insights, even by non-technical analysts and decision-makers in entrepreneurship ecosystems and government organizations.

Keywords Graph analytics · Unicorn companies · Startup ecosystems · Technology management · Entrepreneurship · Innovation management

---

Supplementary Information The online version contains supplementary material available at (https://doi.org/10.1007/978-3-031-61589-4_6).

A. AlKatheeri · S. Abdulmajeed · A. Albedwawi · A. AlSheebany · S. Safi · G. Ertek (✉)
College of Business and Economics, United Arab Emirates University, Al Ain, UAE
e-mail: gurdal@uaeu.ac.ae

A. AlKatheeri
e-mail: 201934081@uaeu.ac.ae

S. Abdulmajeed
e-mail: 201909690@uaeu.ac.ae

A. Albedwawi
e-mail: 201915196@uaeu.ac.ae

A. AlSheebany
e-mail: 201902385@uaeu.ac.ae

S. Safi
e-mail: ssafi@uaeu.ac.ae

© The Author(s), under exclusive license to Springer Nature Switzerland AG 2024
A. Emrouznejad et al. (eds.), *Business Analytics and Decision Making in Practice*, Lecture Notes in Operations Research, https://doi.org/10.1007/978-3-031-61589-4_6

Introduction

As first defined by Aileen Lee in a 2013 TechCrunch article, a unicorn company is a startup that has exceeded a valuation of $1 billion [1]. Unicorns are important for individuals, investors, and nations, and they can bring disproportionate economic value to their founders and investors because of their high valuations and rapid growth. Unicorn companies can also bring economic, technological, and social benefits to the country where they are founded by creating new markets or expanding existing ones, contributing to GDP growth, fostering innovation, advancing technology, creating jobs, and reducing unemployment [2].

Therefore, it is important to analyze the characteristics, valuations, and growth of unicorn companies with the goal of facilitating the creation of startup ecosystems at the micro and macro levels, which can foster unicorns.

While there are multiple research studies that analyze startup ecosystems and unicorn companies [3, 4], only two of these studies have applied graph analytics extensively [5, 6]. However, some of the graph visualization techniques (layout algorithms) that we applied in our study have not been used in [5, 6], or, to the best of our knowledge, any other study. Furthermore, our study aims to develop a systematic methodology focused specifically on graph analytics to analyze the characteristics of unicorn companies. Such a graph-based approach can provide insights that can alternatively be obtained through other methods. However, it can also enable the derivation of novel types of insights, like other research in business and management literature [7] that uses graph analytics to develop business insights. Owing to space limitations, background on graph analytics, graph analytics tools, and graph layout/visualization algorithms are provided in Appendix A of the Supplement [8], which accompanies and complements this paper. Furthermore, a review of the literature on startup and unicorn success, visual analytics, and graph analytics applications for analyzing startups and unicorns is also provided in Appendix A of the Supplement [8].

The methodology is illustrated using a global dataset of unicorn companies, with data on their valuations, country of origin, industry/sector, and valuations [9]. Through the case study, we illustrate how graph visualizations can enable the derivation and understanding of meaningful insights, even by non-technical analysts and decision makers in entrepreneurship ecosystems and government organizations. Enabling and democratizing analytics for nontechnical users is an advantage of graph visualization and analytics over machine learning and other algorithmic approaches.

We acquired an older dataset of 658 unicorn companies around the globe in tabular format, transformed it into a graph dataset using graph analytics techniques and software, and then analyzed the results to extract novel insights into unicorns and their characteristics.

The "Analytic Ascendancy Model" by Gartner [10–12] identifies four levels of data analytics maturity; namely descriptive, diagnostic, predictive, and prescriptive [12]. The primary objectives of the analytics techniques corresponding to these four maturity levels are gaining insights into the data by discovering hidden patterns

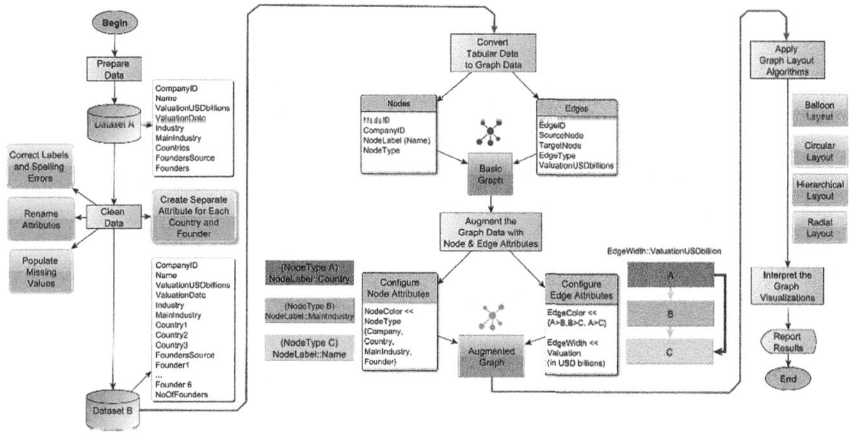

Fig. 6.1 Methodology developed for and applied in the presented study

(descriptive), diagnosing root causes (diagnostic), predicting future outcomes (predictive), and prescribing optimal decisions (prescriptive). Our research scope is descriptive analytics, which is the earliest yet inevitable stage of analytic maturity. The developed methodology aims to enable and support informed investment and strategy decisions for startup ecosystems and startups across industries and sectors. The developed methodology can be applied as is or adopted/extended to the analysis of datasets in selected/multiple industries, sectors, and countries.

The paper is structured as follows: Sect. "Introduction" readily introduced the concept of unicorn companies, why they are important, and the scope and contributions of the study. Appendix A in the supplement [8] provides background on graph analytics and summarizes earlier work in the startup and unicorn literatures. Section "Analysis and Results" in the paper describes the developed methodology and its application using a case study dataset. Section "Conclusion" presents the analysis process as well as the insights and findings gained from the analysis are provided.

Analysis and Results

Four different layouts, namely balloon, circular, hierarchical, and radial, were generated in the case study using yEd software, as displayed in Fig. 6.2. The source *.graphml files for yEd and the parameter settings for all four layouts, as well the *.pdf outputs of the resulting visualizations for easy browsing, are shared in the Supplement package [8]. While many different types of insights are possible, only

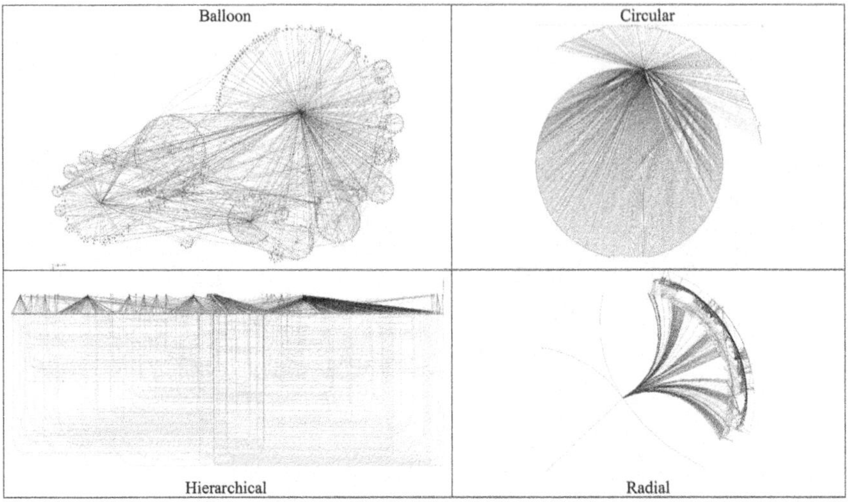

Fig. 6.2 Visualization of the data using different graph layout algorithms

some of the sample types of insights that were identified in the study and only those obtained from balloon layout are as follows[1]:

- *Insight Type A* refers to the identification of companies and/or industries that belong to a specific country or very few countries. This includes the identification of countries with a particular specialization, with all or most of their unicorns starting in a specific industry.
- *Insight Type B* identifies countries that have the largest number of unicorn companies.
- *Insight Type C* is the identification of countries and industries that are proximate to each other in the graph, revealing implicit associations between them.
- *Insight Type D* is the identification of companies that belong to multiple countries.
- *Insight Type E* identifies companies with high valuations within a country or industry.

While the same insights can be obtained through the analysis of tabular data, especially through cross-tabular analysis [13], the generated visualizations were found by the research team to be more visible and effective in communicating the message. However, further research involving a formal usability study is required, as is standard practice in information visualization research [14–16]. Thus, the usability of the generated visualizations and the validity and usefulness of the presented research can be theoretically established.

[1] The files for balloon layout in the Supplement are slightly different than the original layout, whose screenshots are provided in the paper.

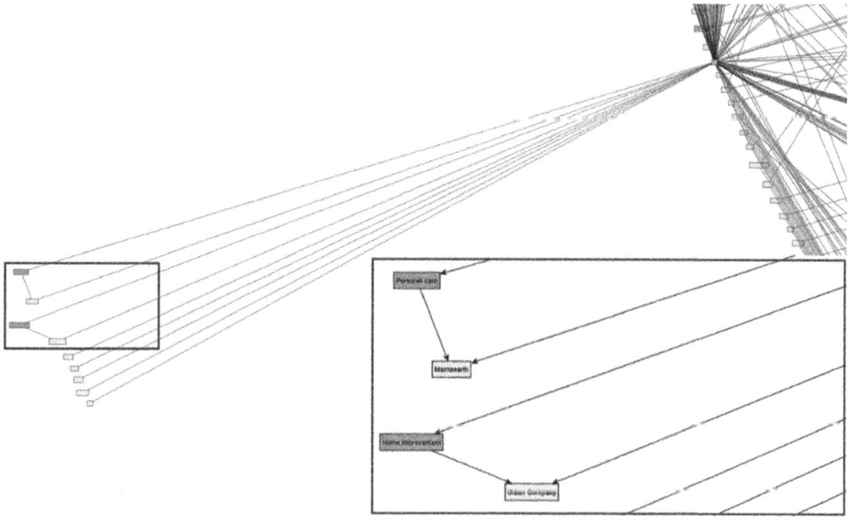

Fig. 6.3 Example of Insight Type A, showing which countries could be best location to grow for certain industries

Insight Type A

Figure 6.3 displays an example of Insight Type A, in which specific industries belong to a specific country. In Fig. 6.2, we have the following two unicorn companies from two industries, both founded in India.

- Mamaearth from Personal care industry
- Urban company from Home improvement industry

Both industries have only one company each, and both industries grew unicorns only in India. This shows the possibility of having more successful companies in India, before any other country, in these two consumer-oriented industries, possibly due to the large population and the corresponding large consumer market in India. While this insight can also be obtained through cross-tabular analysis, for example, using MS Excel's pivot table analysis, the graph visualization can help in detecting this relation more easily and in grasping its meaning. While the rest of the graph (upper right area of the figure) is much denser and more connected, these four nodes stand out, and visually suggest that there is a notable outlier pattern [17].

Insight Types B and C

In Fig. 6.4, the top three countries have unicorn companies with respect to degree (total number of arcs from/to that node).

1. US.
2. China.
3. India.

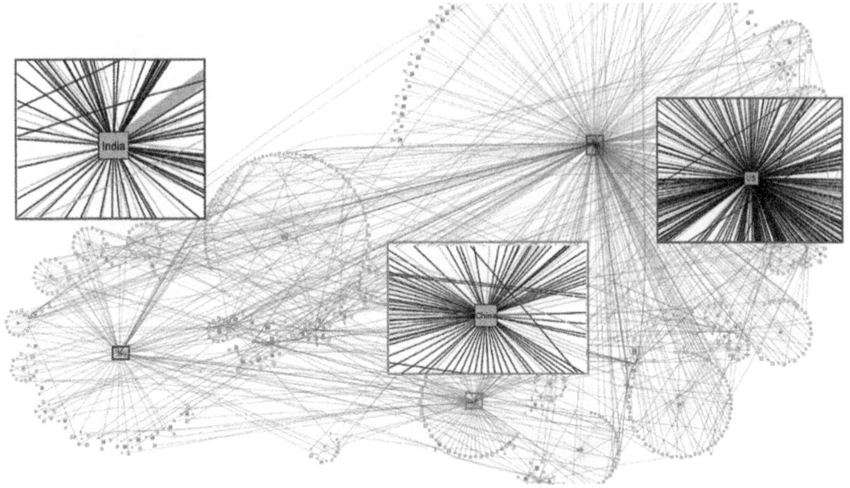

Fig. 6.4 Example for Insight Types B and C

Furthermore, China is closer to the US than India on graph visualization, suggesting that the industries and companies grown outside of the US and China exhibit more similarities than those from the US and India.

Top countries (Insight Type B) can also be obtained through cross-tabular analysis by counting the number of unicorn companies in each country and sorting them. However, graph visualization can help in detecting this relationship more easily and in grasping its meaning; the top three countries are startup hubs, as visually presented. The circles around the countries display the spheres of influence.

The pattern regarding the similarity between the US and China (Insight Type C) can also be revealed through analysis of tabular data, yet the graph visualization enables the detection of this pattern without any further analysis steps, readable directly from the constructed graph visualization.

Insight Type E

Figure 6.5 shows that most video game companies are from the US, and the highest valuation is for Epic Games. This is an example of sight Type E. The visualization in Fig. 6.4 is obtained when the analyst zooms in different areas of the graph visualization, which could be considerably easier than browsing through many rows of tabular data/information. Node colors and arc flows can also help detect patterns. Furthermore, to a degree, graph visualization materializes abstract concepts of country, industry, and company into solid elements that can be visually grasped. For example, it can be more effective, striking, and more intuitive and illustrative to see the high valuation in the video games industry literally flowing into Epic Games.

Further insights obtained for the above five insight types can be found in Appendix D of Supplement [8].

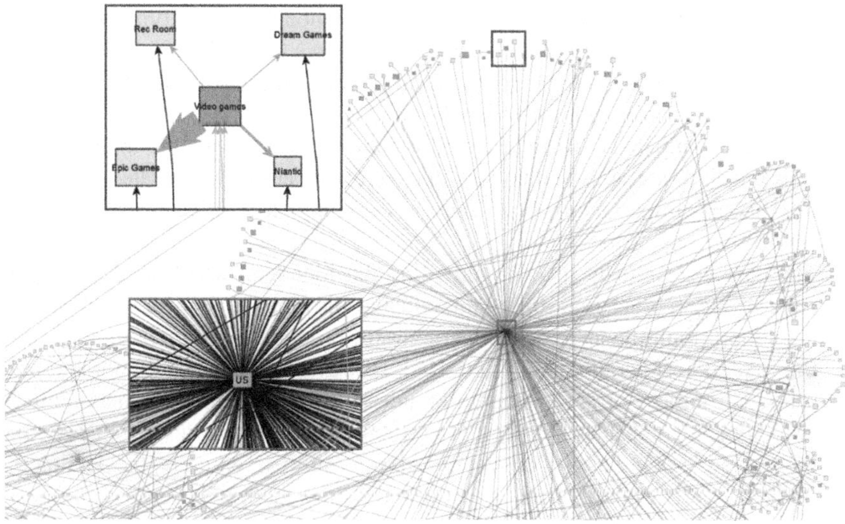

Fig. 6.5 Example analysis for Insight Type E

As mentioned earlier, as a data contribution of the presented research, all the data files, as well as the graph files (Fig. 6.2) and their pdf outputs, have been compiled into a single supplement file, which is readily publicly shared as open data and open analytics [8].

Conclusion

In this study, a custom-developed data analytics methodology was developed and applied for the graph-based analysis of unicorn companies and startup ecosystems. The applicability and practical contributions of the methodology are illustrated through the analysis of a public dataset and discovery of domain insights.

A unique aspect of the developed methodology is that it is based on graph analytics techniques, especially different types of graph layouts. To conduct graph analytics, tabular data must be represented as graph data by mapping properties in the yEd graph analytics software, applying different graph layouts, and analyzing the results. The detailed steps of the methodology and the settings and parameters for the graph layout algorithms are provided in Appendices B and C of Supplement [8].

There are some threats to the validity of the research, as well as possible research avenues that can be addressed in the future, as described below.

1. The dataset used in this study was obtained from Kaggle [9], but its validity and data collection method were not described by the user who posted it. More complete datasets have recently been found under Kaggle with up-to-date list of unicorns, which would be much more valid to analyze. Yet, the focus of our

current research was the development of a methodology and types of insights rather than the accuracy of the insights. Hence, we used a less complete dataset for this case study. Still, there is a need to select the most complete data and validate the source data in future research.

2. For some companies, the industry information was not available in the dataset. This resulted in graphs and layouts that were not perfectly accurate with biased insights. It is important to obtain and analyze complete verified datasets in future studies.
3. The data were mostly missing regarding the founders, which made it impossible to incorporate important and valuable information into the analysis. As a solution, multiple successive queries were prompted in the chat interface of the ChatGPT generative AI system [18]. However, the ChatGPT accepted only a limited number of companies to be queried in each prompt (a maximum of 25 companies in each prompt). Furthermore, the ChatGPT results still need to be verified, making the process/approach infeasible. Eventually, the data on the founders were excluded from the case study; thus, the missing data on founders did not affect our analysis results. However, adding nodes to founders in future research could reveal significant new insights.
4. While the same insights can be obtained through the analysis of tabular data, the visualizations were found to be quite visible and effective in communicating the message to the research team. Furthermore, some of the insights revealed in graph visualizations may require extensive computation if they are achieved through methods that analyze tabular data. In future research, it is important and necessary to conduct usability studies with many users across different backgrounds to test whether graph visualizations are better at understanding and communicating the obtained insights.
5. There are many other possible insight types [7] that can be derived from the readily available four visualizations, as well as many more that could be generated even by yEd software alone. This is topic for future research.

In conclusion, this study presents a novel formal methodology for graph-based analysis of unicorn companies. The methodology enables the derivation of industry and country insights, even by non-technical professionals, as well as effectively and intuitively visualizing insights that can be derived using business intelligence/ analytics tools and methods.

References

1. Lee, A. (2013, November 2). *Welcome to the unicorn club: Learning from billion-dollar startups*. TechCrunch. https://techcrunch.com/2013/11/02/welcome-to-the-unicorn-club/
2. Kenney, M., Zysman, J., & Bearson, N. (2021). Unicorns, Cheshire cats, and the new dilemmas of entrepreneurial finance. *Venture Capital, 23*(1), 73–94. https://doi.org/10.1080/13691066.2020.1783914

3. Jinzhi, Z., & Carrick, J. (2019). The rise of the Chinese unicorn: An exploratory study of unicorn companies in China. *Emerging Markets Finance and Trade, 55*(15), 3371–3385. https://doi.org/10.1080/1540496x.2019.1610877
4. Malyy, M., Tekic, Z., & Podladchikova, T. (2021). The value of big data for analyzing growth dynamics of technology based New Ventures. arXiv.org. https://arxiv.org/abs/2104.03053
5. Bao, Y. (2019). *Data visualization of Chinese internet companies* [Ph.D. Thesis]. Politecnico di Milano. POLITesi–Archivio digitale delle tesi di laurea e di dottorato. Retrieved December 27, 2023, from https://www.politesi.polimi.it/handle/10589/149332
6. Van-Meeteren, M., Tzameret, H. R., & Watson, J. (2021). *The role of start-ups in the insurance knowledge space*. Loughborough University. Report. https://doi.org/10.17028/rd.lboro.14485278.v1. Retrieved December 27, 2023, from https://repository.lboro.ac.uk/articles/report/The_role_of_start-ups_in_the_insurance_knowledge_space/14485278
7. Ertek, G., Tokdemir, G., Sevinç, M., & Tunç, M. M. (2017). New knowledge in strategic management through visually mining semantic networks. *Information Systems Frontiers, 19*, 165–185. https://doi.org/10.1007/s10796-015-9591-0
8. AlKatheeri, A., Abdulmajeed, S. A., Albedwawi, A., AlSheebany, A., Safi, S., & Ertek, G. (2024). *Supplement to "A graph analytics methodology for analyzing startup ecosystems"*. https://ertekprojects.com/ftp/supp/19.zip
9. DeepContractor. *Unicorn companies dataset*. Kaggle. Retrieved December 27, 2023, from https://www.kaggle.com/datasets/deepcontractor/unicorn-companies-dataset
10. Gartner. *What is data and analytics?* Retrieved December 12, 2023, from https://www.gartner.com/en/topics/data-and-analytics
11. O'Leary, D. E. (2009). The impact of Gartner's maturity curve, adoption curve, strategic technologies on information systems research, with applications to artificial intelligence, ERP, BPM, and RFID. *Journal of Emerging Technologies in Accounting, 6*(1), 45–66. https://doi.org/10.2308/jeta.2009.6.1.45
12. Eriksson, T., Bigi, A., & Bonera, M. (2020). Think with me, or think for me? On the future role of artificial intelligence in marketing strategy formulation. *The TQM Journal, 32*(4), 795–814. https://doi.org/10.1108/TQM-12-2019-0303
13. Abdallah, S., Malik, M., & Ertek, G. (2017). A data mining methodology for the analysis of patient arrivals into healthcare centers. In *Proceedings of the 2017 international conference on information technology–ICIT 2017* (pp. 52–61). Singapore, Singapore. https://doi.org/10.1145/3176653.3176740
14. Burch, M., Huang, W., Wakefield, M., Purchase, H. C., Weiskopf, D., & Hua, J. (2020). The state of the art in empirical user evaluation of graph visualizations. *IEEE Access, 9*, 4173–4198.
15. Ventocilla, E., & Riveiro, M. (2020). A comparative user study of visualization techniques for cluster analysis of multidimensional data sets. *Information Visualization, 19*(4), 318–338. https://doi.org/10.1177/1473871620922166
16. Koylu, C., & Guo, D. (2017). Design and evaluation of line symbolizations for origin–destination flow maps. *Information Visualization, 16*(4), 309–331. https://doi.org/10.1177/1473871616681375
17. Cao, N., Lin, Y.-R., Gotz, D., & Du, F. (2018). Z-Glyph: Visualizing outliers in multivariate da-ta. *Information Visualization, 17*(1), 22–40. https://doi.org/10.1177/1473871616686635
18. ChatGPT. Retrieved December 27, 2023, from https://chat.openai.com

Chapter 7
A Visual Analytics Methodology for the "Life Under Water" (SDG14) Sustainable Development Goal (SDG)

Malak AlOmari, Dania Awwad, Hind ElHassan, Rawan Suleiman, Kursad Asdemir, and Gurdal Ertek

Abstract The Ocean Health Index (OHI) is a framework that measures the health of the world's oceans with 10 goals of societal, ecological, and economic importance. The OHI framework is one of the most well-grounded and reliable approaches to the "Life Under Water" (SDG14) Sustainable Development Goal of the United Nations (UN). In this study, data from the OHI were analyzed using a customized visual analytics methodology to reveal the current state of these goals for the Gulf Cooperation Council (GCC) countries, and to investigate the temporal trends of each goal. The methodology was implemented as a collection of public online interactive visual analytics dashboards. The methodology utilizes root cause analysis to delve deeper into the underlying SDG14 subgoals which are the main causes of temporal trends, in particular, the decline in some of the goal indicators. This methodology can be used for benchmarking, as well as strategy and policy development, to aid the sustainable use of marine resources.

Supplementary Information The online version contains supplementary material available at (https://doi.org/10.1007/978-3-031-61589-4_7).

M. AlOmari · D. Awwad · H. ElHassan · R. Suleiman · K. Asdemir · G. Ertek (✉)
United Arab Emirates University, Al Ain, UAE
e-mail: gurdal@uaeu.ac.ae

M. AlOmari
e-mail: 700036897@uaeu.ac.ae

D. Awwad
e-mail: 700036873@uaeu.ac.ae

H. ElHassan
e-mail: 700035593@uaeu.ac.ae

R. Suleiman
e-mail: 700037141@uaeu.ac.ae

K. Asdemir
e-mail: kursad.a@uaeu.ac.ae

© The Author(s), under exclusive license to Springer Nature Switzerland AG 2024

A. Emrouznejad et al. (eds.), *Business Analytics and Decision Making in Practice*, Lecture Notes in Operations Research, https://doi.org/10.1007/978-3-031-61589-4_7

Keywords Life under water · SDG14 · Sustainable development goals · SDG · Marine studies · Visual analytics · Dashboard

Introduction

Research Topic

"Life Under Water" (SDG14) [1], is among the Sustainable Development Goals (SDGs) set by the United Nations (UN) [2]. Sustainable Development Goals (SDG) call for governments and individuals around the world to act to end poverty and inequality, protect the planet, and ensure that all people sustainably enjoy health, justice, and prosperity. SDG14 is the SDG that relates to sustainable protection, management, and enhancement of the Earth's marine resources, including coasts, seas, and oceans [1]. Marine Life also encompasses plants, animals, and other organisms that live in the seas or oceans. A significant level of change is expected in marine resources in the upcoming decades, including a significant decrease in marine animal biomass [3], with potentially disruptive impacts on populations around the world. However, most nations have not been able to accomplish the set goals in SDG14. For example, for the SDG14 targets that expired in 2020 (14.2, 14.4, 14.5, and 14.6), no country on Earth was able to achieve all four targets. Furthermore, most countries scored considerably below the targets [4]. Therefore, there is a need for research to identify the underlying causes of failure to reach the SDG14 targets and to develop solutions in order to progress towards targets as planned. To this end, the application of data analytics, the techniques for the modeling and analysis of data [5, 6] can provide fresh insights that can guide plans and actions. One notable aspect of SDG14 is that it exhibits an insignificant (Pearson) correlation with the other SDGs [7], suggesting that its focused analysis, independent of the other SDGs, is adequate.

The Ocean Health Index (OHI)

The Ocean Health Index (OHI) is a comprehensive framework that measures the health of the world's water resources based on 10 goals related to human benefits and ecological and economic importance [8]. The 10 goals are shown in Fig. 7.1. The goals of biodiversity, (Coastal) Livelihoods and Economies, Food Provision, and Sense of Place are composed of subgoals, as outlined in the OHI web portal [9]. The 10 goals and the associated eight subgoals proposed by the OHI and listed with sample analysis [10] measure the conditions of a suite of benefits or goals that support human well-being and sustainable marine ecosystems. The framework has been applied for monitoring and assessment in various countries and regions across the world, as detailed in the Supplement [11]. The OHI goal and subgoal scores are

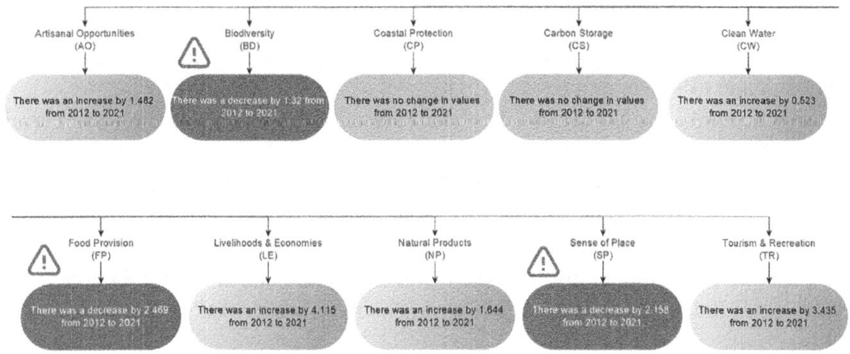

Fig. 7.1 The results of temporal analysis for the GCC

composite indicators, as they combine multiple sub-indicators through aggregation [9].

There are multiple other frameworks that address the monitoring and assessment of marine resources, with four additional frameworks listed in [12]. In this study, the OHI was selected for important reasons. Besides being well-grounded due to a high level of scientific rigor [8], the underlying methods of the OHI are exceptionally well-documented to fine granular detail [9], with the source code publicly available as a toolbox [13, 14]. Thus, the OHI is one of the best frameworks for use in SDG14 research.

Scope of the Study

The United Arab Emirates (UAE) is located on the Arabian Peninsula and is a member of the Gulf Cooperation Council (GCC) [15]. With 1381 km of natural coastlines and abundant marine resources [16], the UAE has a vital role and responsibility in preserving and protecting life underwater. In this paper, answers are sought to the questions regarding the performance of UAE and GCC countries with regard to Life Under Water (SDG14), as well as the best practices in the UAE that serve towards the achievement of SDG14.

Research Focus: GCC and UAE

To achieve SDG 14 through the preservation of healthy marine ecosystems, all countries must overcome several challenges. In this study, the Gulf Cooperation Council (GCC) countries, in particular the United Arab Emirates (UAE), are the focus of the analysis. Alsharhan et al. [17] provides a review of water governance in the UAE. The challenges of achieving SDG, including SDG14, also exist for the selected region and the UAE, which can be facilitated through research on the topic, a clear, systematic roadmap, and effective practices [17, 18].

UAE Federal Law No. 24 of 1999, also referred to as the "Protection and Development of the Environment" law, outright forbids the discharge of pollutants and wastes into aquatic environments [18]. There are multiple other laws in the UAE at the federal and local levels, as well as decrees and international conventions relating to aquatic resources [18]. The UAE authorities aim to develop, implement, monitor, and promote environmentally friendly sustainable practices to ensure the long-term health and vitality of its waters. These practices are described in the Supplement [11].

The OHI website readily includes a sample analysis of OHI data [10]. However, these results are static visualizations (and are not interactive), limited in scope, and do not yield easily interpretable, focused insights for the UAE or the GCC. Therefore, the initial motivation of the current research was to conduct an analysis of the OHI data to discover insights into the GCC through easily interpretable visualizations. However, at later stages of the research, the motivation and objective can be extended to designing interactive dashboards not only for GCC countries, but for any continent, geographic region, or economic union in the world.

Related Work

Kunčič [19] presents an analysis of Arab countries with respect to their similarities based on all SDG metrics. The authors employed the unsupervised machine learning technique of cluster analysis in conjunction with visual analytics to derive insights for the region. The results suggest great variations among the analyzed countries, and that SDG14 is one of the SDGs that has high variance, resulting in clusters of countries with better scores for some SDG14 goals yet with worse scores for other SDG14 goals.

In an earlier study, a research team that included the initial founders of the OHI framework [20] analyzed the OHI dataset as of 2013 to reveal the patterns and emerging trends in global ocean health. The analysis included map visualizations to display annual changes, tabular analysis, regression analysis, and scatter plot matrix. The same analysis, enriched with additional analyses such as flower plots superimposed on maps, and for the U.S. West Coast, can be found in [21]. Most of the analyses in [20] and [21], yet with the recent data from the year 2023 and for the globe, can be found on the OHI website [9]. Daigle et al. [22] conducted a similar analysis with the OHI data for Canada, enhanced with heatmaps and a novel goal, namely Aboriginal Needs. As an alternative to the OHI, [23] presents an integrated framework for coastal management that is based on the concept of a dashboard.

Sachs et al. [24] developed a dashboard, which is mainly a heatmap, not specifically for SDG14, but for all SDG goals. While containing intuitive and effective visualizations, none of the mentioned studies reported a dashboard with interactivity, which is the main contribution of the current research.

Multiple systemically designed dashboards have been reported in the literature relating to water resources and marine management, as detailed in the Supplement [11]. The only interactive dashboard application for marine resource monitoring that the authors of this study have encountered is the Marine Information System for Europe [25]. The web portal provides multiple interactive analytics dashboards, including country-specific interactive dashboards, such as the dashboard for Netherlands [26], all constructed using Tableau software, as in the current study. However, the data source for [25] is not the OHI, but rather a different repository [27] populated by each member country. The literature does not seem to report a functional interactive dashboard for measuring and monitoring ocean health within the context of the OHI framework.

Method

OHI Methodology

The data used in this study were collected from the Ocean Health Index. Hence, understanding the methodology of the OHI team is crucial. According to [28], the OHI is the most comprehensive dataset of ocean health and marine ecosystems. The underlying methodology, including all the formulas, computational methods, and sub-criteria, is well documented [9]. Furthermore, the OHI dataset is used in the assessment of SDG14 metrics by the UN, as documented in Table 2.1 (Page 53) of [29].

Research Scope

This study focuses on the Gulf Cooperation Council (GCC), which is an economic union of the Kingdom of Saudi Arabia (KSA), the United Arab Emirates (KSA), Qatar, Kuwait, Bahrain, and Oman [15]. In this study, data from the OHI were analyzed first for all GCC countries, and then for the United Arab Emirates (UAE) from 2012 to 2021, to uncover changes in the measures of each goal. The analysis provided insights into how the OHI goals in the GCC countries shifted in value over time and highlights areas for improvement to support the sustainable use of marine resources.

Based on the analysis of the OHI goals and subgoals for the GCC, the next stage of the study was to develop a set of visual analytics dashboards that enable the benchmarking of any combination of selected continents, geographical regions, economic unions, or customized sets of countries and/or regions with respect to the

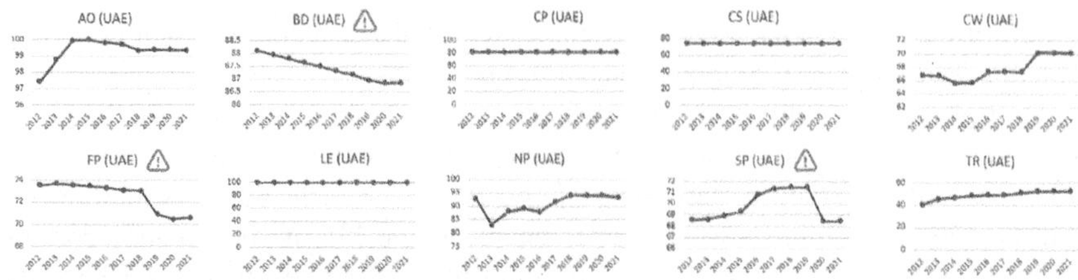

Fig. 7.2 The results of temporal analysis for the UAE

OHI goals and subgoals. The designed and developed dashboards are publicly made available to support stakeholders in SDG14 around the world in their benchmarking, analysis, and planning.

Analysis and Results

Temporal Analysis for the GCC

The first analysis was the overall investigation of how the goals changed from 2012 to 2021 for the entire GCC region. For this analysis, the average for all countries was computed for each goal and year combination. The results of this analysis are shown in Fig. 7.1. As shown in Fig. 7.1, the goals shown in red have experienced drops in value over the past ten years, which is the case for the goals of Biodiversity (BD), Sense of Place (SP) and Food Provision (FP). By contrast, the goals shown in blue indicate a rise in the values of these goals. and the goals shown in gray have experienced no change in values during the past 10 years.

Temporal Analysis for the UAE

The temporal analysis was carried out to track changes in the UAE's goals' values through the selected time range of 2012–2021, as displayed in Fig. 7.2. The problematic trends are marked with a red exclamation icon. Detailed interpretations of Fig. 7.2 are presented in the Supplement [11].

Root Cause Analysis for the UAE

Root cause analysis was conducted to identify the subgoals that could be the underlying cause of the decreasing trend in the values of the goals for the UAE. An earlier

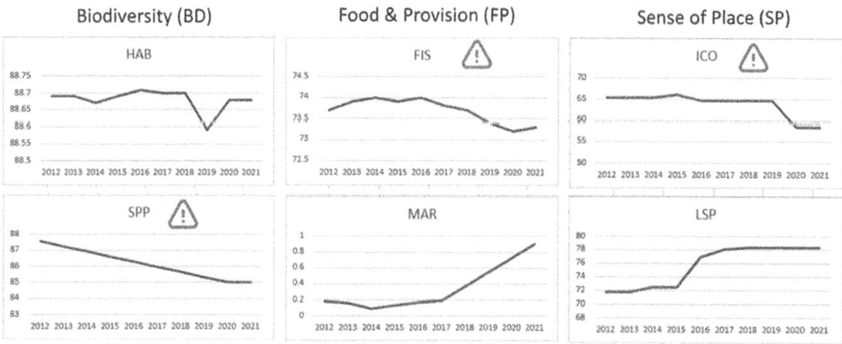

Fig. 7.3 The trend analysis of subgoals of BD, FP, and SP showing the results of root cause analysis for the UAE

analysis showed that The Biodiversity (BD), Food Provision (FP), and Sense of Place (SP) goals exhibited falling values over 2012–2021. Therefore, the root cause analysis focused on these three goals. The subgoals of the three goals were analyzed to identify the root cause of each decline, as shown in Fig. 7.3. As a result, Species Condition (SPP), Fisheries (FIS), and Iconic Species (ICO) subgoals were identified as the root causes of the decline in the score values for Biodiversity (BD), Food Provision (FP), and Sense of Place (SP) goals, respectively. The detailed results are provided in the Supplement [11].

Visual Analytics Dashboards

The analysis conducted earlier revealed the trends in goals and subgoals that are the underlying root causes of deteriorating goals. However, the analysis is limited to the GCC economic union and the countries within them. To extend the benefits to all countries and regions around the world, it is necessary to go at least one step further. To this end, a collection of visual analytics dashboards, Dashboards A-F, has been designed and implemented (Fig. 7.4). Furthermore, the source data for the dashboards were updated to include the values for the year 2022 [30]. The source data have also been augmented to include additional data attributes (fields) for continent, geographical regions, and economic unions. To this end, Help for GPT for Sheets™ and Docs™ extension for Google Sheets [31] has been used to map, through generative AI [32], the countries and regions in the original dataset to the values in the newly created/appended attributes.

The dashboards were developed using Tableau Public [33], an online visual analytics platform that is easy to use and learn [34]. Tableau Public is a free-to-use yet closed-source web service, yet the dashboards created in the free version must be public. This was not an issue, as one objective of the study was to make all dashboards available and accessible to the whole world.

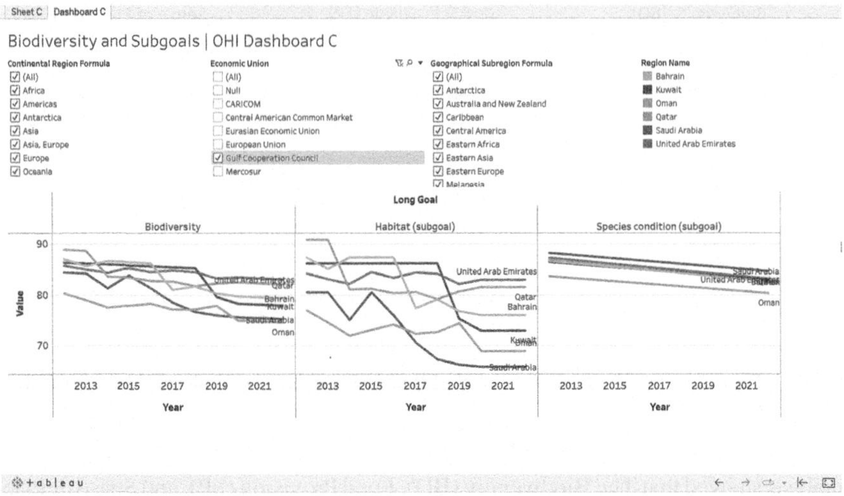

Fig. 7.4 Dashboard C (Biodiversity BD and Subgoals), filtered for Gulf Cooperation Council (GCC)

In this study, only the (Current) Status values were considered, and the data rows relating to Predicted Future Status and Goal Score were excluded [9]. This is because status reflects only the status of the current year, regardless of other dimensions. In the OHI methodology, as documented in Fig. 2.1 of [9], Resilience, Pressures, and Trends constitute the Predicted Future Status, which in turn is used together with Current Status to compute the Goal Score. By focusing on Current Status scores, the confounding of any predictive effects was eliminated. The calculation method for the status scores is well-documented [9] and source code is available online in a public repository [13].

With respect to visual design, the principles proposed by the information visualization pioneer Edward Tufte have been used in designing dashboards [35]. These principles include the placement of data in an appropriate context for assessing cause and effect, the employment of small multiples that offer variations of a central theme, conduct of quantitative comparisons, and overall, design of "information thick" visualizations [35, 36].

Dashboard A provides geographical map visualization for a year for a given goal. Dashboard B provides time series visualization over multiple years for a selected goal. Dashboard C provides time series visualization over multiple years for the goal Biodiversity (BD) (first pane) and its subgoals (second and third panes) (Fig. 7.4). Dashboard D-F are similar to Dashboard C, but they are for the other three goals with subgoals (Food Provision (FP), Livelihood and Economies (LE), and Sense of Place (SP)).

Conclusion

In this paper, first, the ocean health data was analyzed for the GCC and the UAE for identifying trends. Then, a root cause analysis for the UAE revealed the subgoals that are the underlying causes for declines. Finally, a visual analytics dashboard that allows for interactive exploration of the OHI data for the whole world was developed, offering novel ways to thoroughly analyze the OHI data and aiding the discovery of actionable insights that can be translated into policies and initiatives.

One threat to the validity of the present research is the validity of the OHI data, specifically with respect to the UAE. A recent study [37] suggests that for the UAE, only some indicators that compose the OHI goal scores are appropriately measured. If these indicators are not accurate, it would mean that the results obtained for the UAE and the GCC in the current study would not be accurate, and the contributions of the current research in terms of insights for the UAE would be weak. However, even if the source data may have problems, the contribution of the current research to the design and development of analytics dashboards is still applicable.

There are multiple avenues for further research that can be pursued to extend this study, that are discussed in detail in the Supplement [11].

Acknowledgements This research was funded by the "SDGs Research Program" of the United Arab Emirates University (UAEU), with the grant code G00004102.

References

1. United Nations. Retrieved January 12, 2024, from https://sdgs.un.org/goals/goal14
2. United Nations–Department of Economic and Social Affairs Sustainable Development. The 17 Goals. Retrieved January 7, 2024, from https://sdgs.un.org/goals
3. Boyce, D. G., Lotze, H. K., Tittensor, D. P., Carozza, D. A., & Worm, B. (2020). Future ocean biomass losses may widen socioeconomic equity gaps. *Nature Communications, 11*, 2235. https://doi.org/10.1038/s41467-020-15708-9
4. Andriamahefazafy, M., Touron-Gardic, G., March, A., Hosch, G., Palomares, M. L. D., & Failler, P. (2022). Sustainable development goal 14: To what degree have we achieved the 2020 targets for our oceans? *Ocean, Coastal Management, 227*, 106273.
5. Tsai, C. W., Lai, C. F., Chao, H. C., & Vasilikos, A. V. (2015). Big data analytics: a survey. *Journal of Big Data, 2*(21). https://doi.org/10.1186/s40537-015-0030-3
6. Runkler, T. A. (2020). *Data analytics*. Springer Fachmedien Wiesbaden.
7. Lafortune, G., Fuller, G., Moreno, J., Schmidt-Traub, G., & Kroll, C. (2018). SDG index and dashboards detailed methodological paper. *Sustainable Development Solutions Network*, 1–56. Retrieved January 7, 2024, from https://www.mdri.org.vn/vsdg/2019_UN_LNOB_Ethnic_Minority/1.%20SDG%20database/1.SDGINDEX.ORG/2018_SDG_GLobalIndex/2018_SDG_GlobalIndexMethodology.pdf
8. Halpern, B., Longo, C., Hardy, D., & Zeller, D. (2012). An index to assess the health and benefits of the global ocean. *Nature, 488*, 615–620. https://doi.org/10.1038/nature11397
9. Ocean Health Index. (2023). Ocean health index 2023: methods. Retrieved January 7, 2024, from https://oceanhealthindex.org/images/htmls/Supplement.html

10. Ocean Health Index. OHI 2021 results. Retrieved January 7, 2024, from https://ohi-science.org/ohiprep_v2021/Reference/methods_and_results/Supplement_Results.html
11. AlOmari, M., Awwad, D., ElHassan, H., Suleiman, R., Asdemir, K., & Ertek, G. (2024). Supplement to 'A visual analytics methodology for the "life under water" (SDG14) sustainable development goal (SDG)'. https://ertekprojects.com/ftp/supp/22.zip
12. Borja, A., Elliott, M., Andersen, J. H., Berg, T., Carstensen, J., Halpern, B. S., & Rodriguez-Ezpeleta, N. (2016). Overview of integrative assessment of marine systems: The ecosystem approach in practice. *Frontiers in Marine Science, 3*, 20. https://doi.org/10.3389/fmars.2016.00020
13. OHI-Science GitHub Repository. Retrieved January 7, 2024, from https://github.com/OHI-Science/ohi-global/blob/draft/eez/conf/functions.R
14. Lowndes, J. S. (2018). What is the ocean health index toolbox and how can it help you? Retrieved January 7, 2024, from https://swiy.co/KUkM. https://docs.google.com/presentation/d/1WKzbvF-XQl3lGzEc44fp8azssod9BcY2wMaAO0ZhFmk/edit#slide=id.p
15. Secretariat General of the Gulf Cooperation Council. The Cooperation Council for the Arab States of the Gulf. Retrieved January 12, 2024, from https://www.gcc-sg.org/en-us/Pages/default.aspx
16. United Arab Emirates Protocol Department–Dubai. About the United Arab Emirates. Retrieved January 12, 2024, from https://www.protocol.dubai.ae/UAE-Dubai/About-the-United-Arab-Emirates
17. Alsharhan, A. S., Rizk, Z. E., Alsharhan, A. S., & Rizk, Z. E. (2020). Water governance in the UAE. *Water Resources and Integrated Management of the United Arab Emirates*, 725–754. https://doi.org/10.1007/978-3-030-31684-6_26
18. Ben-Romdhane, H., Jabado, R. W., Grandcourt, E. M., Perry, R. J. O., Al Blooshi, A. Y., Marpu, P. R., & Ghedira, H. (2020). Coral reefs of Abu Dhabi, United Arab Emirates: Analysis of management approaches in light of international best practices and a changing climate. *Frontiers in Marine Science, 7*, 541. https://doi.org/10.3389/fmars.2020.00541
19. Kunčič, A. (2018). SDG-specific country groups: Subregional analysis of the Arab Region. *Review of Middle East Economics and Finance, 14*(2), 20170020. https://doi.org/10.1515/rmeef-2017-0020
20. Halpern, B. S., Longo, C., Lowndes, J. S. S., Best, B. D., Frazier, M., Katona, S. K., & Selig, E. R. (2015). Patterns and emerging trends in global ocean health. *PLoS One, 10*(3), e0117863. https://doi.org/10.1371/journal.pone.0117863
21. Halpern, B. S., Longo, C., Scarborough, C., Hardy, D., Best, B. D., Doney, S.C., & Samhouri, J. F. (2014). Assessing the health of the US West coast with a regional-scale application of the ocean health index. *PLoS One, 9*(6), e98995. https://doi.org/10.1371/journal.pone.0098995
22. Daigle, R. M., Archambault, P., Halpern, B. S., Stewart Lowndes, J. S., & Côté, I. M. (2017). Incorporating public priorities in the Ocean Health Index: Canada as a case study. *PLoS One, 12*(5), e0178044. https://doi.org/10.1371/journal.pone.0178044
23. de Alencar, N. M. P., Le Tissier, M., Paterson, S. K., & Newton, A. (2020). Circles of coastal sustainability: A framework for coastal management. *Sustainability, 12*(12), 4886.
24. Sachs, J. D., Schmidt-Traub, G., & Durand-Delacre, D. (2016). Preliminary sustainable development goal (SDG) index and dashboard. Sustainable Development Solutions Network. Retrieved January 7, 2024, from http://www.nax.hla21.de/resources/160215-Prel-SDG-Index-a-SDG-Dashboard.pdf
25. Marine Information System for Europe. Dashboards on marine features under other policies. Retrieved January 7, 2024, from https://swiy.co/KUkL. https://water.europa.eu/marine/data-maps-and-tools/map-viewers-visualization-tools/dashboards-on-marine-features-under-other-policies
26. Marine Information System for Europe. Holland. Retrieved January 7, 2024, from https://water.europa.eu/marine/countries-and-regional-seas/country-profiles/netherlands
27. EIONET Central Data Repository. Retrieved January 7, 2024, from https://cdr.eionet.europa.eu/
28. Gulseven, O. (2020). Dataset on the marine sustainability in the United Arab Emirates. *Data in Brief, 31*, 105742. https://doi.org/10.1016/j.dib.2020.105742

29. Sustainable Development Solutions Network. (2017). Global responsibilities: International spillovers in achieving the goals. PART 2 SDG index and dashboards methodology. Sustainable Development Solutions Network. https://www.jstor.org/stable/resrep15870.5
30. Tableau Public. OHI Dashboards. Retrieved January 12, 2024, from https://public.tableau.com/app/profile/gurdal.ertek/vizzes
31. GPT for Work. Retrieved January 7, 2024, from https://gptforwork.com/
32. Fui-Hoon Nah, F., Zheng, R., Cai, J., Siau, K., & Chen, L. (2023). Generative AI and ChatGPT: Applications, challenges, and AI-human collaboration. *Journal of Information Technology Case and Application Research, 25*(3), 277–304. https://doi.org/10.1080/15228053.2023.2233814
33. Tableau Public. Retrieved January 7, 2024, from https://public.tableau.com/app/discover
34. Tableau. Tableau eLearning for Creator. Retrieved Janaury 7, 2024, from https://elearning.tableau.com/
35. Tufte, E. R. (1997). *Visual explanations: Images and quantities, evidence and narrative*. Graphics Press.
36. Zimmermann, B. B. (1997, October). Applying Tufte's principles of information design to creating effective Web sites. In *Proceedings of the 15th Annual International Conference on Computer Documentation* (pp. 309–317). Retrieved January 7, 2024, from https://doi.org/10.1145/263367.263406
37. Gulseven, O. (2020). Measuring achievements towards SDG 14, life below water, in the United Arab Emirates. *Marine Policy, 117*, 103972. https://doi.org/10.1016/j.marpol.2020.103972

Chapter 8
Benchmarking Machine Learning Algorithms to Predict Profitability Directional Changes

Panagiotis G. Artikis, Nicholas D. Belesis, Georgios A. Papanastasopoulos, and Antonios M. Vasilatos

Abstract This study evaluates machine learning techniques, including Random Forest, Stochastic Gradient Boosting, and AdaBoost, against Logistic Regression in predicting European profitability directional changes. The research addresses the growing need for better prediction models in financial analysis. Focusing on the superiority of machine learning, the study investigates cross-validation strategies, finding that classic methods outperform rolling forward. Results reveal constant high accuracy across predicting horizons, challenging conventional methods. DuPont analysis and raw accounting data are employed, with raw data being as insightful as financial ratios. The research contributes methodologically by demonstrating the robustness of machine learning and pushing for practical computational efficiency. Implications extend beyond academics and industry, directing the design of prediction models and underlining the necessity of different data sources. Future research could explore machine learning for metrics of profitability in levels and assess the value relevance of raw accounting items. This research aligns with literature while providing fresh insights into predictive modeling in financial analysis.

Keywords Profitability · Directional changes · Machine learning

Introduction

The estimation of a firm's future profitability is an essential phase to assess the intrinsic value of a firm. Forecasting profitability is a difficult task, as multiple factor affects firm's profitability [16]. In the relevant stream of the literature, researchers usually use financial ratios and traditional econometric methods. We use both theoretically specified models and a dataset of raw accounting items derived from the financial statements. We use machine learning algorithms, as it is found that they outperform

P. G. Artikis · N. D. Belesis · G. A. Papanastasopoulos · A. M. Vasilatos (✉)
University of Piraeus, 18534 Piraeus, Greece
e-mail: vasilatos@unipi.gr

© The Author(s), under exclusive license to Springer Nature Switzerland AG 2024

A. Emrouznejad et al. (eds.), *Business Analytics and Decision Making in Practice*, Lecture Notes in Operations Research, https://doi.org/10.1007/978-3-031-61589-4_8

traditional regression methods. The purpose of this study is to (a) examine the performance of traditional modes compared to machine learning, (b) benchmark several machine learning algorithms, (c) conduct several sensitivity tests.

As we know from Gerakos and Gramacy [10], the predictions for the level of future profitability cannot beat a random walk. Furthermore, forecasting the direction of future profitability is of equal economic importance [22].

We use three machine learning algorithms that they have been used in classification tasks in the accounting and finance literature, Random Forest, Stochastic Gradient Boosting and AdaBoost. These algorithms are optimized for predictions and out of sample performance. Opposite to the regressions, these algorithms can use a large number of features and can capture more complex, non-linear relationships [17].

Our sample consists of publicly listed firms from 30 European countries. Sample size 526,680 firm years. All data have been downloaded from LSEG-Datastream. For our analysis, we have derived 30 raw accounting items, commonly found in the financial statements.

In our initial approach, we use the factors of the DuPont analysis as features. We use several different cross validation approaches to train our algorithms. We find that no matter the cross validation method, the performance of our algorithms remains the same. The machine learning algorithms outperform the Logistical Regression.

Next, we use the raw accounting data to build our model. We perform a Sequential Feature Selection to find the most important raw accounting items which we will use to fit the algorithms. We find that when we use raw accounting items, the prediction performance is almost equal to the DuPont analysis approach.

A sensitivity analysis is performed. The sample is split in two sub-samples, countries that can be classified as western, developed economies and the rest European countries. Despite of the inter-regional economic differences between the two samples, the algorithms exhibit similar performance throughout the sample. In the next step of our analysis, we choose not to tune the parameters of the algorithms. Our results indicate that even without hyper-tuning, the algorithms perform reasonably well.

Finally, we test several profitability measures, Return on Assets, Return on Net Operating Assets, Free Cash Flows and Operating Cash Flows, using both the DuPont analysis and the raw accounting items. Evidence suggests that raw data items lead to better out of sample performance for all of these profitability measures compared to the DuPont analysis.

This paper makes several contributions to the literature. We contribute to the fundamental analysis literature by using machine learning and raw accounting data. We test several cross validation methods, we test an approach where we do not tune the algorithms' parameters, and thus, the computational efficiency is increased. We examine several profitability measures as Anand et al. [1] and we address their call to investigate more algorithms rather only the Random Forest. Finally, while Chen et al. [6] predict only the directional changes of EPS, we expand their analysis in further five profitability measures (ROE, ROA, RNOA, FCF, CFO).

Methodology

Sample Cross-Validation and Hyper Parameters Tuning

In all the literature that involves machine learning, there is always a sample splitting in training, validation and testing samples. Typically, models are trained in the training sample, they are tested in the validation sample and then the best model selected by this procedure, is used to make forecasts in the testing sample (out-of-sample forecasts).

Further, during the training and validation stage of the models, the algorithm's parameters have to be tuned. A common approach to this task is to use a k-fold cross validation. In cross validation, the sample is split in k-folds while on fold is used as a holdout sample to evaluate model's performance. This split is made randomly and it is repeated several times until all folds have been used once as a holdout sample. As in [4, 6], we use panel data and a random split in k-folds is inappropriate, as we will end up forecasting past values using future values.

In the first steps of our analysis, we follow an approach similar to them. Our test period ranges from 2015 to 2020. We train and validate our models in the three preceding years using a time series split cross validation. For test year 2015, we train and validate in years 2012, 2013, 2014. Cross validation is not performed randomly; in the beginning, we train in 2012 and validate in 2013. Then, we train in 2012 and 2013 and validate in 2014. During the training-validation phase, we use grid search to find the optimal hyper-parameters. Finally, we use the best set of hyper-parameters we found during training-validation, and we train in all years from 2012 to 2014 and we obtain out-of-sample predictions for 2015. We repeat the process for the next year we want to predict (2016) by using the three previous years (2013–2015) for training and validation.

Both Bao et al. and Chen et al. (see [4, 6]), support that they use a few preceding years for validation because by this way they use the most recent available information to evaluate their models. They do not report using any grid search technique or Bayesian optimization to tune the hyper parameters; instead, they use parameters that they empirically know that perform well. As mentioned, we replicate their analysis; however, we additionally perform a sensitivity analysis, in which we use all the available firm years before the first test year (2015) as a training sample and in that way we test if using less and more recent years of data for training and validation, improves forecasting performance. Moreover, we repeat our analysis without hyper tuning the algorithms' parameters in an attempt to examine whether hyper tuning indeed improves forecasting performance.

Performance Evaluation Metrics

The profitability's directional change prediction is a binary classification problem (profitability increase versus profitability decrease), therefore classification evaluation metrics must be used. Common metrics include accuracy score and balanced accuracy score (BAC) [12]. Yet, these two measures come with serious pitfalls when there is class imbalance (refer to Bao et al. [4] for an extensive analysis). For this reason, we follow the suggestions of Fawcett [7] and Larcker and Zakolyukina [13] and we use the Area Under Curve (AUC) as our classification performance metric.

The Out-of-Sample Performance of the DuPont Analysis

DuPont analysis is a simple decomposition of Return on Equity (ROE) into three components, profit margin (PM), total asset turnover(ATR) and equity multiplier(EM). Alternatively, it has been also used as a decomposition of Return on Net Operating Assets (RNOA) into profit margin and total assets turnover [18] . It is widely used in financial statement analysis and it has been found that a change in asset turnover is positively related to future changes in earnings [21].

We use the following specification of DuPont analysis (hereafter DP analysis) in an attempt to predict ROE directional changes.

$$ROE = PM \times ATR \times EM \quad (1)$$

We use the components of DP analysis as features to predict directional changes in ROE (ΔROE). We use the features both in levels and in their first difference format, as [21] has noted that the change in asset turnover is positively related to future earnings. We use the logistic regression (LR) to fit the model.

We notice that when the differentiated DP features are used, there is a 23,5% performance improvement in the AUC score. We have established our benchmark using the logistic regression and the DP analysis. The prediction performance is 63%, which is the average AUC for the test years from 2015 to 2020. This AUC score is very similar to the AUC score (57.96%) Chen et al. have achieved using the DP analysis and the logistic regression.

In the next step of our analysis, we fit more sophisticated/complex machine learning algorithms and we try to beat the benchmark performance. We use Random Forest (RF) [5], Stochastic Grading Boosting (SGB) [9] and AdaBoost (ADB) [8]. The results are summarized in the Table 8.1.

When the more complex algorithms are used, it is noticeable that the prediction performance is vastly improved. The best performing algorithm, SGB, improves the AUC score by 42.4% compared to the AUC score achieved by the LR. Compared to Chen et al., their RF and SGB also performed better than the LR but they did not achieve such an improvement in their AUC score. It has to be noted though that

Table 8.1 Out-of-sample performance metrics for test period 2015–2020

Input features	Algorithm	AUC score averaged over the test period 2015–2020
		ΔROE
DP level	LR	0.51
DP 1st diff.	LR	0.63
	RF	0.889
DP 1st diff.	SGB	0.897
	ADB	0.868

they have used Earnings per Share (EPS) as their profitability measure while in this analysis we have used ROE.

Does the Most Recent Accounting Information Boost Performance?

In the previous analysis, we performed a rolling cross validation. We used the three years before the test year for training and validation and then we repeated that again by rolling one year ahead the sample. The result was that every time the training and validation samples had the same size. In this way, the algorithm was trained only in the three more recent years before the test year.

We use the best performing algorithms of the initial analysis and we alter the cross validation strategy. The sample from 2005 to 2014 is used for training and validation and then we make predictions in the test set, from 2015 to 2020. The algorithm is trained in 2005 and validated in 2006. Then it is trained in 2005 and 2006 and validated in 2007. This process repeats until we have trained the algorithm in $k - 1$ folds, meaning that in the last iteration we train from 2005 to 2013 and validate in 2014. When this process is complete, we test the algorithm in the holdout sample from 2015 to 2020 without performing any further training and validation (Table 8.2).

The incorporation of only the most recent information in our cross validation does not improve prediction performance. Contrariwise,the performance of all algorithms has been improved (except RF's which remained the same), even marginally. Keep in mind though, that the incorporation of older accounting information, not only improved the performance, but also the performance was kept consistent even for five year ahead predictions, thus proving that it is feasible to make successful multi-years ahead predictions.

Motivated by these findings, we have decided to further test the algorithms by once again altering the training- validation sample size. We again performed cross validation with hyper parameter tuning, but this time the training and validation was

Table 8.2 Out-of-sample performance metrics for test period 2015–2020 and train-validation 2005–2014

Input features	Algorithm	AUC score averaged over the test period 2015–2020
		ΔROE
DP 1st diff.	RF	0.889
	SGB	0.903
	ADB	0.874

Table 8.3 Out-of-sample performance metrics for test period 2015–2020 and train-validation 2005–2010

Input features	Algorithm	AUC score averaged over the test period 2015–2020
		ΔROE
DP 1st diff.	RF	0.886
	SGB	0.898
	ADB	0.874

performed from 2005 to 2010. In that way, there is a five year gap between the last training fold (2009) and the first test year (2015) and a 10 year gap with the last test year (2020) (Table 8.3).

The results indicate that even with only five years of information, the algorithms are able to sustain their predictive performance and make successful multi step ahead predictions. That is also an indication, at least regarding this dataset, that cross validation remains effective, no matter the number of iterations ($k - splits$).

The Out-of-Sample Performance of Raw Accounting Data

Raw accounting data have been used by Ou and Penman [19] to predict earnings. Since then, Chen et al. have also replicated this approach to predict directional changes in EPS. Furthermore, Bao et al. used the raw data items of Ou and Penman to detect accounting fraud. The use of raw accounting data has been characterized as a "kitchen sink" approach, as it does not rely on any theoretical approach.

We use 30 accounting items commonly found in the financial statements. Bao et al. suggests, as there is a scale difference between the raw items, each item has to be normalized (see [11]). Ou and Penman use a stepwise regression to select the variables that will include in their model. A model consisting raw data items as features is very large and may lead to over-fitting. Similarly, we use a Sequential Feature Selection technique (SFS). We fit a logistic regression in the training-validation sample (using cross validation) and we include all the raw data items as features. Then, features

Table 8.4 Out-of-sample performance metrics for test period 2015–2020 and train-validation 2005–2014, raw accounting items

Input features	Algorithm	AUC score averaged over the test period 2015–2020
		ΔROE
Raw accounting items, 1st diff, standardized	LR	0.773
	RF	0.833
	SGB	0.839
	ADB	0.835

are removed once at a time, until the removal of further features decreases the AUC score. We then train-validate-test the 19 selected raw data items by the SFS by using all the algorithms we have used in previous analyses.

The results indicate a significant improvement (22.6%) in the prediction performance of the LR compared to the DP analysis. On the other hand, all the other algorithms exhibit a slight decrease in their performance, yet they still achieve a high AUC score (83.3–83.9%). Although this is not the best performing approach, it provides evidence that raw accounting items have high predictive power (Table 8.4).

Sensitivity Analyses

In this section we conduct sensitivity tests to evaluate the robustness of our methods. Therefore, initially we alter the sample selection, by selecting a sub-sample of countries. Artikis et al. [2] have selected a group of countries that share common characteristics. They can be characterized as developed, western economies that follow the same accounting standards (IFRS) and share a common legal culture. Evidence suggests that in developed countries with high investor protection, firms exhibit better quality earnings. We seek to test whether firm characteristics (DP analysis) exhibit greater prediction power in this sub-sample compared to the full sample (Table 8.5).

Table 8.5 Out-of-sample performance metrics for test period 2015–2020 and train-validation 2005–2014, inter-regional comparison

Input features	Algorithm	AUC score averaged over the test period 2015–2020	
		Developed countries	Rest of the countries
DP 1st diff.	LR	0.646	0.649
	RF	0.886	0.916
	SGB	0.895	0.918
	ADB	0.867	0.906

The results of this analysis indicate that using a sub-sample of developed countries does not improve performance. In an attempt to shed more light, we re-fit the models by removing the developed countries from the sample. We find that the prediction performance marginally increased, against all previous analyses. Nonetheless, we cannot provide evidence that inter-regional economic differences affect the performance of the employed algorithms, which exhibit a similar performance in all the samples we have tested.

Is Parameter Tuning Necessary?

Many researchers support that tuning the parameters of an algorithm is of greater importance than the selection of the appropriate algorithm [14, 15]. On the other hand, other researchers are more sceptical about tuning [20], while others provide evidence that in some cases, not tuning at all the parameters does not mean that the algorithm will exhibit inferior performance [23].

We fit the original sample, using the DP analysis, with all the algorithms' parameters set to default values. The algorithms, apart from LR which makes random guesses (AUC = 50.6%), perform reasonably well, yet not on their full potential. The trade-off between model performance and computational efficiency should always been taken under consideration; a large number of parameters that require tuning can vastly increase computation time during the algorithm's fit (Table 8.6).

Out-of-Sample Performance of Further Profitability Measures

Anand et al. [1] have used RF to predict the directional changes of Return on Assets (ROA), Return on Equity (ROE), Return on Net Operating Assets (RNOA), Free Cash Flows (FCF) and Operating Cash Flows (CFO), using a sample of US firms. Although they have used lags and differences of the profitability measures as features, we use the DP analysis and test how it performs in other measures than ROE.

Table 8.6 Out-of-sample performance metrics for test period 2015–2020 and train-validation 2005–2014, no hyper tuning

Input features	Algorithm	AUC score averaged over the test period 2015–2020
		ΔROE
DP 1st diff., no hyper tuning	LR	0.506
	RF	0.838
	SGB	0.840
	ADB	0.814

Table 8.7 Out-of-sample performance metrics for test period 2015–2020 and train-validation 2005–2014, further profitability measures

Input features	Algorithm	AUC score averaged over the test period 2015–2020			
		ROA	RNOA	FCF	CFO
DP analysis, 1st diff. (Raw accounting items, 1st diff., standardized)	LR	0.666 (0.909)	0.566 (0.613)	0.5414 (0.7)	0.655 (0.696)
	RF	0.904 (0.926)	0.707 (0.751)	0.693 (0.699)	0.769 (0.733)
	SGB	0.908 (0.928)	0.711 (0.744)	0.689 (0.713)	0.77 (0.733)
	ADB	0.902 (0.912)	0.6961 (0.735)	0.6961 (0.69)	0.765 (0.727)

The results are summarized in Table 8.7. The DP analysis does not perform well (compared to previous analyses) in the cases of RNOA, FCF and CFO. This can be explained by the fact that the specification of the DP analysis we have used has no theoretical links to the profitability measures we seek to predict. For this reason, we use the raw accounting items approach, which does not rely on theory. Indeed, by using this approach, the performance has been improved across all profitability measures. Anand et al. have found mixed results; in general they find higher prediction accuracy for cash flow-measures, however, in their portfolio analysis approach earnings-based measures outperform the cash-flow measures.

Discussion and Conclusion

In this paper we benchmark several machine learning algorithms, Random Forest, Stochastic Gradient Boosting and AdaBoost and we generate one-step and multiple-step ahead predictions of profitability directional changes. We compare the performance of the algorithms with the traditional Logistic Regression. We use DuPont analysis to predict directional changes in ROE. As expected, machine learning algorithms outperform the Logistic Regression.

We test different cross validation approaches. We find that the approach of rolling forward cross validation, in which the training-validation folds size is kept constant, and always incorporates the most recent information in the training sample, is not superior in performance compared to the classic cross validation approach.It is found that even in multiple-steps ahead forecasting horizons, the algorithms perform equally well.

Following the literature, we refit our data using raw accounting data instead of the DuPont analysis. There is not improvement in the prediction performance, apart from the Logistic Regression, but the forecasting accuracy remains high across all algorithms.

Several sensitivity analyses have been performed to test the robustness of the findings. The algorithms have been fitted to different sub-samples of the dataset. We split the sample in different European regions which consist of countries that are classified as advanced, western economies and the rest of the sample. Evidence supports that inter-regional economic differences do not affect prediction performance.

Next, we fit the algorithms to the data without tuning their parameters. The prediction performance deteriorates a bit, yet it still yields high AUC scores in all algorithms except from the Logistic Regression. In the final sensitivity test, further profitability measures (ROA, RNOA, FCF, CFO) are predicted using the DuPont analysis. As expected, DuPont analysis did not perform equally well in the cash-flow based measures of profitability. The analysis was repeated, but instead of the DuPont analysis, raw accounting data items were used. Prediction performance was boosted for all profitability measures and algorithms. We notice though that earnings-based measures can be predicted more accurately.

Our study contributes to the relevant literature in several ways. In the best of our knowledge, it is the first study that uses machine learning to predict profitability directional changes in the European setting. Further, we have thoroughly examined various cross validation approaches and we provided evidence that our results stand for all the cross validation techniques we have tested. We contribute to the literature that uses raw accounting items for prediction. The results support that raw items are informative and relevant to the prediction of profitability directional changes. Moreover, in certain scenarios, they perform better than financial ratios. Finally, our results indicate that certain algorithms perform adequately without parameter tuning, which is supportive to the relevant stream of literature.

This research has created implications for both academics and professionals. The choice between cross validation techniques or the tuning of the algorithms' parameters should always been made by considering the computational efficiency and not only the prediction accuracy. We show that less computational intensive approaches can still produce accurate predictions. This is quite important, as the continuous increase in the size and complexity of the data raises the computational efficiency. The utilization of variable selection techniques, as the Sequential Feature Selection, which we have used to reduce the raw data we used as features, enhances prediction accuracy and increases computational efficiency. Last but not least, Industry experts should examine multiple profitability measurements, as there is not a single measure that outperforms all the other under all circumstances.

We acknowledge several limitations in our study. First, we do not examine private companies. Given the difference that exists in the earnings quality between public and private companies [3], our results may differ if we examine private companies. Second, more profitability models have to be tested apart from the DuPont analysis.

For future research, we believe that it is worthy to use machine learning in order to forecast profitability measures in levels and not only the directional changes. In addition to that, the value relevance of the raw accounting items can be researched. Finally, our study can be compared to the study of Chen et al., but they predict directional changes of Earnings per Share (EPS).

Funding The research work was supported by the Hellenic Foundation for Research and Innovation (HFRI) under the 4th Call for HFRI PhD Fellowships (Fellowship Number: 009116).

Acknowledgements This work has been partly supported by the University of Piraeus Research Center.

References

1. Anand, V., Brunner, R., Ikegwu, K., & Sougiannis, T. (2019). Predicting profitability using machine learning. Available at SSRN 3466478.
2. Artikis, P., Diamantopoulou, L., Papanastasopoulos, G., & Sorros, J. (2022). Asset growth and stock returns in European equity markets: Implications of investment and accounting distortions. *Journal of Corporate Finance, 73*, 102193.
3. Ball, R., & Shivakumar, L. (2005). Earnings quality in UK private firms: Comparative loss recognition timeliness. *Journal of Accounting and Economics, 39*, 83–128.
4. Bao, Y., Ke, B., Li, B., Yu, Y., & Zhang, J. (2020). Detecting accounting fraud in publicly traded U.S. firms using a machine learning approach. *Journal of Accounting Research, 58*, 199–235. https://onlinelibrary.wiley.com/doi/abs/10.1111/1475-679X.12292
5. Breiman, L. (2001). Random forests. *Machine Learning, 45*, 5–32.
6. Chen, X., Cho, Y., Dou, Y., & Lev, B. (2022). Predicting future earnings changes using machine learning and detailed financial data. *Journal of Accounting Research, 60*, 467–515. https://onlinelibrary.wiley.com/doi/abs/10.1111/1475-679X.12429
7. Fawcett, T. (2006). An introduction to ROC analysis. *Pattern Recognition Letters, 27*, 861–874.
8. Freund, Y., & Schapire, R. (1997). A decision-theoretic generalization of on-line learning and an application to boosting. *Journal of Computer and System Sciences, 55*, 119–139.
9. Friedman, J. (2002). Stochastic gradient boosting. *Computational Statistics & Data Analysis, 38*, 367–378.
10. Gerakos, J., & Gramacy, R. (2013). Regression-based earnings forecasts. Chicago Booth Research Paper.
11. Han, J., Kamber, M., & Mining, D. (2006). *Concepts and techniques* (Vol. 340, pp. 94104–3205). Morgan Kaufmann.
12. He, H., & Ma, Y. (2013). *Imbalanced learning: Foundations, algorithms, and applications* (John Wiley & Sons).
13. Larcker, D., & Zakolyukina, A. (2012). Detecting deceptive discussions in conference calls. *Journal of Accounting Research, 50*, 495–540.
14. Lavesson, N., & Davidsson, P. (2006). Quantifying the impact of learning algorithm parameter tuning. *AAAI, 6*, 395–400.
15. Mantovani, R., Rossi, A., Vanschoren, J., Bischl, B., & Carvalho, A. (2015). To tune or not to tune: Recommending when to adjust SVM hyper-parameters via meta-learning. In *International joint conference on neural networks (IJCNN)* (pp. 1–8).
16. Monahan, S. (2018). Financial statement analysis and earnings forecasting. *Foundations And Trends® In Accounting, 12*, 105–215. https://doi.org/10.1561/1400000036
17. Mullainathan, S., & Spiess, J. (2017). Machine learning: An applied econometric approach. *Journal of Economic Perspectives, 31*, 87–106.

18. Nissim, D., & Penman, S. (2001). Ratio analysis and equity valuation: From research to practice. *Review of Accounting Studies, 6*, 109–154.
19. Ou, J., & Penman, S. (1989). Accounting measurement, price-earnings ratio, and the information content of security prices. *Journal Of Accounting Research*, *27*, 111–144. http://www.jstor.org/stable/2491068
20. Probst, P., Bischl, B., & Boulesteix, A. (2018). Tunability: Importance of hyperparameters of machine learning algorithms. Preprint at ArXiv:1802.09596
21. Soliman, M. (2008). The use of DuPont analysis by market participants. *The Accounting Review, 83*, 823–853. https://doi.org/10.2308/accr.2008.83.3.823
22. Wahlen, J., & Wieland, M. (2011). Can financial statement analysis beat consensus analysts' recommendations? *Review of Accounting Studies, 16*, 89–115.
23. Weerts, H., Mueller, A., & Vanschoren, J. (2020). Importance of tuning hyperparameters of machine learning algorithms. Preprint at ArXiv:2007.07588

Chapter 9
The Role of Institutional Factors in Shaping Environmental Performance: Evidence from Developed and Developing Countries

Jawahir M. Alshehhi and Panagiotis D. Zervopoulos

Abstract The present study employs a novel Bayesian data envelopment analysis (DEA) approach to investigate the impact of institutional factors on the relative environmental performance of developed and developing countries. This study establishes a relationship between formal institutional factors (e.g., government effectiveness and corruption control) and environmental performance. The Bayesian DEA method employed here provides valid estimates, outperforming other existing bias-correction DEA techniques in terms of lower mean square error and mean absolute error. The efficiency estimates are incorporated in a two-step generalized method of moments (GMM) for linear dynamic panel data applied to a balanced panel encompassing 144 countries spanning from 2002 to 2019. Consistent with prior research on economic growth and environmental efficiency, this study reveals an inverse relationship between corruption control and environmental performance in developing countries. This relationship is statistically significant but marginal. Unlike developing countries, the relationship between control of corruption and environmental performance is not statistically significant for developed countries. This result extends the concept of "grease the wheel" theory to environmental efficiency. Moreover, the study highlights the substantial and statistically significant impact of past environmental performance on current environmental performance. This underscores the necessity for countries to reinforce their commitment to regulating energy consumption and its resultant CO_2 emissions.

Keywords Environmental efficiency · Data envelopment analysis · Institutions · Bayesian methods · Generalized method of moments

J. M. Alshehhi
College of Business Administration, University of Sharjah, Sharjah, UAE
e-mail: U19106186@sharjah.ac.ae

P. D. Zervopoulos (✉)
Department of Management, University of Sharjah, Sharjah, UAE
e-mail: pzervopoulos@sharjah.ac.ae

© The Author(s), under exclusive license to Springer Nature Switzerland AG 2024
A. Emrouznejad et al. (eds.), *Business Analytics and Decision Making in Practice*, Lecture Notes in Operations Research, https://doi.org/10.1007/978-3-031-61589-4_9

Introduction

This research explores how institutional factors impact environmental efficiency. The influence of institutions on economic performance is widely acknowledged [1], supported by empirical studies [2], emphasizing their significant role in cross-country income gaps and growth [3]. Institutional economists focus on integrating societal "rules of the game" or "constraints shaping human interaction" into economic theory [1, p.3].

Recent focus has been on environmental institutional quality, particularly in addressing global warming and the impact of human-induced greenhouse gas (GHG) emissions on climate change [4, 5]. Scholars warn that climate change poses the biggest threat to both human and economic development [6–8]. Reduction of greenhouse gas emissions is crucial to meet the goals set by the Paris Climate Accord [9], yet worldwide efforts to cut CO_2 emissions have yielded disappointing results [10].

Expectedly, Countries struggle to lower their GHG emissions without endangering their economic gains [11]. Diverse national priorities and capacities contribute to disparities in climate change goals [12, 13]. Nations dealing with poverty or inequality may deprioritize environmental protection due to limited resources [11]. Researchers suggest strengthening domestic institutional quality to resolve conflicting interests in emission reduction [7, 11]. Efficient institutions can aid nations in mitigating the ecological consequences of elevated economic growth, enabling them to minimize pollution [4]. Local institutions largely shape the efficacy of economic, energy, and environmental policies [11]. Panayotou [14] highlighted the pivotal role of institutional quality in enhancing environmental quality, even in economically challenged countries.

In economic studies, institutions' significance and their link to economic growth have been extensively examined [15–17]. It is argued that institutional settings have an influence on endogenous elements that lead to economic growth and development. These settings play a critical role in fostering and maintaining economic progress [18] or, more frequently, obstructing it [19]. Agostino et al. [16] emphasize that formal institutions, including regulatory quality and the rule of law, drive economic prosperity.

Institutions shape a nation's overall environment and determine its economic success or failure [20]. Control of corruption is a widely studied institutional factor assessing a country's institutional quality [17, 18, 21–23]. Leff [24] introduced the "grease the wheel" theory, suggesting corruption's potential benefits, especially in developing countries, where it can counteract bureaucracy or dysfunctional regulations by facilitating transactions [25].

Besides corruption, the impact of other institutional factors on economic activity has been studied. Afonso et al. [15] highlighted the importance of the rule of law and regulatory quality. Agostino et al. [16] discussed government effectiveness, while [18] focused on voice and accountability. Halkos and Tzeremes [19] and Aparicio et al. [23] argued that corruption is interconnected with a country's regulatory and political system effectiveness, collectively influencing economic development.

Agostino et al. [16] noted varying effects of institutions on economic growth. They found that government effectiveness has the most substantial impact, followed by regulatory quality.

Aparicio et al. [23] noted institutional variations between developed and developing nations, with stronger institutions in developed countries regulating corruption and enforcing stricter laws. Consequently, corruption is anticipated to have a negative impact on economic activity in countries with strong institutions but a positive effect in those with weaker institutions [26]. Méon and Sekkat [21] and Méon and Weill [26] reached the same result, stating that the quality of institutions accounts for variations in economic efficiency.

In existing literature, the impact of corruption on economic growth remains debated [21]. Several studies suggest corruption hinders growth in both developed and developing nations [18, 27]. This perspective is evident when assessing economic performance across countries. Aidt [18] highlighted how practices facilitating quick transactions, like red tape, could negatively affect the broader environment. While beneficial on an individual level, the "greasing" effect might harm society and the economy over time.

Studies often focusing on corruption's institutional impact reveal an indirect correlation with energy or environmental efficiency [28–31]. Tateishi et al. [11] noted that nations with stronger institutions tend to be more efficient environment. In contrast, [32] discovered a varied impact of environmental regulations on energy efficiency.

This study broadens existing research on institutional effects on economic efficiency, emphasizing their influence on environmental efficiency. It employs a two-stage approach considering conventional economic factors and environmental indicators like energy use and CO_2 emissions. Firstly, a Bayesian generalized directional distance function (GDDF) data envelopment analysis (DEA) model is used to estimate environmental efficiency. This model is suitable for handling undesirable variables like CO_2 emissions, with its Bayesian extension minimizing estimation biases. Directional distance function DEA, especially with undesirable variables, is widely used [33, 34]. The novel Bayesian-GDDF DEA approach, drawing on [35] work, significantly reduces efficiency biases.

This study is the first to delve into how institutional factors influence environmental efficiency using Bayesian data envelopment analysis (DEA) for bias-corrected assessments. It explores relationships among various formal institutions and their interactions with country classification in relation to environmental efficiency. Apart from its methodological novelty, this research supports the "grease the wheel" theory from economic growth literature rather than environmental efficiency. It reveals an inverse link between control of corruption and environmental efficiency in developing countries, a statistically significant yet minor effect. In contrast, developed countries' environmental efficiency isn't significantly impacted by formal institutional factors but rather by lagged environmental performance.

The rest of this research is structured as follows: Sect. "Methodology" outlines the methodology, focusing on the Bayesian generalized directional distance function data envelopment analysis approach. Section "Sample and Data Description" describes the sample and outlines the dataset. Section "Empirical Results" analyzes the empirical findings, while Sect. "Conclusion" serves as the conclusion.

Methodology

This study employs a two-stage methodology. Initially, a novel Bayesian generalized directional distance function data envelopment analysis (GDDF DEA) approach yields bias-corrected efficiency estimates. Subsequently, these estimates become the response variable in a regression model, with institutional factors and interaction effects acting as explanatory variables. A two-step generalized method of moments (GMM) approach for linear dynamic panel data, which addresses endogeneity and reverse causality, is applied.

Bias-Corrected Efficiencies (Stage 1)

This study employs the GDDF DEA model [36] to determine relative efficiencies $[\{\theta_j\}_{j=1}^{n} \in [\theta_L, 1]$ where $\theta_L \in (0,1)]$ for sample countries considering inputs $x_i = (x_1, \ldots, x_m) \in \mathbb{R}_+^m$ and desirable outputs $y_r = (y_1, \ldots, y_s) \in \mathbb{R}_+^s$ along with undesirable outputs like CO_2 emissions $b_\eta = (b_1, \ldots, b_\kappa) \in \mathbb{R}_+^\kappa$. The selected model treats undesirable outputs as by-products of the desirable outputs.

The GDDF DEA program used reads as follows:

$$\theta = \min \frac{1 - \frac{1}{m}\sum_{i=1}^{m} \beta g_i / x_{io}}{1 + \frac{1}{s+\kappa}\left(\sum_{r=1}^{s} \beta g_r / y_{ro} + \sum_{\eta=1}^{\kappa} \beta g_\eta / b_{\eta o}\right)}$$

$$s.t. \sum_{j=1}^{n} \lambda_j x_{ij} + \beta g_x \leq x_{io} \quad i = 1, \ldots, m$$

$$\sum_{j=1}^{n} \lambda_j y_{rj} - \beta g_y \geq y_{ro} \quad r = 1, \ldots, s$$

$$\sum_{j=1}^{n} \lambda_j b_{\eta j} - \beta g_b = b_{\eta o} \quad \eta = 1, \ldots, \kappa$$

$$\sum_{j=1}^{n} \lambda_j = 1 \tag{9.1}$$

$$\lambda \geq 0$$

$$g_x = 1, g_y = 1, g_b = -1$$

where λ is the weight allocated to each of these three categories of variables, and g_x, g_y, and g_b are the direction vectors for inputs, desirable outputs, and undesirable outputs, respectively. Additionally, the proportional decreases in inputs and undesirable outputs are expressed by $\beta g_i/x_{io}$ and $\beta g_\eta/b_{\eta o}$, while the proportional increase in the reference country's desirable outputs is expressed by $\beta g_r/y_{ro}$ (identified by the subscript o in model (9.1)).

A Bayesian DEA method, developed by [35], corrects biases in efficiencies. This method combines a uniform likelihood and beta prior to approximate efficiencies obtained from the program (9.1). It is confirmed that the mix of these distributions leads to a posterior mean resembling the parameter estimator [37, pp. 325–326].

Initially, the expression $\{\theta_j\}_{j=1}^{v} \in [\theta_L, 1)$, where $v \subset n$, represents all sample efficiencies excluding ones. Employing the likelihood function of these efficiencies, we derive the expected value (9.2) and the unbiased estimator (9.3) for the parameter θ_L, articulated as follows:

$$E_v\left\{\hat{\theta}_L\right\} = \theta_L + \frac{1-\theta_L}{v+1} \tag{9.2}$$

$$\tilde{\theta}_L = \frac{\hat{\theta}_L(v+1)-1}{v} \tag{9.3}$$

If we presume that the parameter θ_L follows a beta distribution, the Bayesian prior is formulated in the subsequent manner:

$$f_{\theta_L}(\theta_L|\gamma,\delta) = \frac{1}{B(\gamma,\delta)}\theta_L^{\gamma-1}(1-\theta_L)^{\delta-1}, \theta_L \in (0,1) \tag{9.4}$$

with γ and δ representing the beta distribution's shape parameters (B).

Furthermore, equating the unbiased estimator (9.3) to the anticipated value of the prior distribution (i.e., $E\{\theta_L\} = \frac{\gamma}{\gamma+\delta}$) yields expression (9.5):

$$\delta = \frac{\left(1-\tilde{\theta}_L\right)\gamma}{\tilde{\theta}_L} \tag{9.5}$$

The collective probability density function of the vector $\Theta = \{\theta_j\}_{j=1}^{v}$ is expressed as:

$$f_\Theta(\Theta) = \int_0^1 f(\Theta|\theta_L)f_{\theta_L}(\theta_L|\gamma,\delta)\mathrm{d}\theta_L = \frac{B(\gamma,\delta-v)}{B(\gamma,\delta)} \tag{9.6}$$

With the limitation on the parameter δ ($\delta > v$), and considering expression (9.5), parameter γ is constrained from the lower bound:

$$\gamma > v\frac{\tilde{\theta}_L}{1 - \tilde{\theta}_L} \tag{9.7}$$

The resulting beta distribution, derived from the posterior, holds parameters γ and $\delta - v$ in the following manner:

$$f_{\theta_L|\Theta}(\Theta) = \frac{f_{\Theta|\theta_L}(\Theta|\theta_L) f_{\theta_L}(\theta_L|\gamma, \delta)}{f_\Theta(\Theta)} = \frac{1}{B(\gamma, \delta - v)}\theta_L^{\gamma-1}(1 - \theta_L)^{(\delta-v)-1} \tag{9.8}$$

Based on the earlier examination, the posterior reflects the overestimated (biased) efficiencies, whereas the prior signifies the bias-corrected estimations.

$$E_v\{\theta_L|\Theta\} > E_v\{\theta_L\} \text{ given that } \frac{\gamma}{\gamma + \delta - v} > \frac{\gamma}{\gamma + \delta} \tag{9.9}$$

The subsequent parameters are utilized to rectify the bias in the efficiency estimations:

$$\varphi = \frac{\tilde{\theta}_L}{\hat{\theta}_L} < 1 \tag{9.10}$$

Derived from Eqs. (9.5), (9.7), and (9.10), Eqs. (9.11) and (9.12) are formulated:

$$\hat{\gamma} = v\tilde{\theta}_L/(1 - \varphi) \tag{9.11}$$

and

$$\hat{\delta} = (1 - \tilde{\theta}_L)\hat{\gamma}/\tilde{\theta}_L \tag{9.12}$$

The parameters γ and δ values in Eqs. (9.11) and (9.12) are calculated using the MATLAB function *betarnd*.

The MATLAB function *normfit* estimates parameters ($\hat{\mu}, \hat{\sigma}$) for a normal distribution that best fits the ratio (9.10). This approach aids in determining bias-corrected estimates (θ_j^c):

$$\theta_j^c = w^{-1}\sum\nolimits_{\rho=1}^{w} \breve{\theta}_{j\rho} \tag{9.13}$$

Here, w denotes the count of Monte Carlo iterations (($\rho = 1{,}000$), and $\breve{\theta}_{j\rho}$ are randomly derived using the MATLAB function *normrnd*. These values are based on parameters $\theta_j\widehat{\mu}$ and $\theta_j\widehat{\sigma}$, where θ_j are obtained from program (1).

Estimation Procedure (Stage 2)

A panel linear dynamic regression model is used to investigate the institutional effects on environmental efficiency. According to [37], this regression model tackles endogeneity, feedback effects, and unobservable individual heterogeneity. Specifically, this study mitigates feedback effects, which are highlighted in the literature [38, 39], by employing environmental efficiency as the regression model's response variable. This variable differs significantly from conventional economic growth proxies as it ranges between zero and one. The subsequent model generates the two-step GMM estimates for this study:

$$\theta^c_{1,j,t} = a_1\theta^c_{1,j,t-1} + a_2\theta^c_{1,j,t-2} + \beta_1 z_{1,j,t} + \beta_2 z_{1,j,t-1} + \beta_3 z_{2,j,t}$$

$$+\beta_4 z_{1,j,t} z_{2,j,t} + \gamma_2 d_2 + \cdots + \gamma_T d_T + \eta_j + \varepsilon_{j,t} \qquad (9.14)$$

$\theta^c_{1,j,t}$ represents Bayesian DEA estimates derived from Stage 1 (where $j = 1, \ldots, n$ and $t = 2, \ldots, T$ signify the sample countries (individuals) and time, respectively). $\theta^c_{1,j,t-1}$ and $\theta^c_{1,j,t-2}$ stand for the lagged response variables. Additionally, $z_{1,j,t}$ and $z_{1,j,t-1}$ represent control of corruption and its lag, while $z_{2,j,t}$ indicates country classification (country dummy, where 0 denotes developed countries, and 1 denotes developing countries). The $d_2, \ldots, d_T$ are time dummies, capturing event effects that might influence the estimates without biasing them. Moreover, η_j and $\varepsilon_{j,t}$ represent the time-invariant individual-specific effect and random noise, respectively.

According to literature [19, 23, 40], countries in the sample were categorized as developed or developing using the IMF definition. Multiple models involving institutional factors, country classification, and interactions were examined before arriving at model (9.14). Standard tests like [41] specification test, [42]'s J-test for overidentification, and the Wald test for linearity were conducted for each alternative model.

Sample and Data Description

This study investigates a dataset encompassing 144 countries over the period 2002 to 2019, categorized into 35 developed and 109 developing nations based on the IMF World Economic Outlook database. Estimation of environmental efficiency involved employing conventional economic inputs such as capital stock (in millions of dollars, constant 2017) and labor (in millions), alongside a positive output measure, GDP (in millions of dollars, constant 2017), following guidelines set forth by [43, 44]. Additionally, the analysis incorporated energy consumption (in millions of tons) and CO_2 emissions (in kg of oil equivalent per $1,000 GDP, constant 2017) to address environmental perspectives within the input–output framework. The economic information was sourced from the World Bank Development Indicators, whereas data on energy consumption and CO2 emissions were retrieved from the International Energy Agency's Enerdata-Odyssey database.

Based on numerous studies exploring institutional impacts on economic activity, e.g., [10, 23, 26, 45], the institutional factors incorporated in this analysis were (a) government effectiveness, (b) political stability, (c) voice and accountability, (d) regulatory quality, (e) control of corruption and (f) rule of law. These factors were sourced from the Worldwide Governance Indicators (WGI) database, presenting continuous data spanning from –2.5 to 2.5. Lower values indicate weaker governance performance, while higher values signify stronger governance performance.

Empirical Results

Using expressions (9.1)–(9.13), we estimated environmental efficiencies for a panel of 144 countries. The average environmental efficiency of the sample countries declined from 2002 to 2019. Notably, the lowest environmental efficiency was observed in 2019. Despite developed nations displaying higher environmental efficiency compared to developing ones, the latter notably enhanced their performance from 2004 onwards. Consequently, there is a convergence in environmental efficiency between developing and developed countries from 2004 to 2019.

Environmental efficiency is regressed on various combinations of institutional factors and country classifications, along with their interactions, using two-step GMM. Only models (9.14) and (9.15) fulfilled the conditions for specification, overidentification, and linear hypotheses tests, including the Arellano and Bond test, Hansen J-test, and Wald test. Table 9.1 presents the two-step GMM estimates for these models. Model (9.15) was employed to validate the reliability of the findings obtained from model (9.14), and both models yield consistent results.

As per the two-step GMM findings (model (9.14) in Table 9.1), developing countries' environmental efficiency was, on average, 14.52% lower than that of the developed countries during 2002–2019. Past environmental efficiency notably positively influenced the present ($\theta^c_{1,j,t}$) considering the coefficients assigned to $\theta^c_{1,j,t-1}$ (environmental efficiency with a one-year time lag) and $\theta^c_{1,j,t-2}$ (environmental efficiency with a two-year time lag).

Aligned with [24–26] "grease the wheel" theory, increased control of corruption in developing nations negatively impacts environmental efficiency ($z_{1,j,t}z_{2,j,t}$). This indicates corruption benefits environmental efficiency in developing countries with weak institutions, yet these benefits are minor. For developed countries, corruption control lacks statistical significance. Model (9.15) confirms outcomes from the model

Table 9.1 Impact of institutional factors on environmental efficiency (two-step GMM)[1]

	(14)	(15)
$\theta^c_{1,j,t-1}$	0.5421***	0.5371***
	(0.1045)	(0.1106)
$\theta^c_{1,j,t-2}$	0.1407***	0.1408***
	(0.03296)	(0.0349)
$z_{1,j,t}$	0.0102	0.0130
	(0.0078)	(0.0095)
$z_{1,j,t-1}$	−0.0039	−0.0081
	(0.0079)	(0.0102)
$z_{2,j,t}$	−0.1452***	−0.1212***
	(0.0000)	(0.0000)
$z_{1,j,t}z_{2,j,t}$	−0.0084*	−0.0084*
	(0.0037)	(0.0040)
$z_{3,j,t}$		−0.0065
		(0.009)
$z_{3,j,t-1}$		0.0041
		(0.0103)
Year Dummies	Yes	Yes
Observations	2592	2592
Arellano and Bond test	−0.5779	−0.4810
p-value	0.5633	0.6306
Hansen J-test	145.42	157.05
p-value	0.4979	0.2515
Wald test	133.76	133.12
p-value	<10–4	<10–4

** Coefficient is significant at the 0.0001 level (2-tailed)
Values in parentheses are standard errors

(9.14), where government effectiveness, an additional factor, also lacks statistical significance. Alternate models with various institutional combinations failed tests for specification, overidentification, and linearity.

This study's findings do not support the indirect corruption-environmental efficiency link, which is found in some extant literature [28–31]. Our findings are more

---

[1] The alternative two-step GMM model is as follows:

$$\theta^c_{1,j,t} = a_1\theta^c_{1,j,t-1} + a_2\theta^c_{1,j,t-2} + \beta_1 z_{1,j,t} + \beta_2 z_{1,j,t-1} + \beta_3 z_{2,j,t} + \beta_4 z_{1,j,t}z_{2,j,t} + \beta_5 z_{3,j,t} + \beta_6 z_{3,j,t-1} + \gamma_2 d_2 + \cdots + \gamma_T d_T + \eta_j + \varepsilon_{j,t} \quad (9.15)$$

where $z_{3,j,t}$ and $z_{3,j,t-1}$ are the government effectiveness and its lag (j = 1, … n and t = 2, … , T).

closely connected to [11], justifying an indirect relationship between the control of corruption and environmental efficiency for developing countries with weaker institutions than the developed countries (refer to Table 9.1).

Conclusion

This research explored how formal institutional factors and country classification impact environmental efficiency. Utilizing a Bayesian data envelopment analysis (DEA), we estimated cross-country environmental efficiency using variables such as labor, GDP, capital stock, CO_2 emissions, and energy consumption. The Bayesian approach provided superior, unbiased estimates without the limitations of traditional techniques, facilitating regression analysis with efficiency as the response variable. The study applied a two-step GMM to regress environmental efficiency on institutional factors (e.g., voice and accountability, government effectiveness, political stability, rule of law, regulatory quality, and control of corruption).

Findings reveal a convergence in environmental efficiency between developed and developing countries during 2002–2019. Both country types made efforts to reduce CO_2 emissions. Developed nations exhibited stronger institutions than developing ones, emphasizing the positive relationships among institutional factors as previously seen in the literature. Notably, an inverse relationship between corruption control and environmental efficiency was observed in developing countries, statistically significant but minor, unlike developed countries. This supports the "grease the wheel" theory in environmental efficiency. Lagged environmental efficiencies significantly influenced current environmental performance, highlighting the necessity for strict energy regulation and CO_2 emission control for sustainable growth.

References

1. North, D. C. (1990). Institutions, institutional change and economic performance. Cambridge University Press.
2. Efendic, A., Pugh, G., & Adnett, N. (2011). Institutions and economic performance: A meta-regression analysis. *European Journal of Political Economy, 27*(3), 586–599.
3. Ahmad, M., & Hall, S. G. (2017). Economic growth and convergence: Do institutional proximity and spillovers matter? *Journal of Policy Modeling, 39*(6), 1065–1085.
4. Salman, M., Long, X., Dauda, L., & Mensah, C. N. (2019). The impact of institutional quality on economic growth and carbon emissions: Evidence from Indonesia, South Korea and Thailand. *Journal of Cleaner Production, 241*, 118331.
5. Kounetas, K., & Zervopoulos, P. D. (2019). A cross-country evaluation of environmental performance: Is there a convergence-divergence pattern in technology gaps? *European Journal of Operational Research, 273*(3), 1136–1148.
6. Oh, W., & Lee, K. (2004). Energy consumption and economic growth in Korea: Testing the causality relation. *Journal of Policy Modeling, 26*(8–9), 973–981.
7. Acheampong, A. O. (2019). Modelling for insight: Does financial development improve environmental quality? *Energy Economics, 83*, 156–179.

8. Sethi, P., Chakrabarti, D., & Bhattacharjee, S. (2020). Globalization, financial development and economic growth: Perils on the environmental sustainability of an emerging economy. *Journal of Policy Modeling, 42*(3), 520–535.
9. International Energy Agency. (2018). Global energy and CO_2 emissions status report. Retrieved from https://www.iea.org/geco/emissions/.
10. Acheampong, A. O., Dzator, J., & Savage, D. A. (2021). Renewable energy, CO_2 emissions and economic growth in sub-Saharan Africa: Does institutional quality matter? *Journal of Policy Modeling, 43*(5), 1070–1093.
11. Tateishi, H. R., Bragagnolo, C., & de Faria, R. N. (2020). Economic and environmental efficiencies of greenhouse gases' emissions under institutional influence. *Technological Forecasting and Social Change, 161*, 120321.
12. Paavola, J. (2006). Institutions and environmental governance: A reconceptualization. *Ecological Economics, 63*(1), 93–103.
13. Nordhaus, W. (2015). Climate clubs: Overcoming free-riding in international climate policy. *American Economic Review, 105*(4), 1339–1370.
14. Panayotou, T. (1997). Demystifying the environmental Kuznets curve: Turning a black box into a policy tool. *Environment and Development Economics, 2*(4), 465–484.
15. Afonso, A., Schuknecht, L., & Tanzi, V. (2005). Public sector efficiency: An international comparison. *Public Choice, 123*(3), 321–347.
16. Agostino, M., Di Tommaso, M. R., Nifo, A., Rubini, L., & Trivieri, F. (2020). Institutional quality and firms' productivity in European regions. *Regional Studies, 54*(9), 1275–1288.
17. Azam, M., & Emirullah, C. (2014). The role of governance in economic development: Evidence from some selected countries in Asia and the Pacific. *International Journal of Social Economics., 41*(12), 1265–1278.
18. Aidt, T. S. (2009). Corruption, institutions, and economic development. *Oxford Review of Economic Policy, 25*(2), 271–291.
19. Halkos, G. E., & Tzeremes, N. G. (2010). Corruption and economic efficiency: Panel data evidence. *Global Economic Review, 39*(4), 441–454.
20. Hall, R. E., & Jones, C. I. (1999). Why do some countries produce so much more output per worker than others? *The Quarterly Journal of Economics, 114*(1), 83–116.
21. Méon, P. G., & Sekkat, K. (2005). Does corruption grease or sand the wheels of growth? *Public Choice, 122*(1), 69–97.
22. Bentzen, J. S. (2012). How bad is corruption? Cross-country evidence of the impact of corruption on economic prosperity. *Review of Development Economics, 16*(1), 167–184.
23. Aparicio, S., Urbano, D., & Audretsch, D. (2016). Institutional factors, opportunity entrepreneurship and economic growth: Panel data evidence. *Technological Forecasting and Social Change, 102*, 45–61.
24. Leff, N. H. (1964). Economic development through bureaucratic corruption. *American Behavioral Scientist, 8*(3), 8–14.
25. Méon, P. G., & Weill, L. (2010). Is corruption an efficient grease? *World Development, 38*(3), 244–259.
26. Méon, P. G., & Weill, L. (2005). Does better governance foster efficiency? An aggregate frontier analysis. *Economics of Governance, 6*(1), 75–90.
27. Mauro, P. (1995). Corruption and growth. *The Quarterly Journal of Economics, 110*(3), 681–712.
28. Wu, S., Li, B., Nie, Q., & Chen, C. (2017). Government expenditure, corruption and total factor productivity. *Journal of Cleaner Production, 168*, 279–289.
29. Ozturk, I., Al-Mulali, U., & Solarin, S. A. (2019). The control of corruption and energy efficiency relationship: An empirical note. *Environmental Science and Pollution Research, 26*(17), 17277–17283.
30. Hao, Y., Gai, Z., & Wu, H. (2020). How do resource misallocation and government corruption affect green total factor energy efficiency? Evidence from China. *Energy Policy, 143*, 111562.
31. Wang, S., Zhao, D., & Chen, H. (2020). Government corruption, resource misallocation, and ecological efficiency. *Energy Economics, 85*, 104573.

32. Cui, S., Wang, Y., Zhu, Z., Zhu, Z., & Yu, C. (2022). The impact of heterogeneous environmental regulation on the energy eco-efficiency of China's energy-mineral cities. *Journal of Cleaner Production, 350*, 131553.
33. Lozano, S., & Gutiérrez, E. (2011). Slacks-based measure of efficiency of airports with airplanes delays as undesirable outputs. *Computers and Operations Research, 38*(1), 131–139.
34. Podinovski, V. V., & Kuosmanen, T. (2011). Modelling weak disposability in data envelopment analysis under relaxed convexity assumptions. *European Journal of Operational Research, 211*(3), 577–585.
35. Zervopoulos, P. D., Triantis, K., Sklavos, S., & Kanas, A. (2023). An alternative Bayesian data envelopment analysis approach for correcting bias of efficiency estimators. *Journal of the Operational Research Society, 74*(4), 1021–1041.
36. Cheng, G., & Zervopoulos, P. D. (2014). Estimating the technical efficiency of health care systems: A cross-country comparison using the directional distance function. *European Journal of Operational Research, 238*(3), 899–910.
37. Ahn, S. C., & Schmidt, P. (1995). Efficient estimation of models for dynamic panel data. *Journal of Econometrics, 68*, 5–27.
38. Glaeser, E. L., La Porta, R., Lopez-de-Silanes, F., & Shleifer, A. (2004). Do institutions cause growth? *Journal of Economic Growth, 9*, 271–303.
39. Aisen, A., & Veiga, F. J. (2013). How does political instability affect economic growth? *European Journal of Political Economy, 29*, 151–167.
40. Angelopoulos, K., Philippopoulos, A., & Tsionas, E. (2008). Does public sector efficiency matter? Revisiting the relation between fiscal size and economic growth in a world sample. *Public Choice, 137*(1), 245–278.
41. Arellano, M., & Bond, S. (1991). Some tests of specification for panel data: Monte Carlo evidence and an application to employment equations. *The Review of Economic Studies, 58*(2), 277–297.
42. Hansen, L. P. (1982). Large sample properties of generalized method of moments estimators. *Econometrica, 50*(4), 1029–1054.
43. Solow, R. M. (1956). A contribution to the theory of economic growth. *The Quarterly Journal of Economics, 70*(1), 65–94.
44. Swan, T. W. (1956). Economic growth and capital accumulation. *Economic Record, 32*(2), 334–361.
45. Kaufmann, D., Kraay, A., & Zoido-Lobatón, P. (1999). Aggregating governance indicators, vol. 2195. World Bank Publications.
46. Poirier, D. J. Intermediate statistics and econometrics: A comparative approach. The MIT Press.

Chapter 10
UAE Stock Markets Prediction: Machine Learning Application

Randa A. Abdelkarim, Yousif Abdelbagi Abdalla, and Ibrahim Abaker Hashem

Abstract The Effective prediction of stock returns holds immense significance in guiding investment strategies, bolstering risk management protocols, and shaping financial policies. This paper endeavors to explore the relationship and predictive prowess encompassing the stock markets of the Gulf Cooperation Council (GCC) nations (Bahrain, Kuwait, Oman, Qatar, and Saudi Arabia), China, as well as oil and gold prices, concerning their impact on the United Arab Emirates (UAE) stock markets. Employing a robust multiple regression technique, this study analyses weekly data spanning from June 2010 to June 2023. The utilization of the multiple regression methodology not only facilitates the evaluation of GCC countries and China stock markets but also casts a spotlight on their respective contributions to the variance observed in the performance of the UAE stock market. Through a meticulous analysis, this study determines interesting patterns. When dissecting the stock markets in Dubai and Abu Dhabi separately, the outcomes distinctly indicate a distinct dependence on Qatar, Oman, Bahrain, and Kuwait in both markets. Notably, Dubai exhibits a discernible dependence on the Chinese and Saudi Arabian markets, surpassing that of Abu Dhabi. In the backdrop of this investigation, it emerges that the variables encompassing oil and gold do not wield statistically significant impact. A few studies are solely concerned with the UAE markets, but to our best knowledge, none of them explores the relationship between all the previously listed factors and the UAE markets. This study collectively unveils a comprehensive understanding of the intricate dynamics governing stock market interactions, presenting invaluable insights for investors, risk managers, and policy formulators.

Keywords GCC stock market · China · Multiple regression · Machine learning

R. A. Abdelkarim
Faculty of Mathematical Science and Informatics, University of Khartoum, Khartoum, Sudan

Y. A. Abdalla (✉)
College of Business Administration, University of Sharjah, Sharjah, UAE
e-mail: yomer@sharjah.ac.ae

I. A. Hashem
College of Computing and Informatics, University of Sharjah, Sharjah, UAE
e-mail: ihashem@sharjah.ae.ac

© The Author(s), under exclusive license to Springer Nature Switzerland AG 2024

A. Emrouznejad et al. (eds.), *Business Analytics and Decision Making in Practice*, Lecture Notes in Operations Research, https://doi.org/10.1007/978-3-031-61589-4_10

Introduction

Oil, gold, and stock markets are three interconnected markets that reveal close correlation. Various factors account for the potential linkages between crude oil and stock markets. These factors include foreign exchange markets, corporate future cash flows, discount factors, and economic cycles. These elements play a crucial role in establishing the connections and interactions between these two markets [1]. Oil and gold are crucial commodities for the global economy, and they are commonly included in investors' equity portfolios. As a result, accurate modeling of how information is transmitted and shared between oil, gold, and stock markets can significantly assist investors in making informed decisions about portfolio management and risk mitigation. By understanding the dynamics of these markets and their interplay, investors can enhance their strategies and optimize their investment outcomes [1].

Numerous studies have extensively explored the relationship between oil price fluctuations and economic activity in both developed and emerging countries, see for example [2]. They investigated the co-movement of oil prices and stock prices in six countries, namely, China, India, Japan (top oil importers), Saudi Arabia, Russia, and Canada (oil exporters). They found a positive and dynamic relationship between the returns of oil and the stock markets in all six countries. The study highlights that the correlations between oil and stock markets are notably strong and statistically significant for most oil-importing countries, particularly in the period following the global financial crisis, from 2008 to 2011. On the other hand [3] examined the marginal effect of uncertainty regarding stock and oil prices on the returns of the United States, Japan, Korea, and Hong Kong. The study used daily data on stock prices and crude oil prices spanning from 1996 to 2015. The findings of this study indicated that price uncertainty plays a significant role in shaping the relationship between oil prices [3], and stock markets in these countries.

Crude oil is considered one of the most important commodities in all world economies. The Gulf Cooperation Council (GCC) countries are considered major contributors to the global economy as oil suppliers, boasting approximately 50% of the world's oil reserves. Consequently, oil prices have been suggested as a crucial global determinant affecting the stock markets of GCC countries [4].

Abuzayed and Al-Fayoumi [4] investigated the spillover of systemic risk between the crude oil market and the returns of individual stock markets in six GCC countries (Saudi Arabia, the UAE, Kuwait, Qatar, Oman, and Bahrain) both before and during the COVID-19 pandemic. The findings of the study revealed that all GCC markets experienced higher levels of systemic risk emanating from the crude oil market. However, it was observed that Saudi Arabia and the UAE were more vulnerable to oil-related risks compared to the other markets. In the same vein, [5] assessed the in-sample and out-of-sample performance of two GPR indices (global GPR and Saudi GPR) and crude oil returns in predicting GCC stock returns, using different estimators, namely the ordinary least squares (OLS) estimator, the bias-adjusted least squares (ALS) estimator, and the feasible generalized least squares (FGLS) estimator. They found that the predictability of crude oil returns is particularly effective in the stock markets of Kuwait, Oman, Qatar, Saudi Arabia, and the United Arab Emirates (UAE). There are a few studies that address the integration between the GCC markets, including [6] which implement the autoregressive distributed lag methodology to investigate the integration between the GCC stock markets, they used Abu Dhabi stock market to present the UAE market. Based on their empirical study, results indicate the relationship between the six GCC stock indices, and Abu Dhabi, Kuwaiti and Saudi stock markets affected by each other.

Recently, machine learning has gained increasing prominence in the fields of economics and finance, primarily due to its significant advantage of efficiently handling vast quantities of structured and unstructured data. Its rapid decision-making and predictive capabilities further contribute to its widespread adoption in these domains [7]. For instance, study by [8] aimed to assess the predictive ability of oil prices and their impact on the Saudi stock market, considering six international stock markets (China, France, UK, Germany, Japan, and the USA). The study used five machine-learning techniques to analyze the data. The findings of this study provided compelling evidence that oil prices exert the most significant influence on the movements of the Saudi stock market. Additionally, the study revealed that the Chinese stock market emerges as the second most influential factor, likely attributed to the extensive trade relationship between China and Saudi Arabia.

Niu et al. [9] used machine learning to investigate how geopolitical risks influence the prediction performance of the US stock market volatility. They argued that the machine learning models demonstrated superior predictive ability compared to the benchmark model. By employing machine learning methods, more precise forecasts of stock market volatility can be achieved, enabling a deeper understanding of how various geopolitical risks influence the stock market. These findings hold significant value for financial institutions, investors, policymakers, and other stakeholders, offering them more accurate and valuable insights into geopolitical risks and their impact on financial markets.

Therefore, this research investigates the existence of a relationship between UAE stock market indices and the other GCC, China, Oil and Gas Markets based on empirical study using multi regression machine learning technique.

Methodology

This study employs a quantitative research design to predict UAE stock markets using multiple regression analysis. The focus is on identifying the relationship between various independent variables and UAE stock market performance.

Raw data undergoes a comprehensive cleaning process to handle missing values, outliers, and inconsistencies. This ensures the integrity of the dataset and the reliability of subsequent analyses.

Multiple Regression Model

The multiple regression model is formulated based on the selected independent variables. The equation is of the form:

$$Y = \beta_0 + \beta_1 X_1 + \beta_2 X_2 + \ldots + \beta_n X_n + \varepsilon$$

where Y represents the UAE stock market return, X1, X2, …, Xn is the return of the chosen independent variables, n is the number of the independent variables, β0 is the intercept, β1, β2, …, βn are the coefficients, and ϵ is the error term.

The regression model is estimated using the R programming language. The variance inflation factor is used to assess the multicollinearity in the multiple regression model. The data is divided into training and test groups, the training is 75% of the data and the rest of the data is 25% of the test group. This helps evaluate the model's performance in predicting unseen data. The predictive accuracy of the model is assessed using Mean Squared Error (MSE), and Root Mean Squared Error (RMSE).

Data Description

The stock markets in the GCC region are Bahrain (BRN), Kuwait (KWT), Oman (OMN), Qatar (QTR) and Saudi Arabia (KSA) and the UAE which consists of Abu Dhabi (DHB), Dubai (DUB) stock markets. The sample period for this study is from June 2010 to June 2023 a total of 681 observations. We will test the predictability power of Brent Crude Oil, the Shanghai Stock Exchange as a proxy for the Chinese stock market (CHN) and gold price (GLD). The weekly data for each market are downloaded from the Investing.com and finance.yahoo.com websites.

The logarithmic difference of the weekly closing prices is used to compute the returns (r_t) as follows:

$$r_t = \log(P_t/P_{t-1}) * 100$$

where P_t and P_{t-1} are respectively the stock prices at week t and week $t - 1$.

Table 10.1 presents the descriptive statistics of our weekly returns. It is obvious that the distribution of the returns is so far from normality, that the p-value of [10] test is equal to zero. Therefore, all the variables have significantly non-normal distributions at the 1% significance level. In addition, from the table, we can conclude that all the variables are left skewed and leptokurtic, except gold returns (rGLD) is left-skewed and Platykurtic.

Table 10.1 Descriptive statistics of the weekly returns over the period from June 2010 to June 2023 with 681 observations

Variables	Mean	Sd	Min	Max	Skew	Kurtosis	Jarque–Bera
rDUB	0.13	2.92	−19.13	12.60	−0.74	5.25	< 2.2e-16
rDHB	0.20	2.09	−16.87	9.03	−0.80	7.30	< 2.2e-16
rBRN	0.05	1.28	−11.89	4.75	−1.44	12.34	< 2.2e-16
rKWT	0.02	2.38	−36.10	7.44	−6.00	83.66	< 2.2e-16
rOMN	−0.04	1.65	−12.44	12.38	−0.44	10.83	< 2.2e-16
rQTR	0.13	2.31	−11.87	12.45	−0.20	3.83	< 2.2e-16
rKSA	0.09	2.47	−16.27	13.76	−0.70	5.19	< 2.2e-16
rOIL	0.00	4.92	−29.07	31.35	−0.28	6.54	< 2.2e-16
rCHN	0.03	2.67	−14.29	9.11	−0.65	3.48	< 2.2e-16
rGLD	0.06	2.15	−10.13	9.01	−0.31	2.07	< 2.2e-16

Data Analysis

It is common to use a portion of the available data for estimating (or training) the model, and the rest of the data to measure the forecasting performance. Therefore, we divide the data into in-sample training data and out-of-sample testing data. We take 75% of the data as an in-sample with 510 observations, and the remaining 25% is considered out-of-sample with 171 observations.

In-Sample Analysis

We estimate our in-sample dependent variables: Abu Dhabi and Dubai weekly returns using the independent variables (weekly returns of the other GCC, China, Oil and Gold) with 75% of the data (510 observations). First, we will estimate Abu Dhabi returns.

Abu Dhabi Stock Market

Here, multiple regression is used to estimate Abu Dhabi's weekly return (rDHB) using our independent variables for the in-sample period. Table 10.2 presents the estimated coefficients and their significance. The R-square value of 0.38 suggests that approximately 38% of the variance in Abu Dhabi returns can be explained by the selected independent variables.

The results indicate that Abu Dhabi returns are affected by the returns of Qatar, Oman, Kuwait, and Bahrain. In addition, Saudia Arabia, Oil, China, and Gold didn't show a relationship with Abu Dhabi during this study period.

Table 10.2 In-sample estimated coefficients for Abu Dhabi weekly returns using multiple linear regression

	Estimate	Std. Error	t-value	P-value
Intercept	0.16	0.07	2.18	0.03 *
rKSA	0.001	0.04	0.03	0.98
rBRN	0.20	0.07	2.96	0.003 **
rOMN	0.20	0.05	4.18	3.43e-5 ***
rQTR	0.35	0.04	8.39	5.14e-16 ***
rKWT	0.12	0.03	3.61	0.0003 ***
rOil	0.02	0.02	0.90	0.37
rCHN	0.03	0.03	1.08	0.28
rGLD	0.01	0.03	0.39	0.70

Signif. codes: '***' 0.001 '**' 0.01 '*' 0.05 '.' 0.1
Multiple R-squared: 0.3836, F-statistic: 38.97 on 8 and 501 DF, p-value: < 2.2e-16.

The relation with Kuwait consists by [6] results, but they investigate the relation with Saudia Arabia in contrast of our result, which may be due to the difference in the study time, since their period is from 2002 to 2006.

Figure 10.1 shows the variable importance for each independent variable in our Abu Dhabi multiple regression model. We can notice that Qatar is the most important one, then Oman, Kuwait, and Bahrain.

Dubai Stock Market

Again, for the in-sample period, multiple regression is used to estimate the coefficients for our independent variables but this time, Dubai weekly return (rDUB) is considered as the dependent variable. Table 10.3 presents the estimated coefficients

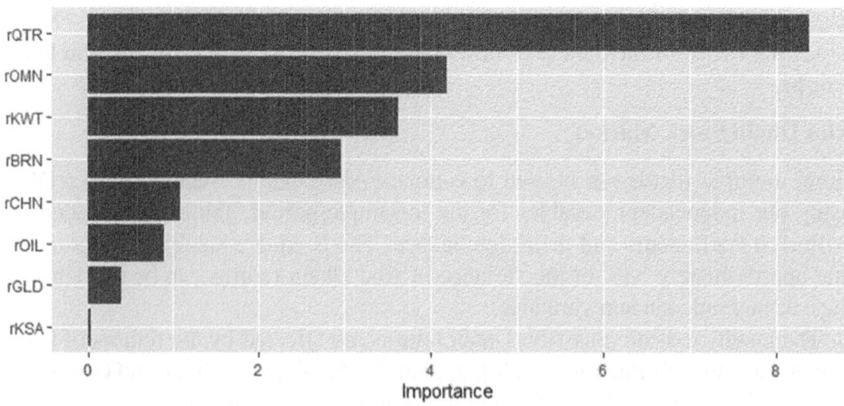

Fig. 10.1 Variable importance for each independent variable in Abu Dhabi multiple regression model

and their significance. The R-square value of 0.5 suggests that approximately 50% of the variance in Dubai returns can be explained by the selected independent variables.

The results show that Dubai returns are affected by the returns of Qatar, Oman, Kuwait, and Saudi Arabia. In addition, China has a significant relationship with Dubai in contrast to Abu Dhabi, due to the trade between them. Same as Abu Dhabi, Oil and Gold don't show a relationship with Dubai. However, [5] indicates that the predictability of oil returns is effective in the stock markets of the United Arab Emirates. Maybe the strong relationship of the UAE returns with the other GCC independent variables erased this relationship. Figure 10.2 shows the variable importance for each independent variable in our Dubai multiple regression model. We can notice that Qatar and Oman are the most important, then Bahrain and Kuwait, and then Saudi Arabia and China.

Table 10.3 In-sample estimated coefficients for Abu Dubai weekly returns using multiple linear regression

	Estimate	Std. Error	t-value	P-value
Intercept	0.08	0.10	0.88	0.38
rKSA	0.08	0.05	1.74	0.08
rBRN	0.37	0.09	4.30	2.07e-5 ***
rOMN	0.45	0.06	7.38	6.86e-13 ***
rQTR	0.48	0.05	8.89	< 2e-16 ***
rKWT	0.14	0.04	3.30	0.001 **
rOil	0.01	0.02	0.51	0.61
rCHN	0.06	0.04	1.68	0.09
rGLD	−0.03	0.04	−0.63	0.53

Signif. codes: '***' 0.001 '**' 0.01 '*' 0.05 '.' 0.1
Multiple R-squared: 0.5001, F-statistic: 62.65 on 8 and 501 DF, p-value: < 2.2e-16.

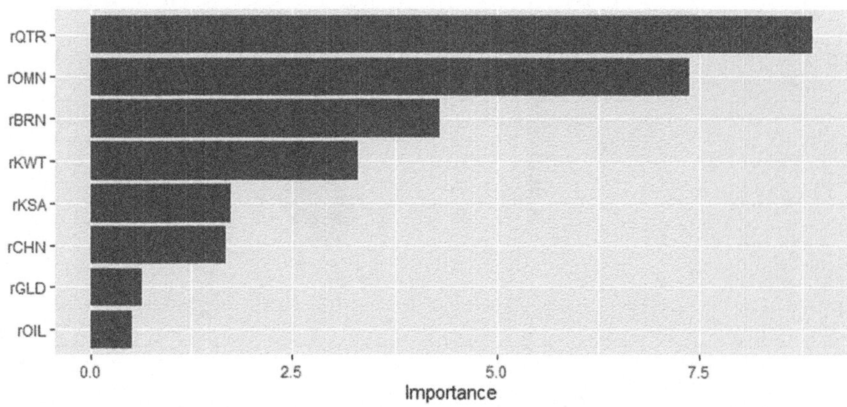

Fig. 10.2 Variable importance for each independent variable in Dubai multiple regression model

Table 10.4 The VIF for the multiple regression model

Variable	rKSA	rBRN	rOMN	rQTR
VIF	1.5	1.3	1.2	1.7
Variable	rKWT	rOIL	rCHN	rGLD
VIF	1.4	1.2	1.0	1.0

To assess the multicollinearity in the multiple regression model, the variance inflation factor (VIF) can be used. Accordingly, the VIF is calculated in Table 10.4 to obtain an idea about the VIF for our models. The findings indicate that all variables in the multiple regression models have a very small VIF (less than 2), indicating there is no multicollinearity.

Out-Of-Sample Analysis

We use the in-sample coefficients' estimations to forecast the out-of-sample dependent variables. Then, we compare the forecasts of the model with the true observed values using the mean absolute error (MAE) and root mean square error (RMSE) forecasting criteria. Figures 10.3 and 10.4 present the scatter plots of the predicted vs actual returns for Abu Dhabi and Dubai, respectively, in which dispersed of the points shows how much the goodness of fit of the model. Here, the points of Fig. 10.3 more closely to the diagonal line, which mean the model of Abu Dhabi better fitted the out-of-sample data than Dubai.

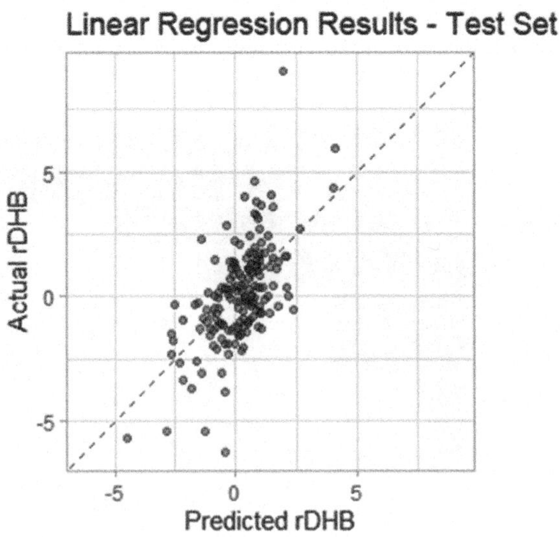

Fig. 10.3 Presents the predicted and actual Abu Dhabi returns for the out-of-sample period

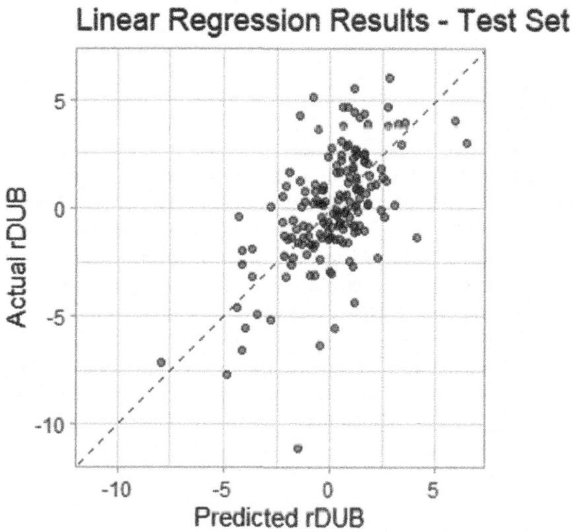

Fig. 10.4 Presents the predicted and actual Dubai returns for the out-of-sample period

Table 10.5 MAE and RMSE for the out-of-sample period

	Abu Dhabi	Dubai
MAE	1.21	1.64
RMSE	1.58	2.17

MAE and RMSE are obtained in Table 10.5 for both Abu Dhabi and Dubai. As expected from Figs. 10.3 and 10.4, MAE and RMSE are smaller for Abu Dhabi, which means the model of Abu Dhabi best detects the relationship between the dependent and independent variables and it has better forecasting accuracy than Dubai.

Conclusion and Recommendation

The paper investigates the relationship of Abu Dhabi and Dubai weekly returns among the other GCC, China, Oil and Gold returns for the period from Jun. 2010–Jun. 2023. Multiple linear regression is implemented for the in-sample and out-of-sample analysis.

The results indicate that both UAE returns indicated a significant relation with Qatar, Oman, Bahrain, and Kuwait. Dubai's returns were significantly affected by Saudi Arabia and China. Oil and Gold returns don't show a significant relationship with the two UAE returns, perhaps the strong relationship with the other independent variables erased this relationship.

The research is important for investors and the government; however, few papers have addressed this topic. Thus, we recommend conducting more research to examine the relation of each GCC stock market to each other or with Western stock markets.

References

1. Mensi, W., Reboredo, J. C., & Ugolini, A. (2021). Price-switching spillovers between gold, oil, and stock markets: Evidence from the USA and China during the COVID-19 pandemic. *Resources Policy 73*(June)
2. Jiang, Z., & Yoon, S. M. (2020). Dynamic co-movement between oil and stock markets in oil-importing and oil-exporting countries: Two types of wavelet analysis. *Energy Economics, 90.*
3. Joo, Y. C., & Park, S. Y. (2017). Oil prices and stock markets: Does the effect of uncertainty change over time? *Energy Economics, 61*, 42–51.
4. Abuzayed, B., & Al-Fayoumi, N. (2021). Risk spillover from crude oil prices to GCC stock market returns: New evidence during the COVID-19 outbreak. *The North American Journal of Economics and Finance, 58.*
5. Alqahtani, A., Bouri, E., & Vo, X. V. (2020). Predictability of GCC stock returns: The role of geopolitical risk and crude oil returns. *Economic Analysis and Policy, 68*, 239–249.
6. Sbeiti, W., & Alshammari, T. (2010). Integration of stock markets in the GCC countries: An application of the ARDL bounds testing model. *European Journal of Economics, Finance and Administrative Sciences, 20.*
7. Al-Maadid, A., Alhazbi, S., & Al-Thelaya, K. (2022). Using machine learning to analyze the impact of coronavirus pandemic news on the stock markets in GCC countries. *Research in International Business and Finance, 61.*
8. Abdou, H., Abedin, M. Z., Elamer, A. A., & Ibrahim, B. A. (2023). The predictability power of oil and global stock markets on Saudi stock market: Application of machine learning techniques
9. Niu, Z., Wang, C., & Zhang, H. (2023). Forecasting stock market volatility with various geopolitical risks categories: New evidence from machine learning models. *International Review of Financial Analysis.*
10. Jarque, C., & Bera, A. (1980). Efficient tests for normality, homoscedasticity and serial independence of regression residuals. *Economics Letters, 6*(3), 255–259.

Chapter 11
Moneyball: Analyzing the Efficiency of English Premier League Strikers Using Data Envelopment Analysis

Aniekan Essien, Marios Kremantzis, Dhiraj Joshi, and Fatema Zaghloul

Abstract In the spirit of the 'moneyball' paradigm, this study introduces the Player Performance Efficiency Index (PEI), designed for English Premier League football players, using an advanced Data Envelopment Analysis (DEA) framework. Our study improves upon traditional DEA models by adding relative weight constraints, positive weight attributions, and cross-efficiency considerations, proposing an index that closely mirrors real-world conditions and aligns with expert opinions. The PEI objectively evaluates players across four key technical dimensions: (i) goal-scoring, (ii) creative playmaking, (iii) dribbling, and (iv) defense. It serves as a valuable decision-making tool for club management, coaches, and players. Empirical findings from our study show a strong correlation between player efficiency and team success, particularly in goal-scoring and playmaking. The PEI enhances existing performance metrics, offering a more comprehensive, objective assessment of player performance.

Keywords Data Envelopment Analysis (DEA) · Player Performance Efficiency Index (PEI) · English Premier League (EPL) · Football analytics

Introduction

In professional football, decision-making has traditionally been based on subjective judgments and human intuitions, often leading to biases and inefficiencies in areas such as player recruitment and fiscal management [1]. The concept of "moneyball" has shown that data-driven analytics can augment these traditional methods by providing objective and precise performance assessments [2]. This need for objective analysis is particularly pressing in the English Premier League (EPL), being a major economic contributor with significant financial expenditures on player salaries

A. Essien (✉) · M. Kremantzis · D. Joshi · F. Zaghloul
Operations and Management Science, Healthcare and Innovation Group, University of Bristol Business School, Bristol, UK
e-mail: aniekan.essien@bristol.ac.uk

© The Author(s), under exclusive license to Springer Nature Switzerland AG 2024

A. Emrouznejad et al. (eds.), *Business Analytics and Decision Making in Practice*, Lecture Notes in Operations Research, https://doi.org/10.1007/978-3-031-61589-4_11

and transfer fees [3]. Certain football clubs, for instance Crystal Palace F.C., often spend heavily on players but fail to achieve corresponding success on the field. This suggests inefficiencies in player performance, salaries, and transfer strategies. Conversely, financially constrained teams such as Brighton, Burnley, and Fulham struggle due to limited resources. To address this, a more objective analytical tool is essential for better player team decision-making.

Current player performance metrics, such as WhoScored ratings, often fail to provide a holistic view as they solely focus on outputs and do not account for varying inputs like playing time, salary, or team-mate ability [4]. These traditional metrics also lack the flexibility to adapt to individual roles and unique playing styles, potentially leading to uninformed decisions on player acquisitions and strategies [5, 6]. Moreover, the homogeneous weightage system applied to player outputs fails to recognize individual roles, responsibilities, and unique playing styles, further diluting the efficacy of such evaluations. In this regard, providing an enhanced performance efficiency index (PEI) could be useful for players and managers as they can pinpoint which of the four technical areas a player can improve upon, as well as identify transfer targets based on the profile of player a manager is looking for.

To address this gap, our research introduces a Performance Efficiency Index (PEI), rooted in the principles of 'moneyball' and operationalized using Data Envelopment Analysis (DEA). The PEI can be used to provide a comprehensive evaluation for EPL players across four key technical dimensions of the game—(i) goal-scoring, (ii) creativity, (iii) dribbling, and (iv) defending. Such an index promises not only to revolutionize management strategies but to also offer actionable insights for coaching staff and players (refer to Table 11.1).

Comparing existing player performance indices in use, our proposed DEA-based framework has the following distinct contributions. Firstly, we conceptualize and design a player evaluation index that allows weights to vary across players, allowing each player to reflect their different playstyle characteristics. Secondly, we establish weight constraints, based on expert consensus, which mirror the evolving dynamics of contemporary football. Our methodology draws inspiration from previous works, such as, Cooper et al. (2007) to align football player assessments with contemporary analytics and player-centric attributes.

The remainder of this study is organized as follows. Section 11.2 reviews the literature and the theory underlying the usage of DEA in developing performance measurement indexes in a sporting context. Section 11.3 outlines the methodology, including DEA and its extensions, along with the process for weight selection and the DEA cross-efficiency. In Sect. 11.4, we apply the methodology to assess the performance and cost-efficiency of EPL football players. The concluding section discusses key findings and highlights future research directions.

Table 11.1 Performance Efficiency Index (PEI) use cases

Entity	Use cases
Managers and coaches	Models can indicate how to improve the performance of inefficient players
	Models help to identify relative strengths and weaknesses of players within the team, as well as for opposition players
	Models provide benchmarks for performances, which coaches can utilize to improve the efficiency of their own players
	Models can inform recruitment policy, contract negotiation and minutes allocated for each player
Players	Models can indicate how inefficient players can improve in order to become efficient
	Models help players understand their strengths and weaknesses, as well as for the opposition players
	Models provide benchmarks for which players can look to learn from and emulate
	Models can inform and provide evidence or justification for an increase in salary
Scouts and transfer policy	Models can identify the most efficient players, under and over-valued players
	Models can inform transfer policy

Literature Review

DEA for Player Performance Evaluation

This study proposes a Performance Efficiency Index (PEI) and establishes a positive correlation between individual player efficiency and overall team success. Unlike prior research that either employed limited metrics or faced challenges due to small sample sizes [7, 8], our approach uses a comprehensive set of performance metrics and incorporates expert opinions in weighting, a methodology validated in other sports [9]. We further adapt DEA cross-efficiency techniques to provide unique player rankings [10, 11], aiding stakeholders in decision-making related to player performance, compensation, and recruitment.

Prior research on team efficiency and success has been mixed. For example, [12] found a strong correlation in the English Premier League and Major League Soccer. [9] applied DEA to Spanish Basketball League player performance, using non-zero weights and relative weight restrictions (RWRs) to address standard DEA limitations [8, 13]. Our study adds to this conversation by calculating a team's weighted-average player efficiency. We find that improved player efficiency positively correlates with team success, endorsing DEA's applicability in evaluating both individual and team performance.

Critiques and Limitations of Existing Performance Indices

The application of data-driven approaches for performance management in sports has garnered considerable attention from analysts, coaches, and stakeholders. Consequently, the application of several metrices for performance measurement is becoming popular. However, the limitations of commonly used evaluative tools cannot be ignored. The use of single metrics such as "goals scored" or "assists" in football, for instance, is often criticized for its one-dimensional nature, as it fails to capture the complexity of player contributions. Even advanced statistics, such as Player Efficiency Rating (PER) in basketball, have been criticized for their limitations. While PER attempts to synthesize a player's positive and negative contributions into one number, it can disproportionately favor offensive players [14].

The "sabermetric" approach, exemplified by metrics like WAR (Wins Above Replacement) in baseball, is not without its shortcomings [15]. These advanced metrics require comprehensive data and complicated calculations, making them inaccessible to the average fan or lower-tier teams with fewer analytical resources [16].

The challenge of creating a universal performance metric is significant, as seen in flawed systems like the FIFA World Cup's Castrol Index [17], and cricket's batting average [18]. These metrics often miss nuanced roles, affecting talent scouting and contract negotiations. For example, the NBA's reliance on PER has limitations in predicting professional success [19] and in football, metrics including "goals scored" undervalue key roles like playmaking [4]. Our study addresses limitations, such as unrealistic DEA weightings, as seen in [6] 5, by introducing relative weight restrictions. We also employ a more detailed set of outputs, representing various technical aspects of football, including goal-scoring, creativity, dribbling, ball carrying, and defense.

Methodology

Data Envelopment Analysis

Standard Data Envelopment Analysis

In general, DEA enables the analysis of the relative efficiency of a set of ***n*** decision-making units (DMUs) that consume ***m*** inputs to generate ***s*** outputs, using the following dual problems (1) and (2):

$$max\phi + \varepsilon\left(\sum_{i=1}^{m} s_{i0}^{-} + \sum_{r=1}^{s} s_{r0}^{+}\right) \quad (11.1)$$

Subject to

$$\sum_{j=1}^{n} \lambda_i x_{ij} = x_{i0} - s_{i0}^{-}, i = 1, \ldots m$$

$$\sum_{j=1}^{n} \lambda_i x_{rj} = \phi_0 y_{r0} + s_{r0}^{+}, r = 1, \ldots s$$

$$\lambda j \geq 0 j = 1, \ldots, n$$

$$s_{i0}^{-} \geq i = 1, \ldots, m$$

$$s_{r0}^{+} \geq r = 1, \ldots, s$$

- Where $\varepsilon > 0$ is an element equal to 0.000001 and helps us incorporate strictly positive weights.
- x_{ij}represents the inputs $i(i = 1, 2, ..., m)$ amounts recorded for $DMU_j, j = 1 \ldots n$.
- y_{rj} represents the inputs $r(r = 1, 2, ..., s)$ amounts recorded for $DMU_j, j = 1 \ldots n$.
- x_{i0} and y_{r0} represents represent the input and output amounts for the DMU_0 being evaluated.

The dual problem:

$$min\omega = \sum_{i=1}^{m} v_i x_{i0} \tag{11.2}$$

Subject to

$$\sum_{r=1}^{s} \mu_r y_{r0} = 1 \tag{11.2}$$

$$-\sum_{i=1}^{m} v_i x_{ij} + \sum_{r=1}^{s} \mu_r y_{rj} \geq 0, j = 1, \ldots, n$$

$$v_i \geq \varepsilon i = 1, \ldots, m$$

$$\mu_r^{-} \geq r = 1, \ldots, s$$

- Where μ_r represents output weights, vi represents input weights.

DEA Extensions

(i) *Incorporating Relative Weight Restrictions (RWRs):* Relative weight restrictions are introduced in (2) to eliminate unrealistic weighting schemes. The relative weight restrictions take the following general form:

$$\mu\text{r} \leq \mu\text{r}'$$

Therefore, these are included as constraints in (2) for our analysis. Where $\mu^{r'}$ and μ^{r} are the weights of outputs r' and r, respectively.

(ii) *Cross-Efficiency (Ranking):* The mathematical formulation for the cross-efficiency evaluation method used in the DEA models provides a unique ranking order of player efficiencies. Additionally, it is worth considering the potential application of the cross-efficiency extension, as outlined in Eqs. (3) and (4). The cross-efficiency score is derived as the mean of these cross-efficiencies, serving the purpose of mitigating any lingering issues associated with impractical weighting schemes, as discussed by Anderson et al. (2002).

$$E_{dj} = \frac{\sum_{r=1}^{s} \mu_r^{d*} y_{rj}}{\sum_{i=1}^{m} v_i^{d*} y_{ij}} \tag{11.3}$$

$$\overline{E_j} = \frac{1}{n}\sum_{d=1}^{n} E_{dj} j = 1, \ldots, n \tag{11.4}$$

$$-\sum_{i=1}^{m} v_i x_{ij} + \sum_{r=1}^{s} \mu_r y_{rj} \geq 0, j = 1, \ldots, n$$

- Where $\mu_r^{d^*}$ and $v_i^{d^*}$ are the weights of outputs and inputs for $DMU_j, j = 1 \ldots n$.

Data and Variable Selection

The analysis uses secondary data from the Premier League 2020/21 season, sourced from FBRef[1] and cross-verified with the official website. Salary data is obtained from Capology,[2] considered the most accurate by researchers [20]. Players with less than 500 min played are excluded for reliability [6, 9, 12]. The study focuses on one position: Strikers, with a final sample size of 55, meeting DEA's validity criteria for sample size [21].

[1] https://fbref.com/en/.

[2] https://www.capology.com.

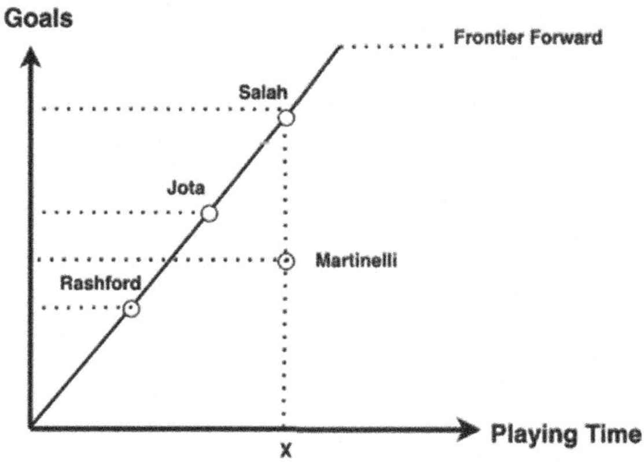

Fig. 11.1 Output-oriented CRS DEA

In Fig. 11.1, we observe an example of DEA in action, comparing the goal-scoring efficiency of forwards including Rashford, Jota, Salah, and Martinelli based on playing time. Rashford, Jota, and Salah are efficient (score goals effectively), receiving an efficiency score of 1 (or 100%), while Martinelli's score is 0.68 (68%), indicating room for improvement.

Inputs and Outputs

In line with [6] 7, our study estimates DEA models for each playing position to maintain homogeneity. Positions are viewed as unique "production technologies" with differing input–output relationships. For instance, forwards and center backs should not be compared on goal-scoring efficiency. The Kruskal–Wallis test confirms statistically significant differences in metrics across positions, supporting position-based and individual analyses. While further segmenting could offer granularity, it may reduce sample sizes and introduce biases.

Primary Data

A survey of 61 amateur footballers and experts informed objective relative weight restrictions for key technical areas of the game. Respondents rated the importance of each metric on a 1–7 scale. If a metric had a statistically significant difference in average rating, it was incorporated as a relative weight restriction in the DEA Models. This ensures that the weights accurately reflect the importance of each metric in its technical area. Summary of these results and weight restrictions can be found in the results section.

Two DEA Models are proposed (refer to Table 11.2): Model-1 focuses on Pure Footballing Efficiency (PFE), providing unique player rankings through cross-efficiency, while Model-2 assesses Cost-Efficiency (CE), considering player salary to identify undervalued or overvalued assets. Both models explore the link between individual efficiency and team success, corroborated by studies such as [6].

Table 11.2 Proposed DEA models

Model	Input	Outputs (4 key areas)
DEA model-1 (Pure footballing efficiency)	Minutes played	1. Goal scoring a. Goals b. SOT c. G-Xg d. GCA e. Big Chances Missed
		2. Creativity (chance creation) a. Assists or XA b. Key passes c. PPA d. SCA
DEA Model-2 (Cost efficiency)	Salary	3. Defensive a. Tackles won b. Blocks and interceptions c. Pressures successful d. Mistake leading to goal
		4. Dribbling and ball carrying a. Dribbles successful b. PrgDist carried c. Carries into final third d. Carries into penalty box e. GCA from dribbling f. SCA from dribbling

Result and Discussion

DEA Model 1: Assessment of Strikers

The sample includes 52 strikers and focuses on two key technical areas: Goal-Scoring and Creativity, as depicted in Fig. 11.2. The study demonstrates how results in these areas can be universally interpreted to identify player strengths and areas for improvement. The columns in the figure represent efficiency in each technical area, as well as cross-efficiency rankings. The optimal weights (color-coded) illustrate player strengths and weaknesses. The last column outlines performance deficiencies, providing a framework for improvement targets. These deficiencies are calculated similarly to [9], offering a metric for gauging a player's potential for enhancement.

Figure 11.2 ranks the Top 25 EPL strikers by goal-scoring efficiency. Harry Kane and Gareth Bale lead as benchmarks, being fully efficient. Kane's efficiency is expected due to his 23 goals and 13 Man Of The Match (MOTM) Awards in 2020/21. Bale's efficiency is notable; despite playing only 920 min, he scored 11 goals and earned 4 MOTM Awards. In contrast, Marcus Rashford and Timo Werner, with

Player	Goal-Scoring Index	Goals	SoT	G-xG	GCA	XE Rank
Harry Kane	1.000	0.00%	0.00%	0.00%	0.00%	4
Gareth Bale	1.000	0.00%	0.00%	0.00%	0.00%	1
Riyad Mahrez	0.945	32.95%	93.65%	0.19%	2.81%	20
Jamie Vardy	0.912	23.53%	78.25%	13.91%	2.07%	17
Timo Werner	0.900	104.86%	68.51%	21.93%	2.27%	42
Marcus Rashford	0.760	77.88%	59.10%	1.36%	4.94%	29
Patrick Bamford	0.731	14.57%	46.46%	3.29%	8.72%	12
Raheem Sterling	0.729	48.19%	75.68%	6.23%	4.21%	27
Mbaye Diagne	0.726	59.30%	176.50%	5.81%	4.92%	43
Mohamed Salah	0.714	39.63%	45.33%	0.94%	15.69%	5
Diogo Jota	0.705	9.00%	80.97%	0.84%	9.34%	2
Callum Wilson	0.693	20.61%	121.10%	2.21%	4.79%	13
Michail Antonio	0.692	28.24%	90.17%	4.80%	4.40%	18
Son Heung-min	0.691	19.09%	60.80%	-1.95%	15.83%	15
Kelechi Iheanacho	0.688	13.72%	83.85%	-0.27%	11.09%	3
RaÃºl JimÃ©nez	0.684	42.27%	165.98%	1.43%	6.85%	19
Anthony Martial	0.666	57.42%	97.36%	6.81%	4.93%	38
Gabriel Jesus	0.657	35.99%	121.10%	2.72%	4.71%	23
Danny Ings	0.640	22.17%	86.92%	0.01%	5.92%	13
Wilfried Zaha	0.618	44.06%	106.09%	-0.61%	7.19%	25
Edinson Cavani	0.608	19.53%	126.97%	-0.11%	8.78%	6
Ollie Watkins	0.567	43.68%	46.46%	4.49%	10.57%	22
Alexandre Lacazette	0.565	14.61%	83.85%	-0.51%	10.30%	7
Sadio ManÃ©	0.554	61.43%	60.80%	9.79%	10.02%	25
Dominic Calvert-Lewin	0.553	16.76%	45.33%	1.36%	17.21%	10
Average Weights		0.1304	0.0221	0.0734	0.2699	
(%) of Overall Weights		25.94%	4.39%	14.60%	53.70%	

Fig. 11.2 Goal-scoring efficiency scores and improvement targets (Strikers)

over 2600 min each, are classified as inefficient, scoring 0.760 and 0.900, respectively. Using optimal weights transformed into percentage deficiencies [9], the analysis pinpoints areas for improvement. Rashford's goal deficiency stands at 77.88%, suggesting he needs about 19 goals per season considering his playing time. Werner's inefficiency lies in Goals (104.86%) and G-xG (18.96%), indicating poor finishing, but he shines in Goal-Creating-Actions (GCA) with a 2.27% deficiency. Werner's 12 assists confirm his playmaking skills.

DEA Model 2: Cost Efficiency Estimate of Strikers

DEA Model 2 evaluates the cost-efficiency of Premier League player wages across four technical areas, drawing on methods similar to [22] in MLB. This model serves as a scouting tool for identifying over- and under-valued players, informing transfer policy. Although valuable in the context of salary deliberations, the model acknowledges that it should not be the exclusive basis for determining player salaries. It recognizes that additional factors, as proposed by [20], play a role in influencing player wages. The primary objective is to highlight disparities in value rather than set rigid salary levels.

The Relationship Between Efficiency and Team Success

The study investigates the correlation between a team's weighted-average-player-efficiency score (TWAPES) and its season-ending Total Points to validate the utility of DEA models in football performance analysis. In line with [6], the hypothesis is that higher TWAPES values lead to higher Total Points. TWAPES is calculated as the average efficiency scores, weighted by player minutes. A strong correlation is found between TWAPES in key areas (goal-scoring, creativity, dribbling) and Total Points, with goal-scoring efficiency showing a significant 0.9087 correlation value (refer to Table 11.3). These results affirm previous findings that attacking efficiency strongly correlates with team success.

This study finds no significant correlation between defensive efficiency and team success, contrasting with [6] and [12]. The discrepancy may arise from different output variables used to measure defensive efficiency, such as omitting goals allowed and clean sheets. Although this does not diminish the significance of our defensive efficiency scores, it underscores the impact of output selection on these metrics. Nonetheless, the research affirms that the chosen DEA Models effectively capture player performance in various aspects such as goal-scoring, creativity, and dribbling, as demonstrated by robust correlations with team success.

Table 11.3 Pairwise Correlations between team success outcomes and TWAPES

	Points	Goals for	Goals against	SoT	Key passes
TWAPES					
Goal scoring efficiency	0.9087**	0.9**	0.7484**	0.7825**	0.6898**
Creativity efficiency	0.5932**	0.5326*	0.7724**	0.5167*	0.7472**
Dribbling and ball carrying efficiency	0.6168**	0.5688**	0.7641**	0.46	0.6712**
Defending efficiency	0.28	0.34	0.30	0.30	0.20
	Final third passes	Passes into penalty area	SCA	SCA drib	GCA
TWAPES					
Goal scoring efficiency	0.738**	0.5672**	0.778**	0.40	0.8851**
Creativity efficiency	0.7887**	0.6896**	0.7184**	0.44	0.4512*
Dribbling and ball carrying efficiency	0.8623**	0.653**	0.6923**	0.53	0.5723**
Defending efficiency	0.36	0.12	0.21	0.34	0.35

Significance levels: * = 5%, ** = 1% or lower

Case Study: Application of DEA Models to Crystal Palace F.C

In the 2020/21 EPL season, Crystal Palace F.C underperformed in relation to their expenses, finishing 14th and struggling in key metrics like goal-scoring and creativity. They were notably poor in statistics that correlate with team success, such as Shots and Key Passes. Their defensive performance was also lackluster, ranking 18th in goals allowed. However, they excelled in dribbling and shot-creating actions via dribbling. To address these issues, the club's management could consider signing a new striker to improve goal-scoring and creativity while also optimizing the efficiency of existing strikers like Christian Benteke, Mitchy Batshuayi, and Wilfred Zaha.

Player Strengths, Weaknesses and Targets for Improvement

DEA Model 1 reveals inefficiencies among Crystal Palace's strikers, confirming wider team analysis (refer to Fig. 11.3). Wilfred Zaha stands out with high efficiency scores in all four game areas. Christian Benteke has a notable premium in goal-scoring cross-efficiency, suggesting an area of strength. Batshuayi needs substantial improvement, with a goal-scoring rate that needs to rise by 58.54% for efficiency and a GCA increase of 9.81%. Zaha could still improve his goal-scoring and GCA by 44.06% and 7.19%, despite already being an efficient finisher. Batshuayi could learn from Zaha to better his own game. Defensively, Zaha excels, while Batshuayi and Benteke lag, especially in creativity.

DEA Model 1 Suggestions

These suggestions are based on the following criteria (**Error! Reference source not found.**):

• Goal scoring efficiency > 0.5	OR	• Creativity XE rank < 20
• Goal scoring XE rank < 20		• Creativity XE rank < 20

DEA Model-1 suggests ideal transfer targets for improving striker efficiency. Kelechi Iheanacho stands out for his high goal-scoring and creativity metrics, coupled with lower wages (£60,000 p/w) compared to current strikers such as Benteke (£120,000 p/w). Patrick Bamford excels in goal-scoring and defending, Adama Traore is a creativity and dribbling specialist, while Anwar El Ghazi offers good value with high cross-efficiency in goal-scoring and creativity. These players present targeted upgrades over current options like Benteke and Batshuayi.

Player	Goal-Scoring Index	Creativity Index	Dribbling and Ball Carrying	Defending
Michy Batshuayi	0.371	0.637	0.636	0.619
Wilfried Zaha	0.618	0.702	0.883	0.829
Christian Benteke	0.542	0.455	0.363	0.581

Player	Goal-Scoring XE Rank	Creativity XE Rank	Dribbling and Ball Carrying XE Rank	Defending XE Rank
Michy Batshuayi	39	32	19	47
Wilfried Zaha	25	15	5	7
Christian Benteke	9	39	35	38

Fig. 11.3 Summary of efficiencies and cross-efficiencies for CPFC Strikers (DEA model 1)

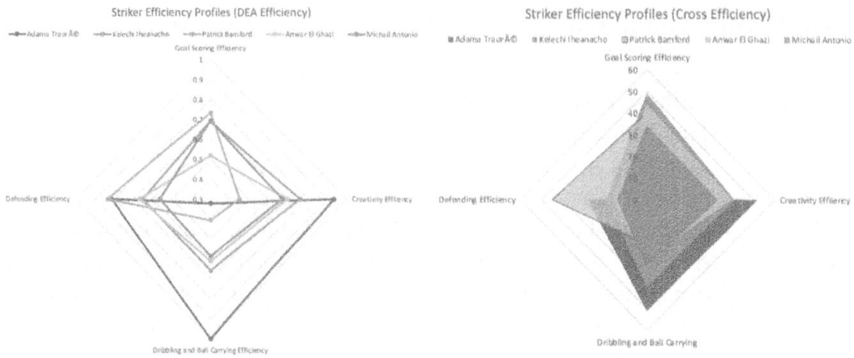

Fig. 11.4 Crystal palace transfer target efficiency and cross-efficiency profile

DEA Model-2 (Cost-Efficiency) Suggestions

DEA Model-1 assesses a player's performance solely in terms of their footballing efficiency, whereas DEA Model-2 is employed to validate whether the recommendations generated by DEA Model-1 are viable from a cost-efficiency perspective. DEA Model-2 can also be utilized to pinpoint players who demonstrate cost-efficiency and are potentially valuable investments. Indeed, Adama Traore, Anwar El Ghazi, Kelechi Iheanacho and Patrick Bamford fit the dual criteria of being footballing and cost-efficient, across all four areas of the game (refer to Fig. 11.4). Michail Antonio stands out as the least remarkable player, given his comparatively lower efficiency, both in terms of performance and cost. DEA Model-2 provides a clear indication that Adama Traore and Patrick Bamford are strong candidates for signing. Traore exhibits full cost-efficiency across all facets of his performance, while Bamford's only cost-related deficiency is associated with his creativity.

Conclusion and Future Research

In line with the objectives of this study, two DEA Models have been implemented which are capable of delivering valuable information regarding player performance to key stakeholders (e.g., managers, coaches and players). Further, in the results and discussion section, it was illustrated how the findings may be utilized to augment decision-making for football clubs, particularly through the hypothetical application of the results to Crystal Palace Football Club. Overall, DEA Model-1 evaluates a player's footballing efficiency by utilizing playing minutes as the only input and serves as the basis for assessing player strengths and weaknesses, in addition to providing objective targets for player improvement. DEA Model-1 is also useful for informing transfer policy through its ability to identify players who fit a specific efficiency profile. In contrast, DEA Model-2 assesses the efficiency of player salaries and reveals the most under- and overvalued players in the English Premier League—particularly useful for reaffirming the suggestions given by DEA Model-1, whilst informing transfer and remuneration policy. We advise decision-makers using the index to take into account the information derived from both the standard efficiency values and the cross-efficiency rankings when assessing potential targets.

Limitations and Future Work

It is important to consider a study of this nature against some limitations. First, our model assumes Constant-Returns-To-Scale (CRS), which may not fully capture player nuances due to factors like fatigue. There's room for incorporating evolving, more precise statistical measures. One way to improve the model is to assess player performance over time, not just in a single season. The Malmquist Productivity Index could provide insights into dynamic player performance. To enhance model accuracy, adding variables like xG-build-up can be beneficial for reflecting specific phases of play. Additionally, adapting the model to account for Decreasing-Returns-To-Scale could provide a more accurate reflection of player performance.

It would be worth attempting to fix some of the limitations outlined earlier. For example, variables such as xG-build-up could be an important addition to the model, as it can reflect the importance of a particular output in relation to a phase of play. This is particularly true, as the accuracy and accessibility of this measure improves. On the other hand, it would be worth exploring how decreasing returns to scale can be incorporated into the current models to allow for a more accurate reflection of reality.

References

1. Baumer, B. S., Matthews, G. J., & Nguyen, Q. (2023). Big ideas in sports analytics and statistical tools for their investigation. *Wiley Interdisciplinary Reviews: Computational Statistics*, Wiley Online Library, p. e1612.
2. Lewis, M. (2004). *Moneyball: The Art of Winning an Unfair Game*, WW Norton & Company.
3. EY. (2019). Premier League Economic and social impact. Retrieved 31 January, 2023, from https://assets.ey.com/content/dam/ey-sites/ey-com/pt_br/topics/ey-economic-advisory-/ey-premier-league-economic-and-social-impact-january-2019.pdf.
4. Brechot, M., & Flepp, R. (2020). Dealing with randomness in match outcomes: How to rethink performance evaluation in European club football using expected goals. *Journal of Sports Economics, Sage Publications Sage CA: Los Angeles, CA, 21*(4), 335–362.
5. Ruiz, J. L., Pastor, D., & Pastor, J. T. (2013). Assessing professional tennis players using data envelopment analysis (DEA). *Journal of Sports Economics, 14*(3), 276–302.
6. Tiedemann, T., Francksen, T., & Latacz-Lohmann, U. (2011). Assessing the performance of German Bundesliga football players: A non-parametric metafrontier approach. *Central European Journal of Operations Research, 19*(4), 571–587.

7. Oukil, A., & Govindaluri, S. M. (2017). A systematic approach for ranking football players within an integrated DEA-OWA framework. *Managerial and Decision Economics, Wiley Online Library, 38*(8), 1125–1136.
8. Thanassoulis, E. (2001). *Introduction to the theory and application of data envelopment analysis.* Kluwer Academic Publishers.
9. Cooper, W. W., Ruiz, J. L., & Sirvent, I. (2009). Selecting non-zero weights to evaluate effectiveness of basketball players with DEA. *European Journal of Operational Research, 195*(2), 563–574.
10. Cooper, W. W., Ramón, N., Ruiz, J. L., & Sirvent, I. (2011). Avoiding large differences in weights in cross-efficiency evaluations: Application to the ranking of basketball players. *Journal of CENTRUM Cathedra: The Business and Economics Research Journal, 4*(2), 197–215.
11. Li, Y., Wang, L., & Li, F. (2021). A data-driven prediction approach for sports team performance and its application to National Basketball Association. *Omega, Elsevier, 98*, 102123.
12. Haas, D. J. (2003). Technical efficiency in the major league soccer. *Journal of Sports Economics, Sage Publications, 4*(3), 203–215.
13. Allen, R., Athanassopoulos, A., Dyson, R. G., & Thanassoulis, E. (1997). Weights restrictions and value judgements in data envelopment analysis: Evolution, development and future directions. *Annals of Operations Research, 73*, 13–34.
14. Berri, D., & Schmidt, M. (2010). *Stumbling on Wins in Basketball*, Pearson Education.
15. Baumer, B. S., Jensen, S. T., & Matthews, G. J. (2015). OpenWAR: An open source system for evaluating overall player performance in major league baseball. *Journal of Quantitative Analysis in Sports, De Gruyter, 11*(2), 69–84.
16. Baumer, B., & Zimbalist, A. (2014). The Sabermetric Revolution: Assessing the Growth of Analytics in Baseball, University of Pennsylvania Press.
17. Chmura, P., Konefał, M., Andrzejewski, M., Kosowski, J., Rokita, A., & Chmura, J. (2017). Physical activity profile of 2014 FIFA World Cup players, with regard to different ranges of air temperature and relative humidity. *International Journal of Biometeorology, Springer, 61*, 677–684.
18. Preston, I., & Thomas, J. (2000). Batting strategy in limited overs cricket. *Journal of the Royal Statistical Society Series D: The Statistician, Oxford University Press, 49*(1), 95–106.
19. Berri, D. J. (2008). A simple model of worker productivity in the national basketball association. *The Business of Sports, Praeger Westport, CT, 3*, 1–40.
20. Li, C., Kampakis, S., & Treleaven, P. (2022). Machine Learning Modelling to Evaluate the Value of Football Players. ArXiv:2207.11361.
21. Nunamaker, T. R. (1985). Using data envelopment analysis to measure the efficiency of non-profit organizations: A critical evaluation. *Managerial and Decision Economics, 6*(1), 50–58.
22. Volz, B. (2009). Minority status and managerial survival in major league baseball. *Journal of Sports Economics, Sage Publications Sage CA: Los Angeles, CA, 10*(5), 522–542.

Chapter 12
The Drivers of Port Productivity for Selected Indian Ocean Ports Using the Malmquist Productivity Index

Adeola Oluwatoyin Osundiran and Makgopa Tshehla

Abstract Efficient ports are pivotal to the health of the nation's economy. For instance, when the port of Seychelles and Port Louis are put into consideration, it is noted that their respective ports are a major source of revenue. In Seychelles, the economy depends largely on the revenue it derives from imports and exports as well as tourism. In Mauritius, the port of Port Louis, handles the entire shipment that comes in to the country via sea. This paper examines the efficiency levels of these ports using the Malmquist based Data Envelopment Analysis. The review of literature lays further credence to the importance of this paper as there has been a paucity of articles on the efficiency level of the port of Port Louis and Port Victoria. Hence, this paper contributes to the growing body of knowledge on the efficiency analysis of ports using Data Envelopment Analysis based Malmquist Productivity Index. The period of examination is from 2008–2018. The results indicated that over the period of 2008 to 2018 the port of Port Louis achieved an annual average productivity gain of 0.91 whilst the Port of Victoria achieved 0.95. The drivers of productivity being tilted more towards technology change.

Keywords Mauritius · Victoria · Data envelopment analysis · Port efficiency · Malmquist productivity index

Introduction

Seychelles lies in the western Indian ocean between Madagascar and India [1]. The main engine of economic growth rests on tourism and fisheries [2]. The country relies heavily on international trade for importing goods as well as for income [3].

A. O. Osundiran · M. Tshehla (✉)
Department of Sustainable Livilihoods, Graduate School of Business Leadership, University of South Africa, Midrand 1686, South Africa
e-mail: tshehlmf@unisa.ac.za

A. O. Osundiran
e-mail: osundo@unisa.ac.za

© The Author(s), under exclusive license to Springer Nature Switzerland AG 2024
A. Emrouznejad et al. (eds.), *Business Analytics and Decision Making in Practice*, Lecture Notes in Operations Research, https://doi.org/10.1007/978-3-031-61589-4_12

The dependence of the country on international trade lends credence to the need to examine the efficiency levels of the port of Victoria, Seychelles.

The port of Port Louis is strategically situated in the Indian Ocean some 2000 kms off the eastern coast of Africa and 890 km to the east of Madagascar [4]. With the upsurge in South-South trade, Port Louis has emerged as a preferred way port for vessels plying on the key maritime routes to Asia, Africa and Latin America [5]. The port handles 99.5% of the country's total volume of external trade, that is almost 7.5 million tonnes of cargo annually.

The body of knowledge has confirmed that to effectively measure port efficiency and port performance, the Data Envelopment Analysis is a veritable tool to provide the necessary results [6]. This study focuses on using the Malmquist Productivity Index to evaluate the efficiency level for Port of Seychelles and Port Louis because these ports are part of the transport system.

In an instance when transport systems are deficient regarding capacity or reliability, they can have an economic cost such as reduced or missed opportunities and lower quality of life [7]. Ports efficiency has become an increasingly important topic because of the connecting links between different transport modes in the global logistics chain, especially the container terminals that are vital to the efficiency of the whole supply chain [8]. When transport systems are efficient, they provide economic and social opportunities and benefits that result in positive multiplier effects such as better accessibility to markets, employment and additional investments [7].

Ports form a vital link in the overall trading chain; consequently, port efficiency is an essential contributor to a nation's international competitiveness [9]. Hence, it is crucial to analyse the effectiveness of transit corridors in other to be able to improve the service of these corridors. The evaluation of efficiency is imperative to sustain a competitive and flourishing global business environment [10].

Data Envelopment Analysis (DEA) has been used frequently to analyze port efficiencies in many part of the world [11]. With respect to this paper, the Data Envelopment Analysis based Malmquist Productivity Index model will be used to establish the drivers of productivity of these selected ports for the period of 2008–2018.

The paper proceeds as follows. In the next section, we will review articles that made use of the DEA model to analyze port efficiency, Sect. 3 discusses the data used, while Sect. 4 deals with the results. Section 5 concludes with discussion on policy implications.

Back Ground of Study

The DEA techniques was developed, notably by [12, 13]. DEA was initially developed to evaluate nonprofit and governmental organisations, but it has subsequently been applied to the service operations of a variety of private companies [3]. The review of literature in DEA indicated its application numerously in ports situated in Asia, America, Latin America, Middle East, Mediterranean, Europe and North America, but very few studies for ports located in Indian Ocean Countries.

The Port of Victoria is a small island developing state and a net importer, its shipping industry is critical for the economy [2]. As the country's primary maritime gateway, the port of Port Louis plays a crucial role in the national economy [4].

The two ports under examination plays a significant role in the boosting of their nation's economy, hence it is important to use the Malmquist productivity index to determine what are the drivers of productivity for these two major ports. The standard approach to the measurement of productivity change over time is the Malmquist index [14–16].

Methodology

The method adopted in this research is exploratory as it is designed to determine the drivers of the productivity for the port of Port Louis and the Port of Victoria that are located Indian Ocean. In this research, the DEA empirical analysis uses two outputs, which are the vessel visits and the container throughput. The four inputs used include the number of berths, the number of cranes, number of tugs and the length of quays. The data sources were obtained from the port authority and also from port websites. Figure 12.1 captures the countries represented in the study.

The Malmquist total factor productivity index was introduced in 1953 before being further developed in the frame of DEA [17]. It was defined as quantity index

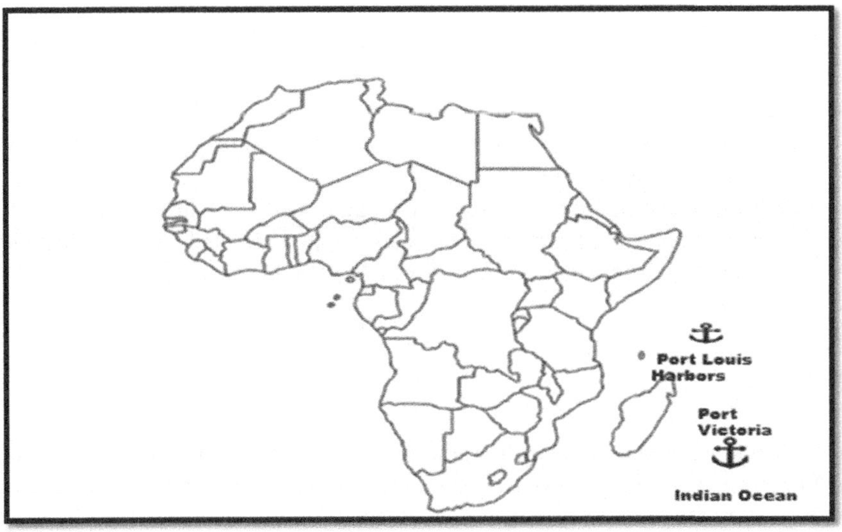

Fig. 12.1 Map of Africa showing Port Louis and Port Victoria. *Source* Osundiran, 2023

$$M\left(Y_{t+1},X_{t+1},Y_t,X_t\right) = \underbrace{\frac{D^t\left(Y_{t+1},X_{t+1}\right)}{D^t\left(Y_t,X_t\right)}}_{\text{Efficiency change}} \underbrace{\left[\frac{D^t\left(Y_{t+1},X_{t+1}\right)}{D^{t+1}\left(Y_{t+1},X_{t+1}\right)} X \frac{D^t\left(Y_t,X_t\right)}{D^{t+1}\left(Y_t,X_t\right)}\right]^{1/2}}_{\text{Technological Change}}$$

Fig. 12.2 Malmquist productivity index formula. *Source* Malmquist (1953)

(QI) as ratios of distance functions where observations were evaluated relative to an indifference curve (IC) [17]. The Malmquist index decomposes productivity change into two components: Catch-up which captures the change in technical efficiency over time, and frontier-shift, which captures the change in technology occurring over time [16]. The Malmquist index is a geometric mean of two indices, evaluated concerning period t and period t + 1 technologies [15]. Figure 12.2, illustrates the Malmquist equation that measures efficiency. The product of efficiency change and the technological change makes up the MPI.

where:

X_t and X_{t+1} input vectors of dimension at time t and t + 1.
Y_t and Y_{t+1} corresponding k- output vectors.
D_t and D_{t+1} denote an input

$$D(x, y) = \max(\rho : (s/\rho s \in L(y))) \tag{12.1}$$

- Where L(y) represents the number of all input vectors with which a certain output vector y can be produced, that is L(y) = (x:y can be produced with x).
- P in Eq. (2) can be understood as a reciprocal value of the factor by which the total inputs could be maximally reduced without reducing output.
- M = measures the productivity change between periods t and t + 1. Productivity declines if M < 1, remains unchanged if M = 1 and improves if M > 1.

Results

The Malmquist model captures the variations in the port performances for the Port of Victoria and port of Port Louis. In the computation of DEA MPI, two significant issues are established, firstly it is the efficiency catch up also known as technical efficiency and the boundary shift technological change, which is also known as the technology change. The Malmquist model allows for the determination of the drivers of productivity which could be efficiency or technology. All the ports reveal several fluctuations in their efficiency levels in the 11-year period examined.

Efficiency Change

The efficiency change (EC) is linked to management's efficiency that causes movement upward or downwards on the production possibility frontier. Hence, ports that have EC = 1 are static regarding efficiency level, therefore revealing no improvement. Ports with EC > 1 indicate improvement in efficiency levels while ports with EC < 1 denote a decline in efficiency level. Figure 12.3 describes the results. The red colored graph represents Port of Port Louis whilst the blue colored graph represents the Port of Victoria. For the eleven years examined the port of Port Louis maintained an EC = 1. The implication of which there has not been any major improvement with regards to efficiency change. On the contrary the Port of Victoria exhibited EC > 1 for the years 2008–2009; 2010–2011; 2013–2014; 2014–2015 and 2016–2017. The implication that these years indicates improvement in efficiency levels.

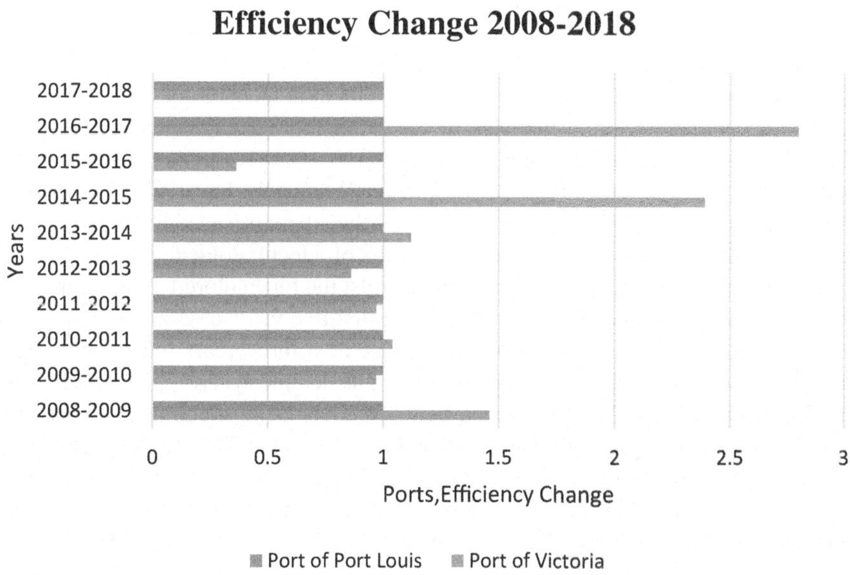

Fig. 12.3 Efficiency change 2008–2018. *Source* Osundiran, 2023

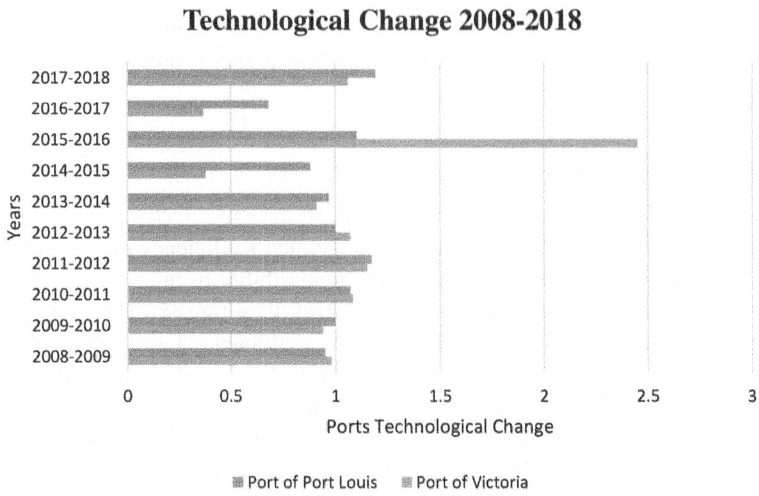

Fig. 12.4 Technological change 2008–2018. *Source* Osundiran, 2023

Technology Change

The technology change is the second aspect of the Malmquist Productivity Index. This is evaluation of the ports based on improvement in technology. With regards to technological change, values above 1 represent technological progress in the sense that more can be produced using fewer resources [18]. Technology change (TC) causes an outward shift in the production frontier. When ports have TC = 1, then there is no improvement in technology. When TC < 1 then there is a need for technological advancement. Figure 12.4, illustrates the results of a technological change. The red colored graph represents Port of Port Louis whilst the blue colored graph represents the Port of Victoria. For the Port Victoria, the first two years of examination shows TC = 1, an indication of no improvement. The next three years indicated a slight improvement where TC > 1. There was a slight decline where TC < 1 for years 2013–2014 and 2014–2015. However, between 2015–2016 a leap in TC where TC > 1. The subsequent years up until 2018 shows a decline in TC. For the port of Port Louis, the first two years shows TC = 1 a static or no improvement. The next two years witnessed another TC > 1 and then a decline for the next three years. The fluctuations continued until 2017–2018 where there was an increase in TC. TC > 1.

Malmquist Productivity Index

The Malmquist productivity index is obtained by the product of TC and EC. Where the MPI > 1 means there is progress in productivity, where MPI < 1 signifies a decline

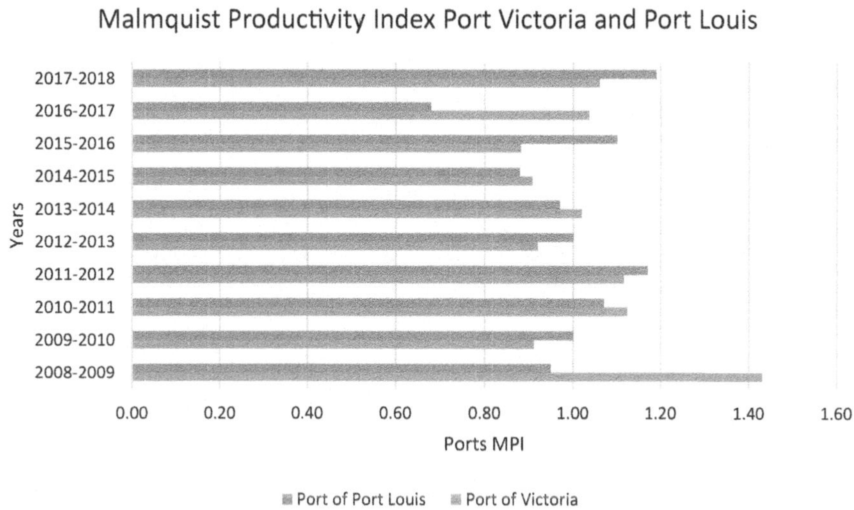

Fig. 12.5 Malmquist productivity index 2008–2018. *Source* Osundiran, 2023

in productivity and where MPI = 1 indicates stagnancy. Figure 12.5 provides the MPI results for the ports. On the average both ports had MPI < 1. Both Ports showed remarkable fluctuations in their MPI.

Conclusion and Recommendation

The results indicated that over the period of 2008 to 2018 the port of Port Louis achieved an annual average productivity gain of 0.91 whilst the Port of Victoria achieved 0.95. The drivers of productivity being tilted more towards technology change. For the Ports of Port Louis and Port Victoria there is need to pay attention to the drivers of productivity which are technology and technical efficiency. This is because the Malmquist productivity Index of the two ports fluctuated. Figure 12.6 indicates the Malmquist Productivity Index for both the Port of Victoria and Port Louis.

The purpose of examining port operations is to ensure that there is improvement both in the short term and in the long term. It is important to evaluate port efficiency because it promotes sustainable port. Port efficiency has become increasingly important because ports are part of the connecting links between different transport modes in the global logistics chain; hence, container terminals are vital to the efficiency of the whole maritime supply chain [8]. Based on the study, it is recommended that the ports continue to build their technological efficiency through engaging state of the art software and applications. This is because development and improvement at ports

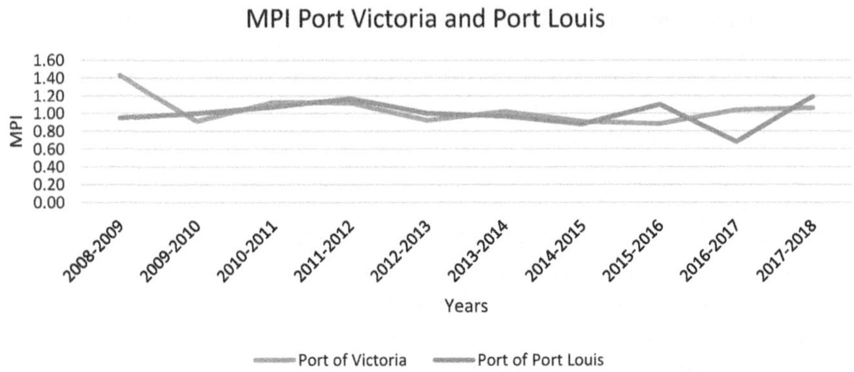

Fig. 12.6 Malmquist productivity index 2008–2018. *Source* Osundiran, 2023

level will have its impact on the economy of the countries involved. When improvement in efficiency level is the focus, port managers can leverage their available resources to ensure that it is achievable. Port authorities should maintain efficiency in all ramifications and at all levels.

References

1. Benoit, Diana. (2021). The Seychelles: A historical perspective. Retrieved 26 September, 2023, from https://www.researchgate.net/profile/Diana-Benoit.
2. UNECA. (2018). United Nations Economic Commission Africa, Country Profile (Seychelles) 2017. Published by the Economic Commission for Africa. Addis Ababa Ethiopia. March 2018. Retrieved 5 April, 2020, from www.uneca.org/publications.
3. Sherman, H. D., & Zhu, J. (2013). Analyzing performance in service organisations. *MIT Sloan Management Review, 54* (4).
4. Mauritius Port Authority. (2023). Retrieved 26 September, 2023, from http://www.mauport.com/information-mauritius.
5. Mauritius Port Authority. (2023). Retrieved 26 September, 2023, from http://www.mauport.com/port_statistics.
6. Zhou, Lele, & Lee, Hyang-Sook. (2023). A study on port (Terminals) Operational Efficiency Analysis Using PCA-DEA Approach: Focusing on Container Terminals on Busan Port. Korean Logistics Research Association. 33, pp. 19–32. https://doi.org/10.17825/klr.2023.33.1.19.
7. Rodrigue, J. P., & Notteboom, T. (2017). |The geography of transport systems. "Transportation and Economic development" Chapter 7 Jean-Paul Rodrigue (2017), New York: Routledge, p. 440. Retrieved 4 July, 2018, from https://transportgeography.org/?page_id=5260. ISBN 978–1138669574
8. Kutin, N., Nguyen, T., & Vallee, T. (2017). Relative efficiencies of ASEAN container ports based on data envelopment analysis. *The Asian Journal of Shipping and Logistics, 33*(2), 67–077. Retrieved 31 October, 2018, from https://doi.org/10.1016/jajsl.2017.06.04.
9. Chudasama, K. M., & Pandya, K. (2008). Measuring efficiency of Indian Ports: An application of DEA. *The ICFAI University Press Journal of Infrastructure, VI*(2) 200, 1–21.
10. Fanou, E. H., & Wang, X. (2018). Assessment of transit transport corridor efficiency of landlocked African countries using data envelopment analysis. *South African Journal of Science,*

114(1/2), 1–7. Retrieved 31 October, 2018, from https://doi.org/10.17159/sajs.2018/2016034 7

11. Perico, A. E., & Riberio da Silva, G. (2020). Port performance in Brazil: A case study using data envelopment analysis. Case Studies on Transport Policy. (Article in Perss). Retrieved 14 February, 2020, from http://doi.org//https://doi.org/10.1016/j.cstp.2020.01.002.
12. Banker, R. D., Charnes, A., & Cooper, W. W. (1984). Some models for estimating technical and scale inefficiencies data envelopement analysis. *Journal of Management Science 30* (9).
13. Charnes, A., Cooper, W. W., & Rhodes, E. (1978). Measuring the efficiency of decision making units. *European Journal of Operations Research, 2*, 429–444.
14. Caves, D. W., Christensen, L. R., & Diewert, W. E. (1982). The economic theory of index numbers and the measurement of input, output and productivity. *Econometrica, 50*, 1393–1414.
15. Färe, R., Grosskopf, S., Norris, M., & Zhang, Z. (1994). Productivity Growth, technical progress, and efficiency change in industrialized countries. *The American Economic Review, 84*(1), 66–83.
16. Huguenin, J. (2012). Data envelopment analysis: A pedagogical guide for decision makers in the public sector. City: Lausanne IDHEAP-Cahier 276/2012. Retrieved 22 August, 2017, from https://serval.unil.ch/resource/serval:BIB_0FC432348A97.P001/REF.
17. Malmquist, S. (1953). Index numbers and indifference surfaces. *Trabjos de Estatistica, 4*(2), 209–242.
18. Bogetoft, P., & Otto, L. (2011). Benchmarking with DEA, SFA, and R. *International Series 1 in Operations Research & Management Science*, Vol. 157. Retrieved 3 July, 2019, from https://doi.org/10.1007/978-1-4419-7961-2.
19. Qiang, Jiang, Ying, Wang, Dong. (2020). Evaluation of port unified efficiency based on RAM-DEA model for port sustainable development. *Journal of Coastal Research, 104*. Retrieved 26 February, 2020, from https://doi.org/10.2112/JCR-SI104-125.1.
20. United Nations Conference on Trade and Development. (2017). Review of Maritime Transport. Retrieved 4 July, 2018, from http://unctad.org/en/PublicationsLibrary/rmt2017_en.pdf.

Chapter 13
The Technical Efficiency of Farms, Its Decomposition into Input Components, and Their Socioeconomic Characteristics at Alahsa Oasis in Saudi Arabia

Ezzeddine B. Mosbah and Amane Iwais

Abstract The current study aims to assess the technical efficiency of farms in the Alahsa region of Saudi Arabia and decompose them into input components. It contrasts the socioeconomic traits of farmers in an attempt to provide an explanation for their efficiency levels. The ID-DEA model was run on a sample of 25 farms. The findings show that farms are categorized into three groups according to technical efficiency scores. With all inputs being efficient (scores = 1), 48% of farms fall into the first efficiency category. Twenty percent of farms fall into the second category of moderately efficient farms (0.5–0.79), meaning that they employ pesticides, irrigation, and production processes at moderately efficient levels (0.89, 0.88, and 085, respectively). The third category consists of inefficient farms (< 0.5), of which only 32% have a moderate efficiency level of their handwork (0.53). Farmers manage their resources differently based on their level of knowledge and understanding of resource utilization, according to the input decomposition. The efficient group seemed to be socioeconomically the youngest, most educated, having the largest percentage of family members, having less experience, and having fewer family members.

Keywords Efficiency · Input decomposition · Saudi Arabia

E. B. Mosbah (✉)
King Faisal University, Alahsa 400, 31892, KSA Hofuf, Saudi Arabia
e-mail: emosbah@kfu.edu.sa; lncs@springer.com

Carthage University, 1054 Amilcar, Tunisia

A. Iwais
University of Tunis, 1007 Tunis, Tunisia

© The Author(s), under exclusive license to Springer Nature Switzerland AG 2024
A. Emrouznejad et al. (eds.), *Business Analytics and Decision Making in Practice*, Lecture Notes in Operations Research, https://doi.org/10.1007/978-3-031-61589-4_13

Introduction

The Kingdom of Saudi Arabia (KSA) has a desert climate and landscape. Its agricultural land area is 9 million ha, accounting for 22.7% of the total land area, of which only 3.22% is cultivated. Many challenges have hampered agricultural integrated development, including a lack of clean water, desertification, low investment, low agricultural worker wages, and climate change negative effect. Primarily increasing agricultural productivity [1]. Farm efficiency, as an important tool, leads to increased agricultural income, food security, and the preservation of a sustainable environment. In this context, the current study seeks to assess the efficiency of farms in Al-Ahsa Oasis, which has historically played a significant role in agricultural production in the Kingdom of Saudi Arabia. Furthermore, in order to investigate the source of the inefficiency, the overall efficiency must be decomposed into its input components in accordance with an input orientation that aims to minimize resource use. The theory of production economics provides a set of theories and tools that resolve and explain inefficiency problems. Efficiency has become an important tool for supporting production managers in their decision-making. Since [2] formulated the first efficiency ratio, many developments have occurred on this topic. The recent ones talk about the nuance and obscurity of the efficiency indicator. For that, they have been oriented towards the decomposition of the overall efficiency in two ways: either according to different stages of DEA processes or measuring inputs, or outputs components.

According to the first way, many researchers are interested in the evaluation of the different stages of component efficiencies (two stages, three stages, or multiple stages). The idea was argued by the fact that some outputs from a given stage may leave the system while others become inputs to the next stage, and new inputs can enter at any stage [3]. These authors proceeded by decomposing the overall efficiency, under additive weighted average form, into multiple stages of components of a structure Conducting a two-stage DEA process decomposition with non-splitable shared inputs and outputs, the same approach was applied by [4] on 17 city branches of China Construction Bank in Anhui Province. Choosing an approach of two-stage process decomposition with slack-based measures, [5] made a frontier DEA projection and proved that the system inefficiency is equivalent to the sum of the inefficiencies of the two stages of US commercial banks.

Kao and Hwang [6] studied efficiency decomposition in two-stage data envelopment analysis applied to Taiwanese non-life insurance companies. They used an additive efficiency decomposition approach wherein the overall efficiency is expressed as a (weighted) sum of the efficiencies of the individual stages. With a two-stage decomposition approach with fixed sum outputs, [7] assessed the efficiency of decision-making units (DMU). Application on 24 non-life insurance companies, the results provided reliable ranking results for DMUs and rich information to assist decision-making [8] tried to decompose the general multistage system into parallel stages. According to him, each stage consumes exogenous inputs and intermediate products to produce exogenous outputs and intermediate products [9] developed a set of DEA models to treat a two-stage system comprised of three sub-DMUs in hybrid form with additional inputs to the second stage, simulating a system's parallel-series internal structure and employing synthetically additive and multiplicative DEA approaches [10] used a three-stage network efficiency decomposition on the chipboard and furniture industries of wood lumber, with fuzzy acceptable and undesirable output and fuzzy input in DEA.

The weakness of this way of decomposition is that it includes intermediate measures under real-world conditions, including desirable and undesirable data. The outcomes of these approaches give efficiency scores for the stages of the process but not the input scores.

Unlike the last method of decomposition, according to this method, researchers don't try to change the structure of the DEA process. Based on the DEA conventional process, researchers aimed to decompose the overall efficiency into its input or output components. For instance, [11] aimed to decompose the technical efficiency of Slovak urban transport companies under fixed proportion technologies (FPTs) to various sources. Three main components were observed: Farrell efficiency, input mix efficiency, and output mix efficiency. The empirical work confirmed the outcomes of the technical inefficiency in the overall input–output ratio. In the same orientation, [12] presented a study of the impact of an input–output specification on the efficiency scores of commercial banks in Slovakia. The paper compared technical efficiency scores for nine input–output specifications of the intermediation approach, nine specifications of the production-like approaches, and 3 network integrated specifications. Results highlighted that input–output sets are a critical judgmental input to efficiency measurement since there is vast diversity in the efficiency scores of input–output sets coming from different approaches, as well as input–output sets associated with the same approach.

The works of [13] didn't make changes in the structure of the original model of [14] in measuring the overall efficiency of the unit, while they extended this model to involve what they call virtual DMU*, which is the projection of the original one on the frontier curve. Many researchers cited this work in order to show that the implementation of weights would lead to more reliable results [15], or could be used in agriculture [16, 17], and improve resource management decisions [18]. In reality, the decomposition of the overall efficiency into input components (or output components) is more than a simple effective measure of technical efficiency or an essay on improving resource management; it constitutes a deep correction of the analysis of results given by the overall efficiency score of the DMU. We can't attribute the global score of DMU inefficiency to all inputs (or outputs) at the same level.

Method and Materials

The appropriate model applied for this analysis was initiated by [19] and tested by [13, 20, 21] in the agricultural sector. The model, taking its origin from the model CCR78 version of input orientation (IO), the derivative ID-DEA represents a decomposition of the overall technical efficiency issued by the conventional CCR78 model into input components. For instance, the ID-DEA involves a supplementary constraint added to the conventional one resulting from the idea stipulating that each DMU of the data envelopment (V) has a peer; the virtual DMU* is situated on the frontier curve and represents the efficient level of the original DMU. The DMU* and DUMs constitute, all together, a new data envelope (V*).

The process of estimation will take the form of one operation, but internal work will be done in two stages. First, the run of CCR78 model on the (V) model, which gives the overall efficiency score of the DMU, will be evaluated, as well as the parameters (u_r and v_i). The second stage will be the run of ID-DEA on the data set of (V*). So the process is composed of two runs of models in the same operation of evaluation.

Assuming that we have n DMUs using 1 input to produce 1 output, In order to measure the global technical efficiency of an evaluated DMU_e (x,y), we will run the CCR78 model on the data set of (V), results give the score of efficiency, θ. Conventional analysis of this result suggests that the DMU_e will be efficient when its input is reduced to the level (θx), where θ is the efficiency score of the original DMUe. In this case of one input, θ is also considered the efficiency score of the input (x).

Now suppose that the virtual DMU_e* is identified by (θx,y) will be considered in the data envelopment. A new constraint will be added to other constraints in the CCR78 model form as follows:

$$\frac{\sum_{r=1}^{s} u_r Y_{rj}}{\sum_{i=1}^{m} v_{i\theta_j} x_{ij}} \leq 1.withj = 1.\dots.n \tag{13.1}$$

For the DMU_e* with 1 input and 1 output, the constraint gives:

$$\frac{uy1}{v\theta_j x1} \leq 1.or\frac{uy1}{vx1} \leq \theta_j.withj = 1.\dots.n \tag{13.2}$$

Now suppose in an input orientation process, that we have $i = 1, \ldots, m$ inputs and $r = 1, \ldots., s$ outputs, a general case (mainly for $i > 1$), we will take into consideration this constraint and rewrite the hypothesis. Suppose that DMU_k will be evaluated, the new constraint will be applied on DMU_k is written as:

$$\frac{\sum_{r=1}^{s} u_r Y_{rk}}{\sum_{i=1}^{m} v_i \theta_{ik} x_{ik}} \leq 1. \, with \, j = k, \tag{13.3}$$

So, the model, which we call Input decomposition of DEA or ID-DEA, will be rewritten as:

$$\max_{u_r, v_i} \sum_{r=1}^{s} u_r Y_{rk},$$

$$s.t. \tag{13.4}$$

$$\sum_{i=1}^{m} v_i x_{ik} = 1$$

$$\sum_{r=1}^{s} u_r Y_{rj} - \sum_{i=1}^{m} v_i x_{ij} \leq 0 j = 1 \ldots . . n$$

$$\sum_{r=1}^{s} u_r Y_{rk} - \sum_{i=1}^{m} v_i \theta_{ik} x_{ik} \leq 0. \, j = k, \, i = 1, \ldots, m$$

The results give coefficients of u_r, v_i, θ_k, θ_{ik}, $i = 1, \ldots., m$.

Data for the current research were collected using a survey on a sample of 25 farmers with 7 inputs (area, handwork, equipment, fertilizer, irrigation, pesticide, and other activities related to production activities (such as harvesting, plowing, pruning, etc.)) and 1 output (production quantity of dates).

Results and Discussion

Running the model with GAMS, obtained results give the global efficiency of the farms, and the 7 related input efficiency scores participated in the production process. Results are presented in Table 13.1 as follows.

The decomposition of the global efficiency gives scores for each input's efficiency. For instance, the farm size efficiency shows that 56% of farms are highly efficient,

Table 13.1 Results of global efficiency and its inputs components

Farm	Global	Size	Hand-work	Materials	Fert-liser	Irrigation	Pesticide	Other activities
25	1	1						1
23	1	1	1	1				
18	1	1	1	1			1	1
16	1	1		1		1	1	
15	1	1						1
14	1		1					1
11	1	1				1	1	
9	1	1					1	1
7	1	1	1					1
6	1	1	1		1			1
4	1						1	1
3	1		1		1		1	1
2	0.92	0.942					0.948	0.924
10	0.737	1				0.7	1	1
12	0.731	0.731		1				0.749
13	0.69	0.915				0.69	0.963	0.749
5	0.644	0.899	0.372			0.237	0.621	
24	0.498	1						0.498
1	0.44	0.44				1		
19	0.37	0.475					1	0.009
17	0.188	0.267		0.514		0.066	0.072	
22	0.161	0.146		0.474	0.2	0.057	0.037	
21	0.16	0.00017	1	0.00026				0.16
20	0.13	0.13		0.415			0.052	
8	0.088	0.102	0.067				0.242	

The classification of these farms by global technical efficiency gives three groups:

- The group of highly efficient farms includes units with scores superior to 0.8; they account for 52% of the total farms.
- The group of mildly-efficient farms includes units with scores between 0.5 and 0.79. The percentage of these farms represents 16%.
- Group of inefficient farms that have a score of efficiency inferior to 0.5. They represent 32%.

The decomposition of the global efficiency gives scores for each input's efficiency. For instance, the farm size efficiency shows that 56% of farms are highly efficient, with scores superior to 0.8; farms that are middle-efficient account for 4%; and farms that are lower-efficient represent 28%. While it appears to be indeterminate in 10% of cases, Concerning the handwork, farms with an efficient score represent 28%; the lowest scores represent 20%. 52% of farms have indeterminate scores. The material efficiency scores are higher for 16%, middle for 4%, and lower for 12%. For 68% of total farms, the materials are indeterminate. The scores of irrigation efficiency are similar to material efficiency scores. Concerning indeterminate scores and inefficiency, although the highest and middle scores are 12 and 8%, respectively. Fertilizer scores are indeterminate for 84%, and higher and lower scores represent each other by 8%. The pesticide efficiency score shows 44% of farms are highly efficient, 4% are middle-efficient, 16% are inefficient, and for 36% of total farms, pesticide efficiency is indeterminate. Finally, other actives (related to soil preparation, harvesting, etc.) with high, middle, and low efficiency scores represent 48%, 8%, and 20% of farms, respectively. Only 24% of farms with indeterminate scores.

The first observation given by these results is that the global score appears different from efficiency scores across inputs. This result confirms the hypothesis stipulating that global efficiency could be the same for all input scores as well, as supposed by the conventional analysis, only when the DMU efficiency equals 1. Whereas, it denies the hypothesis when DMUs are inefficient and score < 1. Then, the global score differs from the inputs efficiency scores, and inputs have their own proper efficiency score, except some of them may have the same scores.

The evidence is that farm number 1 has a score of global technical efficiency equal to 0.44. The decomposition of this score shows that only the farm area explains this inefficiency, which means that the use of the farm land is not optimal, while irrigation efficiency is used at an optimal level because its efficiency score equals 1. The result of the decomposition of the global efficiency score of farm number 22, which is 0.161, gives different scores for inputs 'efficiency such as the farm area, material, fertilizer, irrigation, and pesticides at different levels of 0.146, 0.474, 0.2, 0.057, and 0.037, respectively. While handwork and other activities are indeterminate.

Farms exhibit varying social characteristics based on their efficiency scores (high, midrange, or poor). For example, the youngest farmers, aged under forty, make up 54% of the efficient group, whereas the intermediate efficient group includes 75% of farmers over forty, and the oldest group has 62% of farmers over forty. According to socioeconomic status, 87.5% of members of the lowest efficient group are married, compared to 100% of members of the high and intermediate efficient groups. The group with the lowest efficiency has the highest level of education; 50% of its members have completed college or higher; the group with the highest efficiency has 46%; and the medium group has a level of education of 25. Additionally, there may be differences in the number of families among groups. For example, the average family size in the high group (38%), compared to the average family size in the middle and lowest groups (0 and 12.5%), is less than 5. This indicates that the inefficient group has a larger family. 30.8% of the most efficient group live in rural areas on farm sites, compared to an average of 25% for intermediate and lower efficiency groups. Because 75% of them have more than ten years of experience, the lowest may seem to have the most, while 69.2% of the highest group and 50% of the middle efficient have more than 10 years. With regard to labour by kind, 54% of the most productive group employs family members, while 62.5% of the least efficient group employs salaried workers. By nature of labour, 61% of the most efficient group has full-time employees, while only 50% of the least productive are (Table 13.2).

In terms of economic characterization, it demonstrates that the more efficient farms possess a smaller land area than the less efficient farms. The average size of their farms is 12873 square meters, compared to 18,637 for the least productive. Despite having more acreage available, the most productive group produces more than the next two taken together: 38,798 tonnes, 10,716 tonnes, and 7645 tonnes, respectively. Concerning the other inputs, the group with the highest efficiency has less machinery and fewer annual production activities than the groups with the intermediate efficiency and the lowest efficiency. On the other hand, the most productive group utilizes greater amount than the medium and lowest groups when it comes to inputs like hand labour, irrigation, pesticide, and fertilizer (chemical and organic) (Table 13.3).

Table 13.2 Social characteristics

Group	Age (< 40 years) (%)	Social status (married) (%)	Education (University level) (%)	Family < 5 members (%)	Living home (Rural) (%)	Expe-rience (> 10 years) (%)	Type of labour (familial) (%)	Nature of labour (full time) (%)
GHE	54	100	46	38	30.8	69.2	54	61
GME	25	100	25	0	25	50	25	100
GLE	38	87.5	50	12.5	25	75	37.5	50

Table 13.3 Economic characteristics of farms

Group	Area	Total Prod	Machinery	Org anic	Chemical	Handwork	Activities Annual	Irrigation	Insecticides
GHE	12,873	38,798	183	4654	902.4	1833	5004	5716	1894
GME	9375	10,716	525	1575	575	200	1437.5	200	700
GLE	18,637	7645	2025	2750	230	737.5	6576	4362.5	902.5

Conclusion

In Saudi Arabia, in addition to the climate change effect, the agricultural sector faces a lot of challenges, mainly the low level of factors' productivity. In order to realize the objectives fixed on the horizon of 2030, farmers may adopt the best practices of the farming system and change their management model based on a global vision of production unity and change their management resources one by one. For that, the authors tried not only estimating the global efficiency but also decomposing it into input components.

The sources of any inefficiency may be related to the socioeconomic characteristics of farmers. The approach to estimation is based on the model of ID-DEA, which decomposes the global efficiency score of a farm into its input components. The model is run on a sample of 25 farms. The main results highlight three group farms:

- A group of farms is highly efficient, having scores equal to more than 0.8 to 1. It represents 48% of total farms.
- The group of farms that were mildly efficient scored between 0.5 and 0.79. It includes 20% of farms.
- A group of farms having low efficiency inferior than 0.5. it involves 32%.

References

1. Ministry of Environment, Water, and Agriculture, Saudi Arabia: Annual report 2021). Retrieved 2 November, 2024, from https://www.mewa.gov.sa/ar/InformationCenter/DocsCenter/YearlyReport/Pages/default.aspx.
2. Farell, M. J. (1957). Measurement of productive efficiency. *Journal of the Royal Statistical Society, 120*(3), 253–281.
3. Cook, W. D., Zhu, J., Bi, G., & Yang, F. (2010). Network DEA: Additive efficiency decomposition. *European Journal of Operational Research, 207*, 1122–1129.
4. Li, L., Dai, Q., Huang, H., & Wang, S. (2016). Efficiency decomposition with shared inputs and outputs in two stage DEA. *Journal of Systems Science and Systems Engineering, 25*(1), 23–38.
5. Chen, Y., Li, Y., Liang, L., Salo, A., & Wu, H. (2015). Frontier projection and efficiency decomposition in two-stage processes with slacks-based measures. *European Journal of Operational Research, 250*(2), 543–554.
6. Kao, C., & Hwang, S. N. (2008). Efficiency decomposition in two-stage data envelopment analysis: An application to non-life insurance companies in Taiwan. *European Journal of Operational Research, 185*(1), 418–429.
7. Zhang, X., Xia, Q., & Wei, F. (2023). Efficiency evaluation of two-stage parallel-series structures with fixed-sum outputs: An approach based on SMAA and DEA. *Expert Systems with Applications, 227*, 120264.
8. Kao, C. (2014). Efficiency decomposition for general multi-stage systems in data envelopment analysis. *European Journal of Operational Research, 232*(1), 117–124.
9. Jianfeng, M., Linan, Q., & Lizhi, D. (2017). Efficiency measurement and decomposition in hybrid two-stage DEA with additional inputs. *Expert Systems with Applications, 79*, 348–357.

10. Noughabi, F. S. A., Mohammadi, N. M., Lotfi, F. H., & Razavyan, S. (2022). Efficiency decomposition in two stages network in data envelopment analysis with undesirable intermediate measures and fuzzy input and output. *Fuzzy Information and Engineering, 14*(2), 123–142.
11. Mendelová, V. (2022). Decomposition of technical efficiency under fixed proportion technologies: An application of data envelopment analysis. *Journal of Productivity Analysis, 57*, 1–22.
12. Boda, M., & Piklova, Z. (2021). Impact of an input-Output specification on efficiency scores in data envelopment analysis: A banking case study. RAIRO-Operation Research 55 S1551S1583
13. Mosbah, E., Zaibet, L., & Dharmapala, P. S. (2020). A new methodology to measure efficiencies of inputs (outputs) of decision-making units in data envelopment analysis with application to agriculture. Socio-economic Planning Sciences 100857.
14. Charnes, A., Cooper, W. W., & Rhodes, E. (1978). Measuring the efficiency of decision-making units. *European Journal of Operational Research, 2*(6), 429–444.
15. Kyrgiakos, L. S., Kleftodimos, G., & Vlontzos, G. et al. (2023). A systematic literature review of data envelopment analysis implementation in agriculture under the prism of sustainability. *Operational Research International Journal, 23*(7).
16. Pan, Z., Tang, D., Kong, H., & He, J. (2022). An analysis of agricultural production efficiency of yangtze river economic belt based on a three-stage DEA Malmquist model. *International Journal of Environmental Research and Public Health, 19*(2), 958.
17. Oukil, A. (2023). Investigating prospective gains from mergers in the agricultural sector through Inverse DEA. *IMA Journal of Management Mathematics, 34*(3), 465–490.
18. Zhang, H., & Zhong, Z. (2022). How does environmental regulation affect the green growth of China's citrus industry? The mediating role of technological innovation. *International Journal of Environmental Research and Public Health, 19*(20), 13234.
19. Mosbah, E. (2007). Mesure et analyse de l'efficacité technique et allocative du dispositif de la formation professionnelle en Tunisie dans le cadre d'application du programme MANFORM [Measure and Analyze of technical and allocative efficiency of Tunisian training system in frame of MANFORM]. Doctoral dissertation, National Institute of Agriculture Tunis, University of Tunis Carthage.
20. Mosbah, E., & Bachta, M. S. (2009). Mesure et Analyse de l'Efficacité des Etablissements de la Formation Professionnelle en Tunisie [Measure and Analysis of vocational Training establishments Effectiveness in Tunisia. *Canadian and International Education Review, 38*, 45–60.
21. Mosbah, E., & Bachta, M. S. (2015). Data envelopment analysis: A Simultaneous measure of technical and allocative efficiency. In: Business Performance Measurement and Management, pp. 487–516. Cambridge Scholars publishing, United Kingdom.

Chapter 14
Synergizing Deep Belief Networks and Arithmetic Optimization for Stock Market Price Prediction: A Hybrid Approach

Noura Metawa, Hussein Al Tamimi, and Rania Itani

Abstract The expectation of stock market prices has become a vital and stimulating mission for both academic and practitioners' in financial study. The unpredictable type of stock market represents that forecasting stock market prices is a stimulating mission. Generally, standard time-series predicting approaches dependent upon static development; therefore, evaluating stock value is a basic issue. Moreover, predicting the stock trend is an important concern because of the comprised variables. Consequently, between income and the economic downturn. However, a recent advancement in the machine learning, particularly from deep learning approaches are developed for researchers to practically utilize such approaches for predicting future stock trends dependent upon historical financial information, financial news, social media broadcast, and stock technical indicators (STIs). This article introduces an arithmetic optimization algorithm with deep belief network-based stock market price prediction (AOADBN-SMPP) model. The proposed technique focuses on the forecasting of stock prices in a long term. To accomplish this, the proposed model follows two key procedures namely prediction and factor optimization. In the first stage, utilizing the DNB model for forecasting stock prices. In the second stage, AOA can be leveraged to optimally fine tune the hyperparameters pertaining to the DBN technique and thereby boosts the classification execution. The AOA's layout is beneficial to optimally adjust the hyperparameters associated to the DBN version. The investigational

N. Metawa (✉) · H. Al Tamimi
College of Business Administration, University of Sharjah, Sharjah, UAE
e-mail: nmetawa@sharjah.ac.ae

H. Al Tamimi
e-mail: hussein@sharjah.ac.ae; tshehlmf@unisa.ac.za

N. Metawa
Faculty of Commerce, Mansoura University, Mansoura, Egypt

R. Itani
College of Business, Murdoch University Dubai, Dubai, UAE
e-mail: rania.itanI@murdoch.edu.au

© The Author(s), under exclusive license to Springer Nature Switzerland AG 2024

A. Emrouznejad et al. (eds.), *Business Analytics and Decision Making in Practice*,
Lecture Notes in Operations Research, https://doi.org/10.1007/978-3-031-61589-4_14

outcome analysis of the proposed type is tested as well as the results are evaluated beneath specialized prospects. The comparative research reported a enhanced effectiveness of the suggested method over recent state of art approaches.

Keywords Stock price expectation · Stock market · Deep learning · Machine learning · Prediction models

JEL Classification G10 · G15 · G17 · D53

Introduction

Background and Context of the Study

Artificial Neural Network (ANNs) become utilized extensively for the prediction of time series throughout the 1980s. Recurrent neural network, radial basis functions networks, and multi-layer perceptron are illustrated of ANNs of forecasters that are frequently programmed using gradient descent techniques to identify appropriate parameters [1]. The dynamic, high volatility and complicated nonlinear nature of stock price period series rendering them extremely difficult to predict due to their erratic movements. Numerous factors including industrial expansion, political unpredictability, legislation, the overall conditions of the world economy, investors psychological research, and many more have an impact on the stock market's performance. The main goal of time series prediction of the price of stock is to anticipate prospective value of shares using historical data from the data set [2]. The necessity for developing an intelligent system that can replicate the nonlinearity and unpredictability associated with the price of stock time series is expanding. Predicting stock prices is a crucial undertaking for both the nation's growth in the economy and investor reduction of risk. The most potent technology is the use of machine learning (ML), that integrates several algorithms to improve efficiency on a given case study.

Many people think that machine learning (ML) is very good at finding reliable information as well as trends in the collection of data. The recently emerged Deep belief Network (DBN) and its variations have eclipsed the conventional Neural Network (NN) regarding the precision of predicting in domains such as language recognition and the processing of images. NN and DBN are both considered an AI algorithm that is used as price indicator for predicting prices [3]. A recent development in deep learning (DL) can identify a deep nonlinear topology in its particular framework and has a great potential for extracting pertinent data from financial time series. The outcomes of the investigation demonstrated that unidirectional LSTM was capable of predicting the price of stocks for investment decisions, and the technique outperformed other forecasting approaches [4].

The stock market's price forecasting is a technique by which mathematical methods are applied to forecast the upcoming inclination of the stock market using data from the past, present, and regulations governing the stock market, utilizing precise survey statistics and stock information about the market [5]. Deep learning selects characteristics dynamically from input with minimal input from humans. Thus, our method allows us to use a huge variety of raw-level information as inputs and without requiring knowledge of predictors such macroeconomic factors. Particular of deep learning is its capacity to recognize abstraction characteristics from data as well as uncover latent nonlinear correlations without depending on economic models and personal experience [6].

From the above research concern, clearly, all the estimations can really handle stock prediction issues. The predictive outcomes are not solely affected by the depiction of the data, yet likewise rely upon the prediction procedure. Additionally, including obvious features and separating them as data rather than all components could discernibly encourage the precision of the prediction models.

This article introduces an arithmetic optimization algorithm with deep belief network-based stock market price prediction (AOADBN-SMPP) model. The proposed technique focuses on the forecasting of stock prices in a long term. To accomplish this, this technique follows two main methods specifically prediction and factor optimization. In the first stage, the DBN model is exploited for forecasting stock prices. Next, the AOA can be used to optimally adjust the hyperparameters linked to the DBN algorithm and thereby boost the classification performance. The design of the AOA helps to optimally Improve the DBN model's hyperparameters. The suggested approach's experimental outcome assessment has been put to the test, and the results are evaluated according to several criteria.

Literature Review

Chong [6] defined Stock market (SM) as dynamic, strange, and volatile nature setting. Forecasting the cost of stocks (SP) is a difficult job as it hinges on several components involving however not confined to political conditions, overall economy, association's monetary reports, execution, etc. Likewise, [7] explained that to enhance the advantage and cut off the disasters, procedures to expect potential gains of the stock early on by examining the example all through late years, could turn out to be uncommonly significant for making SM improvements.

The following is an attempt to highlight the methods used by some researchers in predicting the market stock price.

Machine Learning

By applying machine learning and statistical methods, Kumar et al. [8] conducted a methodical review of stock market prediction. They believe that for investors to make more profits and use their money properly, predicting the stock price is a big challenge for them. A novel hybrid deep learning method has been used by [9] in stock price prediction of China's commercial bank. The authors proposed an enhanced K-means grouping algorithm to group out banks with comparable price movements. They concluded that the use of this approach, more accurate prediction of stock price can be achieved which eventually will help investors to make more profitable decisions. [10] also support the use of hybrid deep learning, to be it has the ability to forecast both linear and non-linear time-series components of the price movement of stocks.

Other methods of market stock price prediction have been attempted by [11] who proposed the ML to forecast stock price by introducing the general (3 by 3) forecasting model to include three types of stocks, overpriced, underpriced, and fairly priced. [12] in their article about how to predict directional changes of stock market prices by employing unplanned forests and LSTM systems. The experimental findings indicate that the multi-feature set offers a day-to-day gain, prior to transaction costs, of 0.54% utilizing random forests and 0.64% using LSTM networks.

Abualigah et al. [13] clarified that in the past, SMs were consistently expected by money related specialists. In any case, data specialists had initiated handling prediction issues with a progression of learning approaches. Moreover, [14] described that PC specialists have begun using AI procedures to deal with the introduction of prediction models and overhaul the precision of prediction. Using significant learning is an accompanying stage in additional creating prediction models with better execution. In SM prediction structures, AI and significant learning methods could aid monetary sponsors and vendors via its decisions. Such systems hope to normally see and study plans between enormous proportions of information. The estimations could be efficiently self-learning and might deal with forecasting mission of price instabilities and additionally create trading approaches [11].

Technical Analysis

Estimating the stock value utilizing the relationship between Tobin's Q variation by section as the composite index, and research was carried out conducted on the Columbia stock market's volatility in prices by [15]. The authors indicated that their methodology is essential to all investors because of the availability of vital information. Li et al. (2020) suggested new approaches to forecast stock market values by incorporating numerical stock prices data (technical analysis) and news sentiments and concluded that these models outperform those that use only technical indicators or news sentiment.

Convolutional Neural Network

A convolutional neural network (CNN)-based SP prediction framework was presented by Sim et al. (2019) to validate the applicability of innovative methods of learning in SMs. While utilizing CNN, 9 technical values have been picked as indicators of the prediction method, and such values were switched over completely to pictures of the time series chart. Moreover, Li and Wu (2022) proposed an improved echo state network (ESN) and deep belief network (DBN) method were built in view of DBN method to investigate LSTM method and the exchange estimation of blockchain (BC) virtual cash in the DBN and to further develop the exchange prediction exactness of the BC virtual money. Furthermore, the boundaries of DBN method were upgraded utilizing particle PSO system, which was confirmed with the stock traded data and The digital currencies of BC as well as and other money methods for investigation.

Zang, et al. (2020) identified two essential strategies have been proposed for expecting the SP of an affiliation. Special assessment technique uses obvious stock price such as opening and closing price, volume exchanged, adjoining closing characteristics, etc. of a stock for expecting the future price of a stock. Currently, significant level sagacious techniques considering either particular or key assessment are used at forecasting stock expenses. Particularly, for SM examination, the data size was enormous and non-straight. To handle with this arrangement of data powerful method was required that can perceive the mysterious models and complex relations in this gigantic educational record. Computer based intelligence systems in this space have exhibited further foster efficiencies by 60 to 86% when appeared differently in relation to the past methods [16].

Bidirectional LSTMs (Long Short-Term Memory)

Yadav et al. suggested two methods for various purposes. Rapid Recurrent Neural Networks (Fast RNNs) are necessary for the first. For SP projections, this approach is used for the first time in this paper. A mixed DEEP LEARNING approach was developed for the following iteration by using the elite elements of FastRNNs, Convolutional Neural Networks (CNN), and Bi-Directional LSTM techniques to foresee sudden variations in the SPs of an organization. Jing et al. propose a hybrid research method through application of LSTM Neural Network approach for examining the method pointers from the SM and the sentiment analysis (SA) outcome of the initial step.

In the research work of [17], utilize the habitually utilized calculations LSTM, Extreme Gradient Boosting (XGBoost), Moving Average, and Last Value technique on over a year of verifiable stock information to develop a prediction model for determining SP [18] identified a new closing SP determining system is suggested, which contains a higher prediction than conventional methods. The machine learning indication portion, a data processing component, and the indicator portion of the method of optimization are the parts of this deep half and half system. Information handling incorporates practical wavelet transform (EWT) based pre-processing and outlier robust extreme learning machine (ORELM) method related post-handling. LSTM network-based deep learning network indicator, as the fundamental piece of the blended casing, is together upgraded by dropout technique and PSO technique. In their work, [19] employ optimum LSTM (O-LSTM) deep learning with flexible stock indicators of value to predict the stock market's never-ending price trend. (STIs). Additionally, it evaluated the procedure for mandating a buy-sell decision at the end of the day. Correlation-Tensor theory is applied with appropriate STIs in order to improve machine learning challenge.

Nabipour et al. [16] stated that SM prediction provides an impression of being a confusing issue since there are numerous factors that really can't be tended to, and it doesn't have all the earmarks of being verifiable during every step. Nonetheless, by genuine use of AI techniques, one can relate past data to the continuous data and train the machine to acquire from it and make reasonable doubts. Simulated intelligence has many models, yet this paper revolves around two, among the most part critical and made the predictions using them.

Deep Learning for Blockchain Financial Products

According to [20] and Lui and Metawa [21], through study, they create a deep learning-based intelligent return-on-investment prediction approach for blockchain financial services (RRP-DLBFP) and they found out that the simulation results emphasized the superiority of the RRP-DLBFP method over the existing state-of-the-art methods in terms of various valuation factors.

Shilpa and Shambhavi [22] noted that SM prediction is overflowing with troubles, and data scientists generally resist issues when they endeavour to cultivate a judicious method. Difficulty and nonlinearity are two essential tasks achieved by the frailty of SM and the connection among hypothesis cerebrum science and market direct. Clearly, there are by and large unpredictable variables, for instance, the society image of associations or political condition of nations, that impact the SMs design. In this way, expecting the data obtained from stock characteristics are pre-processed and suitable estimations to be used, the example of stock characteristics and record is expected.

Hypothesis

The hypotheses of this study have been framed to address gaps and inconsistencies in current literature regarding the difficulties encountered by the investors market in predicting stock market prices.

In this paper, the authors attempt to shed light on a new technique that give better outcome in long term stock market prices predictions.

H1: AOADBN-SMPP follows the "prediction" method.

Proposition 1: This first hypothesis tests the step 1 of our model using the DBN.

H2: AOADBN-SMPP follows the "factor optimization" method.

Proposition 2: This second hypothesis tests the step 2 by optimally fine tuning the hyperparameters.

H3: *AOADBN-SMPP works in combing 2 methods "predication and "factor optimization".*

Proposition 3: The third hypothesis test the combination of hypotheses 1 and 2 to confirm that the new model of AOADBN-SMPP generates better outcomes from current available approaches.

In summary, each hypothesis of this study is intensely rooted in prevailing academic dissertation nevertheless purposes to fill gaps or resolve incongruities within that discourse.

Methodology and Methods

Data Source and Sample

The data for this study is not a real data. It is a simulation data from Shanghai Market hina.

Specifications of the Variables

In this paper, evaluation parameters were used to assess the performance of different methods in stock prices' predication and compare them with our suggested model. Hence, we consider these parameters and methods as our study's variable, and we categorize them as follows:

Evaluation Parameters:

- True Positive Rate (TPR)
- True Negative Rate (TNR)
- Accuracy (ACC)
- F-Measure:

Methods:

- Linear Regression (LR)
- Support Vector Machine (VM)
- Naïve Bayes (NB)
- Multi-layer Perception (MLP)
- Randon Forest (RAF)
- Arithmetic Optimization Algorithm with Deep Belief Network-Based Stock Market Price Prediction (AOADBN-SMPP)

Variables Employed for Comparison Purposes:

- Training accuracy (TA)
- Training Loss (TL)
- Validation accuracy (VA)
- Validation Loss (VL)

The Suggested Model

In this article, a recent method has devised for forecasting stock prices in a long term. To accomplish this, the AOADBN-SMPP technique follows two main methods specifically prediction and factor optimization. In the first step, the DBN model is exploited for forecasting stock prices. Next, the AOA was employed for the best possible hyperparameter adjustments associated to the DBN approach and thereby boost the classification performance. Figure 14.1 depicts the overall block diagram of AOADBN-SMPP approach.

DBN Based Predictive Model

At the first stage, the DBN model is exploited for forecasting stock prices. The network structure is comprised of output layer, noticeable layer V, and concealed layer H (multiple RBM); all the layers are comprised of different activation functions (namely sigmoid), and the layer is interconnected bi-directionally via w linking weight, a noticeable layer bias, and b hidden layer bias. The DBN fault identification method created through neuron node that is activated through energy depending on the RBM architecture is well-suited for data processing [23]. Figure 14.2 depicts the infrastructure of DBN.

(1) Data feature learning. Input the dataset from the V1 visible layer to the H1 hidden layer of RBM1. Depending on the characteristics that the influence energy comprised in the abovementioned dataset is larger when compared to

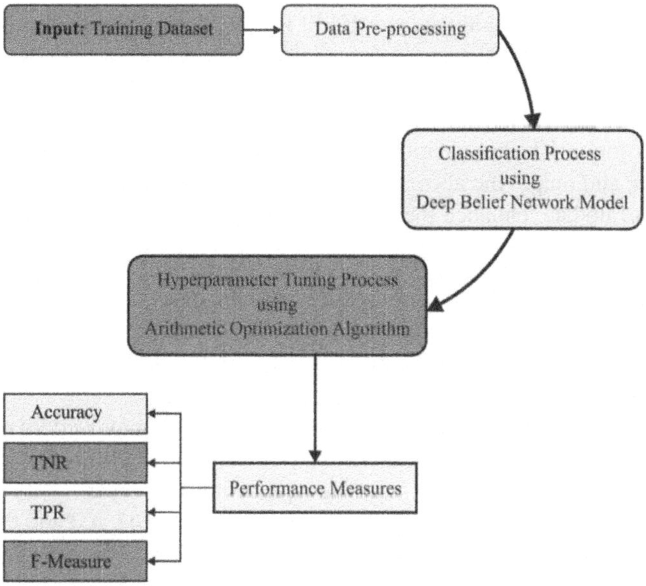

Fig. 14.1 Overall block diagram of the proposed model

Fig. 14.2 Architecture of DBN

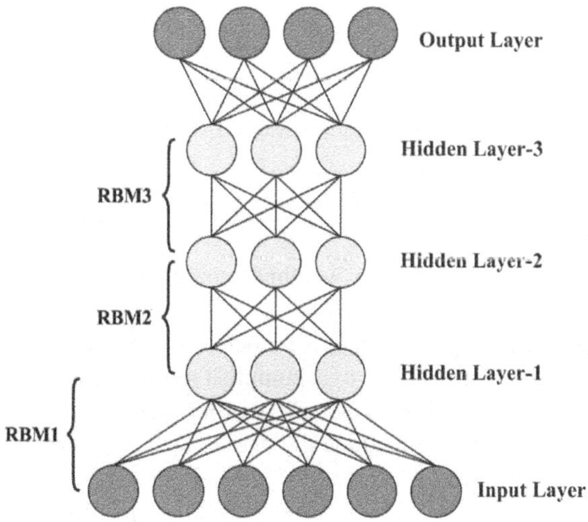

the other mechanisms, the energy computation of H1 hidden layer neuron node of the RBM1 is activated. Based on the upgrade of parameters namely offset bH1 of H1, weights w1, and offset aV1 of V1 afterward every neuron node in the H1 hidden layer of RBM1 functions the dataset, the inherent nature of the abovementioned vibration dataset is learned.

(2) Weight update. The hidden layer H1 of RBM1 is employed as the visible layer V2 of RBM2, and the principle of (1) is utilized to implement self-learning on the dataset for updating the offset bH2 of H2, the connection weight w2, and the offset aV2 of V2.
(3) Result output. The last operation of the dataset in the hidden layer Hn of the final RBMn, and output the abovementioned outcome through the O output layer.
(4) Reverse update, enhance the detection rate. Backpropagation the outcome of (3).

From the back to the front, the output layer O upgrades the w connection weight of all the neurons, the bias a of V noticeable layer, and b bias of the H unknown layer based on the feedback data of the preceding layers for additionally improving the fault detection rate of classifier method.

RBM concentrations on a particular layer in their learning procedure and ignored others. Consider that DBN with L levels, with Wi being the RBM weighted matrixes at ith layers. Still, the hidden unit at ith layer procedures the layer $(i+1)$ input units. During this case, the gathered of hidden layers as $h = \{h1, h2, h3\}$ and the gathered of weighted matrixes are assigned as $W = \{W1, W2, W3\}$. The weighted matrixes amongst layers ith and $(i + 1)$ that is demonstrated as Wi, but jth hidden layer is demonstrated as hj. The energy function has been employed to compute the joint distribution of visible as well as hidden layers:

$$p(v, h) = \frac{e^{-E(v,h)}}{\Sigma_{v,h} e^{-E(v,h)}} \tag{14.1}$$

whereas $E(v, h)$ represents the RBM energy function,

$$E(v, h) = -\sum\nolimits_{i=1} a_i v_i - \sum\nolimits_{j=1} b_j h_j - \sum\nolimits_{i,j} v_i h_j W_{ij} \tag{14.2}$$

W_{ij} represents the weighted amongst the visible as well as hidden layers, a_i and b_j defines the visible as well as hidden layers coefficients. The projected approach creates utilizes of stochastic gradient descent (SGD) method succeeding log-likelihood (L.L.) to accomplish optimal trained. It is obtained by improving the RBM parameters b, and wij. The derivative of $\log p(v, h)$ interms of W_{ij}, a_i, and b_j are measured to upgrade the bias as well as weight as follows.

$$W^{t+1} = W^t + \eta\big(p(h|v)v^T - p(h|v)v^T\big) - \lambda W^t + \alpha \Delta W^{t-1} \tag{14.3}$$

$$a^{t+1} = a^t + \eta(v - \tilde{v}) + \alpha \Delta a^{t-1} \tag{14.4}$$

$$b^{t+1} = b^t + \eta\big(p(h|v) - p(\tilde{h}|\tilde{v})\big) + \alpha \Delta b^{t-1} \tag{14.5}$$

In the expression, $(h_j = 1|v) = \sigma(\Sigma_{i=1}^{m} w_{ij} v_i + b_j)$, $p(v_i = 1|h) = \sigma(\Sigma_{j=1}^{n} w_{ij} h_j + a_i)$, and σ represent the logistic sigmoid function. $\tilde{v}$ and $\tilde{h}$ implies the reconstructed v and h, correspondingly. N denotes the hidden node count, η represents the learning ratio, the momentum weighted can be denoted as α, and the weight decay is indicated by λ.

AOA Based Hyperparameter Tuning Model

Khatir et al. [24] determined in their study that the AOA can be used to optimally fine tune the hyperparameters related to the DBN technique and thereby boost the classification performance. In this section, the concept of AOA is defined. The major motivation of AOA derives from the usage like Subtraction (S), Addition (A), Division (D), and Multiplication (M) arithmetical operators to resolve optimization problems. Next, we discuss the dissimilar effects of those operators on optimization problems and the searching methodology of AOA.

Initialization Phase

As meta-heuristic optimization algorithm, AOA is depending on population behaviour. The set of population X comprising N searching agent is demonstrated in the following expression. Here, all the rows indicate a searching agent.

$$X = \begin{bmatrix} x_{1,1} & \cdots & x_{1,j} & x_{1,n-1} & x_{1},n \\ x_{2,1} & \cdots & x_{2,j} & \cdots & x_{2,n} \\ \cdots & \cdots & \cdots & \cdots & \cdots \\ \vdots & \vdots & \vdots & \vdots & \vdots \\ x_{N-1,1} & \cdots x_{N-1,j} & & \cdots & x_{N-1,n} \\ x_{N,1} & \cdots & x_{N,j} & x_{N,n-1} & x_{N,n} \end{bmatrix} \quad (14.6)$$

Afterward producing the population, the fitness of searching agent is calculated, and the optimal one is defined. Then, AOA performs exploration or exploitation via the Math Optimizer Accelerated (MOA) values:

$$MOA(t) = \text{Min} + t \times \left(\frac{\text{Max} - \text{Min}}{T} \right) \quad (14.7)$$

In Eq. (14.7), $MOA(t)$ denotes the value of MOA at t-th iterations. Min and Max denote the minimal and maximal values of the accelerating function, correspondingly. t indicates the existing iteration, and T represents the maximal iteration. The searching agent implements the exploration stage if $r_1 > MOA$, or else the exploitation stage is implemented.

Exploration Phase

In this section, the exploration stage of AOA is defined. Based on the motivation, the Multiplication (M) and Division (D) operators are presented to accomplish higher distributed value or decision. Both operators are arithmetically defined in the following:

$$X_{i,j}(t+1) = \begin{cases} X_{besi,j}(t) \times MOP \times \left(\left(UB_j - LB_j\right) \times \mu + LB_j\right), r_2 < 0.5 \\ X_{best,j}(t) \div (MOP + \varepsilon) \times \left(\left(UB_j - LB_j\right) \times \mu + LB_j\right), orelse \end{cases} \tag{14.8}$$

In Eq. (14.8) $X_{i,j}(t+1)$ signifies the *j-th* location of the *i-th* solution in the following iteration. $X_{best,j}(t)$ signifies the optimal solution in the *j-th* position. LB_j & UB_j denotes the lower and upper bounds, correspondingly. ε denotes a smaller integer number, and r_2 indicates the arbitrary number lies in zero and one. $\mu = 0.5$, that indicates the control function. Furthermore, the Math Optimizer is evaluated by:

$$MOP(t) = 1 - \frac{t^{\frac{1}{\alpha}}}{T^{\frac{1}{\alpha}}}$$

In Eq. (9), $\alpha = 0.5$ indicates the dynamic variable that defines the accuracy of the exploitation stage.

Exploitation Phase

In this section, we discuss the exploitation stage of AOA. In comparison with *D* and *M* operators, AOA exploits the Subtraction (S), and Addition (A) operators for deriving higher density solutions as (*S* & *A*) is simple technique the target region because of the lower dispersion. The arithmetical equation is defined by:

$$X_{i,j}(t+1) = \begin{cases} X_{best,j}(t) - MOP \times \left(\left(UB_j - LB_j\right) \times \mu + LB_j\right), r_3 < 0.5 \\ X_{besi,j}(t) + MOP \times \left(\left(UB_j - LB_j\right) \times \mu + LB_j\right), orelse \end{cases} \tag{14.10}$$

In Eq. (14.10) r_3 signifies an arbitrary number lies within zero and one.

Findings and Analysis

In this section, the simulation outcomes offered by the AOADBN-SMPP model are clearly examined. The training accuracy (TA) and validation accuracy (VA) obtained by the AOADBN-SMPP method on test dataset is shown in Fig. 14.3. The investigational result inferred that the AOADBN-SMPP technique has achieved maximal values of TA and VA. Particularly the VA is higher than TA.

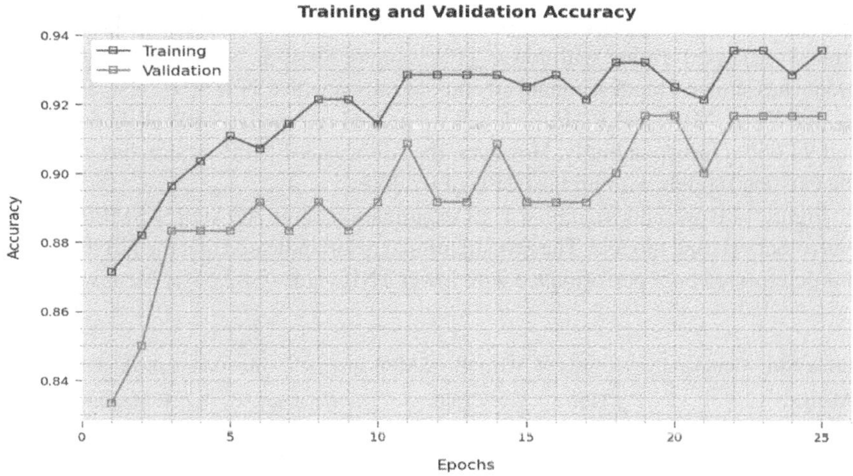

Fig. 14.3 TA and VA analysis of AOADBN-SMPP approach

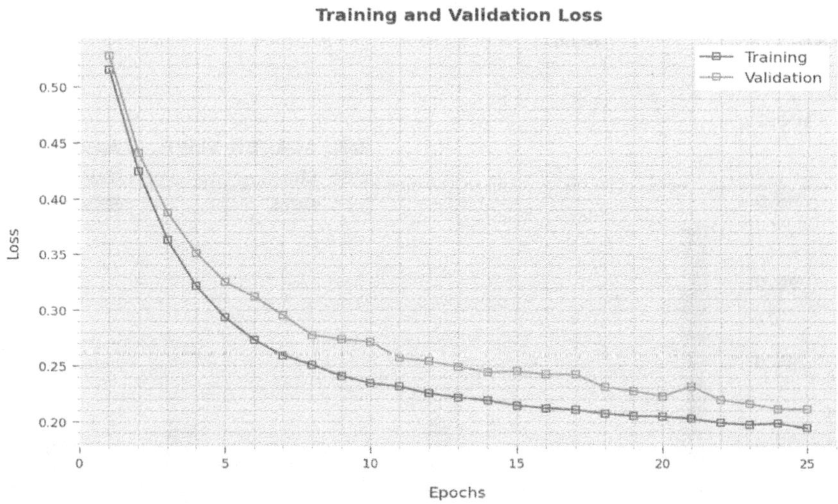

Fig. 14.4 TL and VL analysis of AOADBN-SMPP approach

The training loss (TL) and validation loss (VL) gained by the AOADBN-SMPP approach on test dataset are recognized in Fig. 14.4. The investigational outcome implied that the AOADBN-SMPP algorithm has achieved minimal values of TL and VL. In specific, the VL seemed to be lesser than TL.

Table 14.1 demonstrates the comparison analysis of AOADBN-SMPP approach under distinct measures with existing algorithms [25]. Figure 14.5 provides a comprehensive comparative TPR examination of the AOADBN-SMPP type with other current types. The experimental outcome implied that the NB model has demonstrated weak performing with lower TPR value of 89.49%. Simultaneously, the LR model has shown a modest enhanced outcome along with a TPR of 91.53%. Followed by, the SVM, MLP, and RAF models have proved reasonable TPR of 92.66, 92.34, and 94.66% correspondingly. However, the AOADBN-SMPP version has exceeded the existing models with highest TPR of 97.36%.

Figure 14.6 presents a detailed TNR examination of the AOADBN-SMPP method with other existing models. The investigational result indicated that the NB technique has determined poor performance with lower TNR value of 87.65%. Meanwhile, the

Table 14.1 Comparative analysis of AOADBN-SMPP approach with existing methodologies

Methods	TPR	TNR	Accuracy	F-Measure
AOADBN-SMPP	97.36	91.07	93.26	93.72
LR	91.53	88.38	90.51	89.51
SVM	92.66	88.51	90.59	90.64
NB	89.49	87.65	89.61	89.07
MLP	92.34	88.59	90.29	89.52
RAF	94.66	80.77	88.56	87.98

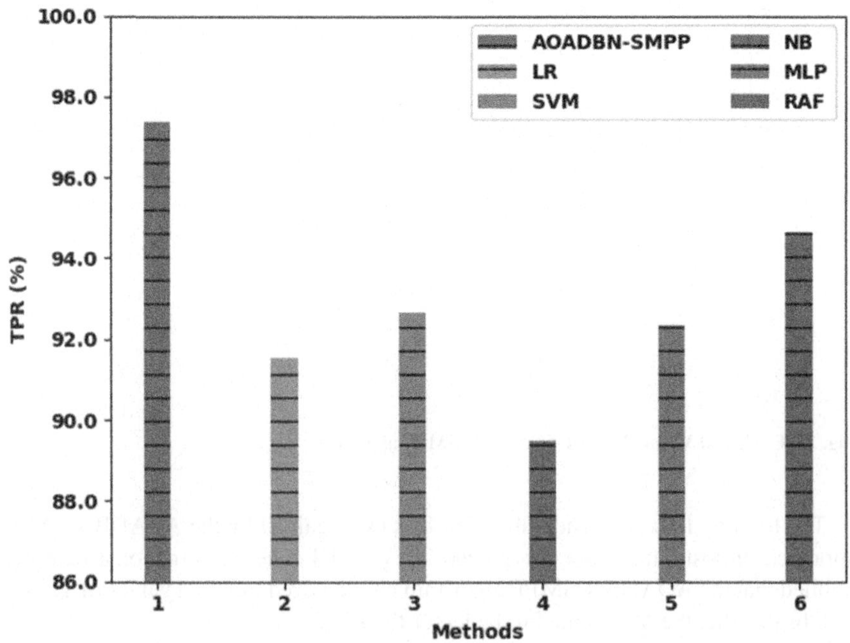

Fig. 14.5 TPR analysis of AOADBN-SMPP approach with existing algorithms

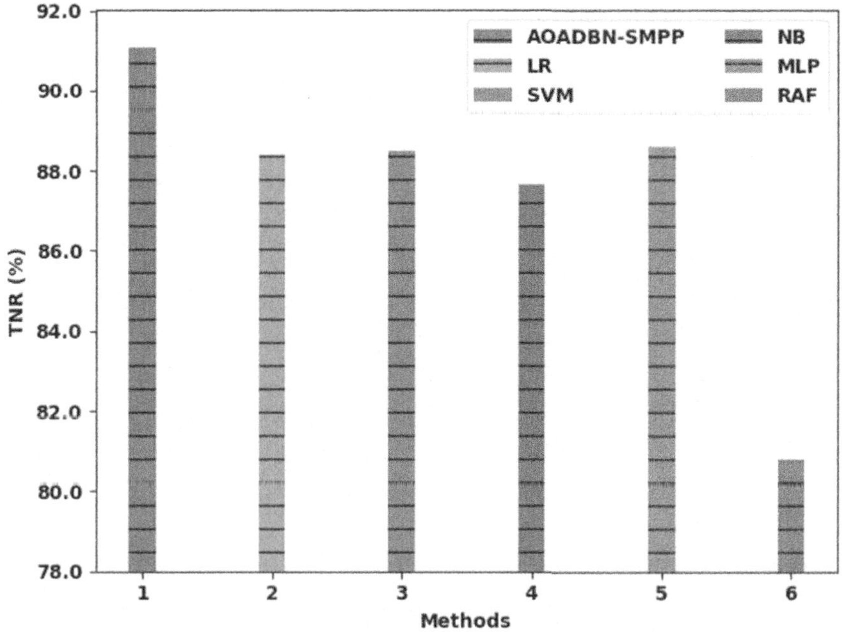

Fig. 14.6 TNR analysis of AOADBN-SMPP approach with existing algorithms

LR algorithm has shown slightly enhanced outcomes with a TNR of 88.51%. Subsequently, the SVM, MLP, and RAF techniques have demonstrated reasonable TNR of 88.51, 88.59, and 80.77% correspondingly. But, the AOADBN-SMPP methodology has surpassed the existing models with higher TNR of 91.07%.

Figure 14.7 offers a comparative $accu_y$ assessment of the AOADBN-SMPP method with other current standards. The experimental outcome inferred that the NB methodology has illustrated poor performance with lower $accu_y$ value of 89.61%. Meanwhile, the LR technique has shown slightly enhanced outcome with the $accu_y$ of 90.51%. Then, the SVM, MLP, and RAF models have demonstrated reasonable $accu_y$ of 90.59, 90.29, and 88.56% correspondingly. But, the AOADBN-SMPP methodology has surpassed the existing models with increased $accu_y$ of 93.26%.

Figure 14.8 renders a detailed comparative $F_{measure}$ examination of the AOADBN-SMPP technique with other existing types. The experimental outcome denoted the NB method has demonstrated poor performance with lower $F_{measure}$ value of 89.07%. simultaneously, the LR method has shown slightly enhanced outcomes with the $F_{measure}$ of 89.51%. Afterward, the SVM, MLP, and RAF models have established reasonable $F_{measure}$ of 90.64, 89.52, and 87.98% correspondingly. But, the AOADBN-SMPP technique has surpassed the existing models with maximal $F_{measure}$ of 93.72%.

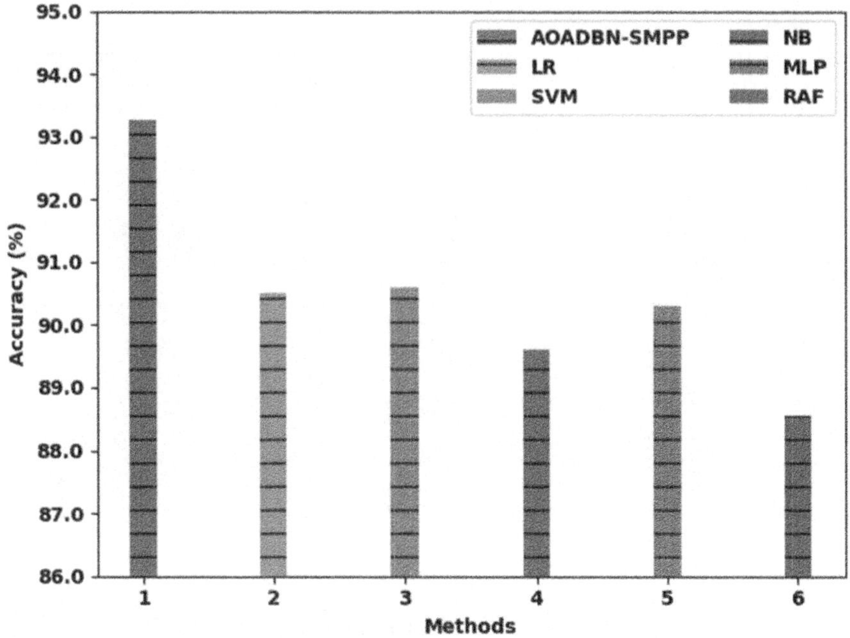

Fig. 14.7 Analysis of AOADBN-SMPP approach with current standards

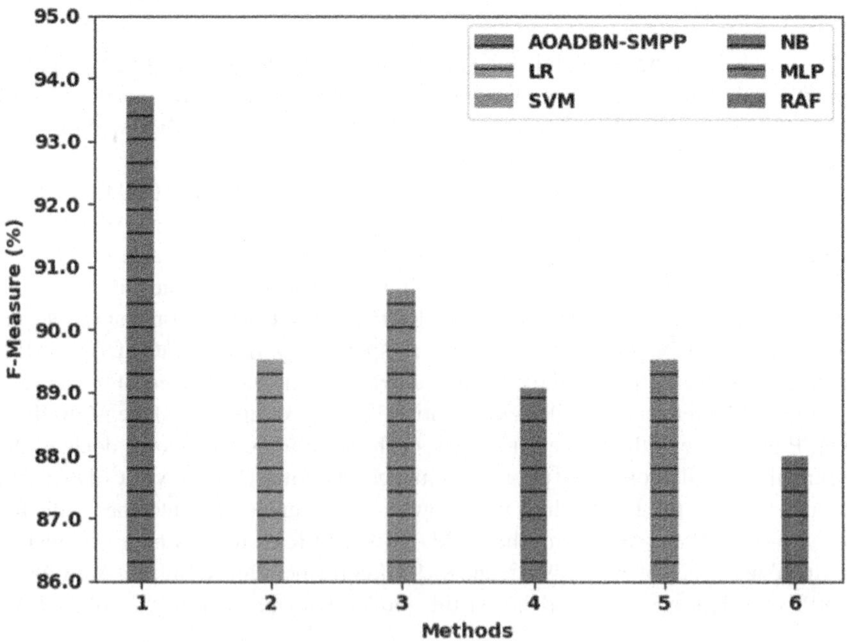

Fig. 14.8 $F_{measure}$ analysis of AOADBN-SMPP approach with existing algorithms

Research Limitations

Few limitations were assessed from this study. First, a major limitation is addressed from the application of the model in the first step, through out the infrastructure of the DBN steps. More specifically, in the reverse update, where the actual result will turn to be different from the expected result and hence generates a loss. Second, an impactful limitation might be due from unsystematic type of risk associated with the stock performance in the market when markets become very volatile due to external economic reason; such us political and/or unexpected events.

Research Contribution

The study contributes to the academic field as it adds new knowledge for study for the literature review. From the other hand, since this study is practical research, its outcome can be used by financial professionals, such and the financial analysts and other agencies and financial institutions that deal with market studies. Additionally, investors (individuals or companies) might benefit.

The novelty of this study replies on its customization of the suggested model. DBN and AOA are already available models exploited in the academic literature and employed in the industry for stock predication and various other reasons. What makes this study unique is that it has combined 2 steps "predication" and "factor optimization" to create a new model for long term stock prices' predictions.

The Study offers opportunities for future research. For instance, further research could be exploited around other methods used, such as the K-Nearest Neighbors (KNN) algorithm or Gradient boosting (GB).

Conclusion

In this article, a novel AOADBN-SMPP method was predicted for the forecasting of stock prices in a long term. To accomplish this, the AOADBN-SMPP technique follows two main methods specifically prediction and factor optimization. At the first step, the DBN model is exploited for forecasting stock prices. Next, the AOA has been leveraged to optimally adjust the hyperparameters associated to the DBN technique and thereby boost the classification performance. The AOA's architecture facilitates the best possible modification of the DBN standard's hyperparameters. The examination of the outcomes of experiments of the AOADBN-SMPP algorithm is tested and the outcomes were evaluated with distinct prospects. The comparison analysis found that the AOADBN-SMPP algorithm outperformed more contemporary state-of-the-art techniques.

The above research emphasizes the significance of stock price prediction in both academic study and real -world financial industry applications. Recognizing that stock markets are inherently unpredictable; the research presents a novel method called the Arithmetic Optimization Algorithm with Deep Belief Network-based Stock Market Price Prediction. The study highlights how machine learning methods—in particular, deep learning algorithms like DBN—are becoming into effective instruments for predicting stock prices. The research findings correspond with the wider pattern observed in the finance industry, wherein sophisticated technologies such as deep learning and machine learning are being utilized more and more to improve decision-making and prediction accuracy. Overall, highlights the significance of ongoing innovation in predicting stock markets models, taking into account the variety of factors affecting stock prices also the opportunity for cutting-edge algorithms to increase forecasting accuracy.

References

1. Kuremoto, T., Kimura, S., Kobayashi, K., & Obayashi, M. (2014). Time series forecasting using a deep belief network with restricted Boltzmann machines. *Neurocomputing, 137*, 47–56.
2. Kumar, D., Sarangi, P., & Verma, R. (2022). A systematic review of stock market prediction using machine learning and statistical techniques. *Materials Today Proceedings 49* (A), 3187–3191.
3. Chen, X. (2015). Stock price prediction via deep belief networks.
4. Nabipour, M., Nayyeri, P., Jabani, H., Mosavi, A., & Salwana, E. (2020). Deep learning for stock market prediction. *Entropy, 22*(8), 840.
5. Aghaeefar, N., Mohammad Pourzarandi, M. E., Afshar Kazemi, M. A., & Minoie, M. (2019). Applying optimized mathematical algorithms to forecast stock price average accredited banks in Tehran STOCK EXCHANGE and Iran Fara Bourse. *Advances in Mathematical Finance and Applications, 4*(3), 77–94.
6. Chong, E., Han, C., & Park, F. C. (2017). Deep learning networks for stock market analysis and prediction: Methodology, data representations, and case studies. *Expert Systems with Applications, 83*, 187–205.
7. Eapen, J., Bein, D., & Verma, A. (2019, January). Novel deep learning model with CNN and bi-directional LSTM for improved stock market index prediction. In 2019 IEEE 9th annual computing and communication workshop and conference (CCWC) (pp. 0264–0270). IEEE.
8. Kumar, G., Singh, U. P., & Jain, S. (2022). An adaptive particle swarm optimization-based hybrid long short-term memory model for stock price time series forecasting. *Soft Computing, 26*(22), 12115–12135.
9. Chen, Y., Wu, J., & Wu, Z. (2022). China's commercial bank stock price prediction using a novel K-means-LSTM hybrid approach. *Expert Systems with Applications, 202* (C), 117370.
10. Zhao, Y., & Chen, Z. (2021). Forecasting stock price movement: New evidence from a novel hybrid deep learning model. *Journal of Asian Business and Economic Studies, 29*(2), 91–104.
11. Harel, A., & Harpaz, G. (2021). Forecasting stock prices. *International Review of Economics & Finance, 73*, 249–256.
12. Ghosh, P., Neufeld, A., & Sahoo, J. K. (2022). Forecasting directional movements of stock prices for intraday trading using LSTM and random forests. *Finance Research Letters, 46* (A), 102280.
13. Abualigah, L., Zitar, R. A., Almotairi, K. H., Hussein, A. M., Abd Elaziz, M., Nikoo, M. R., & Gandomi, A. H. (2022). Wind, solar, and photovoltaic renewable energy systems with

and without energy storage optimization: A survey of advanced machine learning and deep learning techniques. *Energies, 15*(2), 578.
14. Shi, Y., Zheng, Y., Guo, K., & Ren, X. (2021). Stock movement prediction with sentiment analysis based on deep learning networks. *Concurrency and Computation: Practice and Experience, 33*(6), e6076.
15. Leonardo, M. P., Eduardo, C. M. C., María, T. H. A., Luis, T. G. J., & Elias, L. R. C. (2022). Formalization of a new stock trend prediction methodology based on the sector price book value for the Colombian market. *Heliyon, 8* (4), e.09210.
16. Nabipour, M., Nayyeri, P., Jabani, H., Shahab, S., & Mosavi, A. (2020). Predicting stock market trends using machine learning and deep learning algorithms via continuous and binary data; a comparative analysis. *IEEE Access, 8*, 150199–150212.
17. Biswas, M., Shome, A., Islam, M. A., Nova, A. J., & Ahmed, S. (2021, April). Predicting stock market price: A logical strategy using deep learning. In 2021 IEEE 11th IEEE Symposium on Computer Applications & Industrial Electronics (ISCAIE) (pp. 218–223). IEEE.
18. Liu, H., & Long, Z. (2020). An improved deep learning model for predicting stock market price time series. *Digital Signal Processing, 102*, 102741.
19. Agrawal, M., Khan, A. U., & Shukla, P. K. (2019). Stock price prediction using technical indicators: A predictive model using optimal deep learning. *Learning, 6*(2), 7.
20. Metawa, N., et al. (2021). Return rate prediction in Blockchain financial products using deep learning. *Sustainability, 13*(21), 11901.
21. Mingzhong Liu,& N. Metawa. (2021). An optimization model for stock market direction prediction. *American Journal of Business and Operations Research, 6*(1), 23-35.
22. Shilpa, B. L., & Shambhavi, B. R. (2021). Combined deep learning classifiers for stock market prediction: integrating stock price and news sentiments. Kybernetes.
23. Niu, G., Yi, X., Chen, C., Li, X., Han, D., Yan, B., Huang, M., & Ying, G. (2020). A novel effluent quality predicting model based on genetic-deep belief network algorithm for cleaner production in a full-scale paper-making wastewater treatment. *Journal of Cleaner Production, 265*, 121787.
24. Khatir, S., Tiachacht, S., Le Thanh, C., Ghandourah, E., Mirjalili, S., & Wahab, M. A. (2021). An improved artificial neural network using arithmetic optimization algorithm for damage assessment in FGM composite plates. *Composite Structures, 273*, 114287.
25. Shen, J., & Shafiq, M. O. (2020). Short-term stock market price trend prediction using a comprehensive deep learning system. *Journal of Big Data, 7*(1), 1–33.
26. Jing, N., Wu, Z., & Wang, H. (2021). A hybrid model integrating deep learning with investor sentiment analysis for stock price prediction. *Expert Systems with Applications, 178*, 115019.
27. Li, X., Liu, Q., & Wu, Y. (2022). Prediction on blockchain virtual currency transaction under long short-term memory model and deep belief network. *Applied Soft Computing, 116*, 108349.
28. Market Prediction. *Entropy, 22*(8), 840.
29. Nabipour, M., Nayyeri, P., Jabani, H,. Mosavi, A., & Salwans, E. (2020). Deep learning for stock market prediction. *Entropy, 22* (8), 840.
30. Shahi, T. B., Shrestha, A., Neupane, A., & Guo, W. (2020). Stock price forecasting with deep learning: A comparative study. *Mathematics, 8*(9), 1441.
31. Sim, H. S., Kim, H. I., & Ahn, J. J. (2019). Is deep learning for image recognition applicable to stock market prediction? Complexity
32. Xie, M., Li, H., & Zhao, Y. (2020). Blockchain financial investment based on deep learning network algorithm. *Journal of Computational and Applied Mathematics, 372*, 112723.
33. Xie, W., & Metawa, N. (2021). Financial stock market forecasting based on intelligent support vector machine regression model [J]. *Journal of Intelligent and Fuzzy Systems, 2021*(2), 1–10.
34. Yadav, K., Yadav, M., & Saini, S. (2022). Stock values predictions using deep learning-based hybrid models. *CAAI Transactions on Intelligence Technology, 7*(1), 107–116.

Chapter 15
A Kernel Bayesian Data Envelopment Analysis Approach for Bias Correction of Efficiencies

Constantinos Zacharias, Panagiotis D. Zervopoulos, Ali Emrouznejad, Konstantinos Triantis, and Gang Cheng

Abstract This study introduces a kernel Bayesian approach to correct the bias of data envelopment analysis (DEA) efficiency estimates. This approach yields consistent estimates for convex sets. The prior distribution of this Bayesian method is "non-informative" in a relative sense as no distributional assumptions are made, like in theoretical Bayesian approaches, and the parameters of DEA efficiency distributions are not used to obtain bias-corrected estimates, as in alternative computational or hybrid Bayesian techniques for statistical inference to efficiencies. Specifically, various kernel distributions, such as Epanechnikov, Biweight, Triweight, and Gaussian, are tested for the prior distribution. In addition, we deploy least cross validation (LCV), rule of thumb (RoT), and least-squares cross validation (LSCV) as bandwidth selection methodologies for every kernel distribution function. Bias correction draws on the ratio of a posterior truncated normal distribution, with μ and σ the respective kernel values, and the above prior kernel distributions with LCV, RoT, and LSCV as bandwidth selection mechanisms. Using scaled samples of 30, 50, 80, and 100 units, the mean square error (MSE) and mean absolute error (MAE) of this Bayesian

C. Zacharias
EY, Financial Services & Risk Management, Athens, Greece
e-mail: Constantinos.Zacharias@gr.ey.com

P. D. Zervopoulos (✉)
Department of Management, University of Sharjah, Sharjah, UAE
e-mail: pzervopoulos@sharjah.ac.ae

A. Emrouznejad
Surrey Business School, University of Surrey, Guilford, UK
e-mail: a.emrouznejad@surrey.ac.uk

K. Triantis
Grado Department of Industrial and Systems Engineering, Virginia Tech, Northern Virginia Center, Blacksburg, VA, USA
e-mail: triantis@vt.edu

G. Cheng
China Center for Health Development Studies, Peking University, Beijing, China
e-mail: maxdea@qq.com

© The Author(s), under exclusive license to Springer Nature Switzerland AG 2024

A. Emrouznejad et al. (eds.), *Business Analytics and Decision Making in Practice*, Lecture Notes in Operations Research, https://doi.org/10.1007/978-3-031-61589-4_15

approach's estimates are as low as 6.45 × 10–3 and 6.4 × 10–2, respectively. Based on real-world data, we show that the new Bayesian method performs better than extant computational bias-correction techniques for DEA efficiencies. At the same time, the MSE and MAE decrease gradually as the sample size increases.

Keywords Data envelopment analysis · Bayesian methods · Kernel estimation · Bias correction · Statistical inference

Introduction

Data envelopment analysis (DEA) is extensively employed to estimate relative efficiency. After the seminal paper introducing DEA [1], a rich set of research papers has been published [2]. The statistical characteristics of the DEA efficiencies have been studied by [3–5].

Regarding DEA as a "frontier" method, any frontier variations can impact the validity of DEA estimators. It has been proven that DEA efficiencies are overestimated for finite samples [4] and are consistent for convex sets [4, 6, 7]. Further research on the shape of the production frontier and the fashion in which the inefficiency is related to the noise can be found in [6, 8–11].

As far as the sensitivity of the efficiency estimators is concerned, there are four primary ways to handle: (a) chance-constrained DEA (CCDEA), (b) two-stage DEA approaches, (c) bootstrap DEA, (d) and Bayesian DEA.

The CCDEA [12–14] supplants the input–output data with randomly distributed to prescribe the stochastic production frontiers. The CCDEA methodology is suitable for addressing the noise in the data, but there are limitations on its statistical background. In [15], a retrospection of the methodology can be found.

A combination of both DEA and stochastic methods is present in two-stage approaches to improve the DEA efficiency estimators' stochastic properties. The [5] and [16–18] are indicative of two-stage DEA approaches for statistical inference to efficiencies.

The bootstrap DEA, also known as the smoothed bootstrap DEA, is mainly used for bias correction and the construction of the DEA efficiency estimators' confidence intervals. The development of the smoothed bootstrap DEA is attributed to [7, 19–23], and [26]. The estimators of the methodology are consistent [7, 23].

Both computational and theoretical Bayesian methods are involved in the statistical inference of the efficiency estimators. As part of the computational approach, in [24], the DEA efficiencies are projected by an iterative technique to a stochastic frontier [25] advanced the aforementioned computational approach. For theoretical Bayesian DEA methods, the selection of prior distribution is mandatory. Numerous prior distributions are studied in [27, 28]. On top, [30] introduced a Bayesian approach that requires prior distribution assumption [29] noted that inappropriate prior selection can deteriorate efficiency estimators.

From a non-parametric standpoint, [31] highlighted the advantages of the application of kernel non-parametric distributions along with [32–42]. The application of the non-parametric methodology on the DEA efficiencies estimation has been studied in [23, 43–45].

This work contributes to enhancing the efficiency estimates validity. A two-stage Bayesian approach is developed. In the first stage, the efficiencies lying in the interval (0, 1] are obtained from a DEA program. In the second stage, a non-informative prior is used, drawing on non-parametric kernel distributions and various bandwidths, while the bias-corrected efficiencies are obtained from a simulated shifted normal distribution. Unlike extant methods, our approach does not make any distributional assumptions about the prior distribution and does not depend on the sample size, while it yields consistent efficiency estimates. In addition, its estimates are strongly correlated with the smoothed bootstrap and the conventional DEA estimates.

The rest of the paper unfolds as follows. In Sect. 2, we present the kernel Bayesian efficiency DEA approach. Section 15.3 discusses the real-world data sets used in this study. Section 15.4 analyzes the empirical results. Section 5 concludes the analysis.

Methodology

Data Envelopment Analysis (DEA)

In our study, a Variable Returns to Scale (VRS) DEA program is used as in [46] to obtain relative efficiencies $\Theta = \{\theta_j\}_{j=1}^{n}$, where $\theta_j \in (\theta_L, 1]$ for any $j = 1, \ldots, n$, and $\theta_L \in (0, 1)$. The VRS DEA model reads as follows:

$$min\theta s.t. \sum_{j=1}^{n} \lambda_j x_{ij} < \theta x_{io}, i = 1, \ldots, m$$

$$\sum_{j=1}^{n} \lambda_j y_{rj} \geq y_{ro}, r = 1, \ldots, s$$

$$\sum_{j=1}^{n} \lambda_j = 1 \tag{15.1}$$

$$\lambda_j \geq 0, j = 1, \ldots, n$$

where, x_{ij}, y_{rj} denote the i th input and r th output of the j th Decision Making Unit (DMU), respectively, and λ_j represents the intensity weights. The Bayesian approach discussed later in this section is applicable to any DEA program yielding efficiencies ranging between zero and unity, such as (1).

Bayesian Statistical Inference

The Prior

A generic formula for the kernel density estimator can be used

$$\hat{f}(\theta) = \frac{1}{nh}\sum_{j=1}^{n} k\left(\frac{\theta - \theta_j}{h}\right) \tag{15.2}$$

where $k(\psi)$ corresponds to the kernel function, and h is called the kernel bandwidth. Let

$$\kappa_j(k) = \int_{-\infty}^{\infty} \psi^j k(\psi) d\psi \tag{15.3}$$

If $\kappa_0(k) = 1$, $\kappa_1(k) = 0$ and $\kappa_2(k) < \infty$ the kernel is of second order. By imposing $\kappa_0(k) = 1$ entails that any weighting function integrates into unity, $\kappa_1(k) = 0$ requires symmetric kernels around 0, and $\kappa_2(k) < \infty$ returns meaningful variance expressions. Table 15.1 highlights the most popular kernels with their second-order properties.

By using basic kernel properties the bias of the kernel density estimator can be formulated as

$$Bias\left(\hat{f}(\theta)\right) = E\left[\hat{f}(\theta)\right] - f(\theta) \approx \frac{h^2}{2} f^{(2)}(\theta)\kappa_2(k) \tag{15.4}$$

Table 15.1 The kernel density formulas with variance condition

Kernel density	Formula	Variance condition
Epanechnikov	$k_1(\psi) = \frac{3}{4}(1 - \psi^2)1\{\lvert\psi\rvert \leq 1\}$	$\frac{1}{5}$
Biweight	$k_2(\psi) = \frac{15}{16}(1 - \psi^2)^2 1\{\lvert\psi\rvert \leq 1\}$	$\frac{1}{7}$
Triweight	$k_3(\psi) = \frac{35}{32}(1 - \psi^2)^3 1\{\lvert\psi\rvert \leq 1\}$	$\frac{1}{9}$
Gaussian	$k_\varphi(\psi) = \frac{1}{\sqrt{2\pi}} e^{-\left(\frac{1}{2}\right)\psi^2}$	1

Table 15.2 Kernel densities with scaling factors

Kernel density	Scaling factor
Epanechnikov	2.345
Biweight	2.778
Triweight	3.154
Gaussian	1.059

where $f^{(2)}(\theta)$ is defined as the second derivative of $f(\theta)$, and a second-order Taylor expansion is used. The bias is dependent on the local design, the kernel, and the bandwidth. Hence, the bandwidth manipulation can result in bias adjustments.

The impact of bandwidth selection can be more significant than the choice of the kernel density estimator. A plethora of bandwidth selection approaches have been proposed, which are broken down into two fundamental classes: rule-of-thumb and data-driven methods.

In this study, the kernel density estimates are deployed using Silverman rule-of-thumb bandwidths for the different kernels along with their scaling factors (see Table 15.2).

The rule-of-thumb bandwidths possess optimal properties only in case the underlying density is correctly assumed. That is the reason why data-driven methods should be explored. The direct plug-in method (DPI) has also been applied as an alternative to the Silverman rule of thumb.

The least squares cross-validation (LSCV) is widely used as a data-driven bandwidth selection approach that aims to derive the optimal bandwidth by minimising the distance between the estimated density and the "true" density. According to [31], the LSCV criterion can be written as

$$LSCV(h) = \frac{1}{n^2 h} \sum_{j=1}^{n} \sum_{\tau=1}^{n} \bar{k}\left(\frac{\theta_j - \theta_\tau}{h}\right) \tag{15.5}$$

where $\bar{k}$ combines three different kernel types.

The Likelihood Cross-Validation (LCV) can be easily interpreted intuitively and is more efficient than the LSCV. The maximisation of the Kullback–Leibler criterion is the objective of the LCV to obtain the optimal bandwidth. The Kullback–Leibler can be formulated as.

$$KL\left(f, \hat{f}\right) = \int f(\theta) \log\left\{\frac{f(\theta)}{\hat{f}(\theta)}\right\} d\theta \tag{15.6}$$

On top of the (6) an average over each datum can be obtained.

$$LCV(h) = n^{-1} \sum\nolimits_{j=1}^{n} \log \hat{f}_{-j}(\theta_j) \tag{15.7}$$

where the $\widehat{f}_{-j}(\theta_j)$ is the leave-one-out estimator.

The Posterior

In [27] and [28], a truncated normal distribution is used. The probability density function of a truncated normal distribution is

$$f(\theta;\mu,\sigma,a,b) = \begin{cases} \frac{1}{\sigma}\frac{\varphi\left(\frac{\theta-\mu}{\sigma}\right)}{\Phi\left(\frac{b-\mu}{\sigma}\right)-\Phi\left(\frac{a-\mu}{\sigma}\right)}, a \leq x \leq b \\ 0, \textit{otherwise} \end{cases}$$

where μ is the average, σ is the standard deviation, $a = min\Theta$, and $b = max\Theta$.

Bias Correction

Let a correction parameter:

$$\xi = \frac{\widetilde{\theta_L}}{\widehat{\theta_L}} < 1,$$

where $\theta \in (0, 1]$ and $\widetilde{\theta_L}$ and $\widehat{\theta_L}$ resemble the efficiencies from a posterior distribution and from a prior distribution, respectively.

The ratio of the posterior to prior distribution is best fitted by a normal distribution with μ and σ the mean and standard deviation, respectively.

The bias-corrected efficiencies θ_j^c are estimated iteratively using Monte-Carlo simulation from a normal distribution with μ_{θ^c} and σ_{θ^c} the mean and standard deviation, respectively.

Dataset

In this study, four real-world data sets are used. The scaled samples consist of 30, 50, 80, and 100 banks located in the European Union. The performance of the bias-corrected efficiencies methodology, as described in the previous section, is examined in the four real-world samples.

A fixed number and type of input and output variables are used for all real-world samples. Specifically, the three inputs (i.e., deposits & short-term funding, equity, and fixed assets) and the two outputs (i.e., gross loans and other earning assets) are selected to estimate the efficiency of the sample banks.

Empirical Results

As part of the statistical analysis of the DEA efficiencies, the kernel density estimates (Epanechnikov, Biweight, Triweight, and Gaussian) with different bandwidths (RoT, LSCV, LCV, and DPI) are estimated. In Figs. 15.1 and 15.2, the kernel densities are estimated along with their bandwidths for 30 and 100 banks, respectively.

As seen in Tables 15.3, 15.4 and 15.5, the performance of our method is measured by calculating the MSE in every sample of 30, 50, 80, and 100 units.

In Tables 15.3, 15.4 and 15.5, the MSE of the kernel Bayesian approach estimates drops gradually, justifying their consistency and the appropriateness of our method in correcting bias. In an extended version of this study, we show that the MSE and MAE of the kernel Bayesian bias-correction method's estimates gradually decrease while the sample size increases. These findings have been tested on samples of up to 500 units.

The bias-corrected efficiencies of our method share common statistical properties with the efficiencies derived from DEA, Eq. (15.1), and the smoothed bootstrap [23]. Specifically, the range of the correlation of the bias-corrected efficiencies with the DEA efficiencies for the kernels applied with their bandwidths across samples can be seen in Table 15.6.

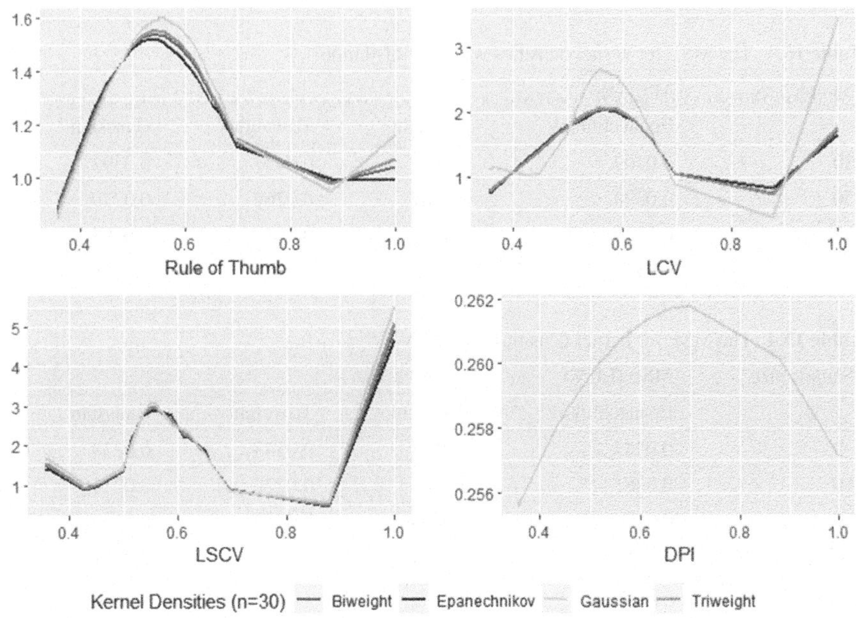

Fig. 15.1 The kernel densities with their bandwidths for 30 banks

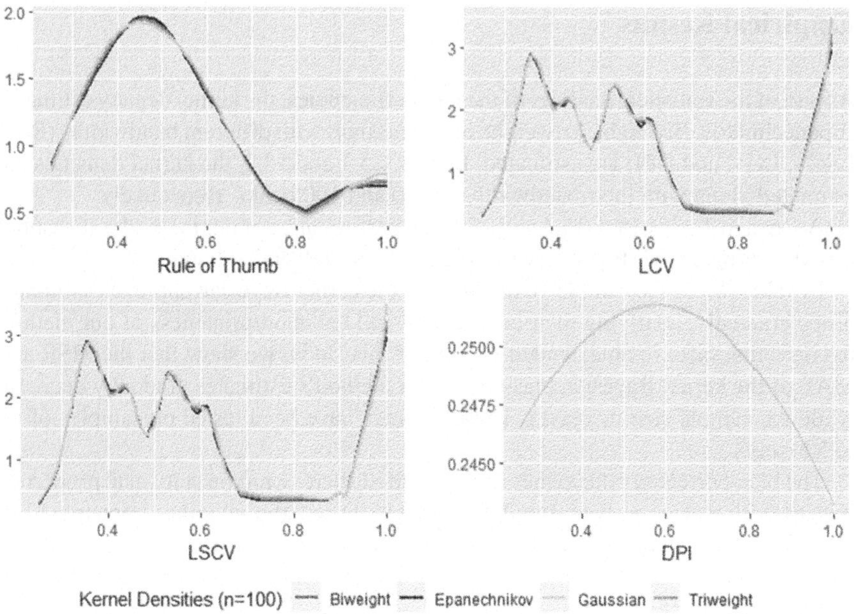

Fig. 15.2 The kernel densities with their bandwidths for 100 banks

Table 15.3 The MSE for kernel densities with rule-of-thumb

Sample size	MSE (RoT)			
	Epanechnikov	Biweight	Triweight	Gaussian
30	0.0409	0.0457	0.0477	0.0393
50	0.0290	0.0267	0.0269	0.0279
80	0.0196	0.0187	0.0180	0.0193
100	0.0191	0.0167	0.0168	0.0178

Table 15.4 The MSE for kernel densities with LCV

Sample size	MSE (LCV)			
	Epanechnikov	Biweight	Triweight	Gaussian
30	0.0347	0.0345	0.0343	0.0238
50	0.0261	0.0208	0.0222	0.0211
80	0.0174	0.0164	0.0159	0.0165
100	0.0153	0.0154	0.0143	0.0151

Table 15.5 The MSE for kernel densities with LSCV

Sample size	MSE (LSCV)			
	Epanechnikov	Biweight	Triweight	Gaussian
30	0.0264	0.0262	0.0260	0.0257
50	0.0202	0.0216	0.0201	0.0197
80	0.0178	0.0166	0.0160	0.0149
100	0.0153	0.0154	0.0143	0.0143

Table 15.6 Correlations of bias-corrected efficiencies with the DEA and smoothed bootstrap

Kernel (Bandwidth)	DEA	Smoothed bootstrap
Epanechnikov (RoT)	[0.9815 - 0.9983]	[0.9350 - 0.9608]
Biweight (RoT)	[0.9815 - 0.9983]	[0.9319 - 0.9606]
Triweight (RoT)	[0.9815 - 0.9983]	[0.9319 - 0.9608]
Gaussian (RoT)	[0.9815 - 0.9983]	[0.9350 - 0.9608]
Epanechnikov (LCV)	[0.9815 - 0.9983]	[0.9350 - 0.9606]
Biweight (LCV)	[0.9815 - 0.9983]	[0.9350 - 0.9606]
Triweight (LCV)	[0.9815 - 0.9982]	[0.9350 - 0.9606]
Gaussian (LCV)	[0.9815 - 0.9982]	[0.9350 - 0.9606]
Epanechnikov (LSCV)	[0.9815 - 0.9983]	[0.9350 - 0.9606]
Biweight (LSCV)	[0.9815 - 0.9983]	[0.9350 - 0.9606]
Triweight (LSCV)	[0.9815 - 0.9982]	[0.9350 - 0.9606]
Gaussian (LSCV)	[0.9815 - 0.9982]	[0.9350 - 0.9609]

Concluding Remarks

The purpose of this study was to develop a Bayesian DEA approach for correcting the bias of efficiencies without distributional assumptions. Instead of making distributional assumptions, non-parametric kernel densities were used along with their bandwidths.

The empirical results showed a gradual decrease in MSE while the sample size increased. Given that DEA efficiencies are asymptotically unbiased [4], high correlations of the bias-corrected efficiencies with the efficiencies obtained from the DEA and the smoothed bootstrap estimates ensure sound statistical properties of our methodology. We acknowledge that the MSE of our paper is higher than the MSE of the Bayesian DEA estimates of [46]. However, no distributional assumptions are required here, unlike the Bayesian DEA approach of [46], which drew on a mix of uniform and beta distributions.

References

1. Charnes, A., Cooper, W. W., & Rhodes, E. (1978). Measuring the efficiency of decision making units. *European Journal of Operational Research, 2*(6), 429–444.
2. Emrouznejad, A., & Yang, G. L. (2018). A survey and analysis of the first 40 years of scholarly literature in DEA: 1978–2016. *Socio-Economic Planning Sciences, 61*, 4–8.
3. Banker, R. D., & Maindiratta, A. (1992). Maximum likelihood estimation of monotone and concave production frontiers. *Journal of Productivity Analysis, 3*(4), 401–415.
4. Banker, R. D. (1993). Maximum likelihood, consistency and data envelopment analysis: A statistical foundation. *Management Science, 39*(10), 1265–1273.
5. Banker, R. D., & Natarajan, R. (2008). Evaluating contextual variables affecting productivity using data envelopment analysis. *Operations Research, 56*(1), 48–58.
6. Kneip, A., Park, B. U., & Simar, L. (1998). A note on the convergence of nonparametric DEA estimators for production efficiency scores. *Econometric Theory, 14*(6), 783–793.
7. Kneip, A., Simar, L., & Wilson, P. W. (2008). Asymptotics and consistent bootstraps for DEA estimators in nonparametric frontier models. *Econometric Theory, 24*(6), 1663–1697.
8. Simar, L. (2007). How to improve the performances of DEA/FDH estimators in the presence of noise? *Journal of Productivity Analysis, 28*, 183–201.
9. Kuosmanen, T., Johnson, A., & Saastamoinen, A. (2015). Stochastic nonparametric approach to efficiency analysis: A unified framework. Data Envelopment Analysis: A Handbook of Models and Methods, 191–244.
10. Coelli, T. J., Rao, D. S. P., O'Donnell, C. J., & Battese, G. E. (2005). An introduction to efficiency and productivity analysis. springer science & business media.
11. Meeusen, W., & van Den Broeck, J. (1977). Efficiency estimation from Cobb-Douglas production functions with composed error. *International Economic Review,* 435–444.
12. Charnes, A., & Cooper, W. W. (1963). Deterministic equivalents for optimizing and satisficing under chance constraints. *Operations research, 11*(1), 18–39.
13. Land, K. C., Lovell, C. K., & Thore, S. (1993). Chance-constrained data envelopment analysis. *Managerial and Decision Economics, 14*(6), 541–554.
14. Olesen, O. B., & Petersen, N. C. (1995). Chance constrained efficiency evaluation. *Management Science, 41*(3), 442–457.
15. Olesen, O. B., & Petersen, N. C. (2016). Stochastic data envelopment analysis—A review. *European Journal of Operational Research, 251*(1), 2–21.
16. Kuosmanen, T., & Kortelainen, M. (2007). Stochastic nonparametric envelopment of data: Cross-sectional frontier estimation subject to shape constraints. Available at SSRN 983882.
17. Zervopoulos, P. D., Sklavos, S., Kanas, A., & Cheng, G. (2019). A multi-parametric method for bias correction of DEA efficiency estimators. *Journal of the Operational Research Society, 70*(4), 655–674.
18. Kuosmanen, T., & Kortelainen, M. (2012). Stochastic non-smooth envelopment of data: Semi-parametric frontier estimation subject to shape constraints. *Journal of Productivity Analysis, 38*, 11–28.
19. Simar, L., & Wilson, P. W. (1998). Sensitivity analysis of efficiency scores: How to bootstrap in nonparametric frontier models. *Management Science, 44*(1), 49–61.
20. Simar, L., & Wilson, P. W. (1999). Estimating and bootstrapping Malmquist indices. *European Journal of Operational Research, 115*(3), 459–471.
21. Simar, L., & Wilson, P. W. (2000). A general methodology for bootstrapping in non-parametric frontier models. *Journal of Applied Statistics, 27*(6), 779–802.
22. Simar, L., & Wilson, P. W. (2000). Statistical inference in nonparametric frontier models: The state of the art. *Journal of Productivity Analysis, 13*, 49–78.
23. Kneip, A., Simar, L., & Wilson, P. W. (2011). A computationally efficient, consistent bootstrap for inference with non-parametric DEA estimators. *Computational Economics, 38*, 483–515.
24. Tsionas, E. G. (2003). Combining DEA and stochastic frontier models: An empirical Bayes approach. *European Journal of Operational Research, 147*(3), 499–510.

25. Liu, J., Sickles, R. C., & Tsionas, E. G. (2017). Bayesian treatments for panel data stochastic frontier models with time varying heterogeneity. *Econometrics, 5*(3), 33.
26. Simar, L., Vanhems, A., & Wilson, P. W. (2012). Statistical inference for DEA estimators of directional distances. *European Journal of Operational Research, 220*(3), 853–864.
27. Chilingerian, J. A. (1995). Evaluating physician efficiency in hospitals: A multivariate analysis of best practices. *European Journal of Operational Research, 80*(3), 548–574.
28. Rosenman, R., & Friesner, D. (2004). Scope and scale inefficiencies in physician practices. *Health Economics, 13*(11), 1091–1116.
29. Friesner, D., Mittelhammer, R., & Rosenman, R. (2013). Inferring the incidence of industry inefficiency from DEA estimates. *European Journal of Operational Research, 224*(2), 414–424.
30. Tsionas, M. G. (2020). A coherent approach to Bayesian data envelopment analysis. *European Journal of Operational Research, 281*(2), 439–448.
31. Henderson, D. J., & Parmeter, C. F. (2015). *Applied nonparametric econometrics*. Cambridge University Press.
32. Jones, M. C., & Sheather, S. J. (1991). Using non-stochastic terms to advantage in kernel-based estimation of integrated squared density derivatives. *Statistics & Probability Letters, 11*(6), 511–514.
33. Abadir, K. M., & Lawford, S. (2004). Optimal asymmetric kernels. *Economics Letters, 83*(1), 61–68.
34. Duin. (1976). On the choice of smoothing parameters for Parzen estimators of probability density functions. *IEEE Transactions on Computers, 100*(11), 1175-1179
35. Epanechnikov, V. (1969). Nonparametric estimation of a multidimensional probability density. *Teoriya Veroyatnostei i ee Primeneniya, 14*(1), 156–161.
36. Hall, P. (1983). Fast rates of convergence in the central limit theorem. *Zeitschrift für Wahrscheinlichkeitstheorie und Verwandte Gebiete, 62*(4), 491–507.
37. Deheuvels, P. (1977). Estimation non paramétrique de la densité par histogrammes généralisés. *Revue de statistique appliquée, 25*(3), 5–42.
38. Rudemo, M. (1982). Empirical choice of histograms and kernel density estimators. Scandinavian Journal of Statistics, 65–78.
39. Loader, C. R. (1999). Bandwidth selection: Classical or plug-in? *The Annals of Statistics, 27*(2), 415–438.
40. Stone, C. J. (1984). An asymptotically optimal window selection rule for kernel density estimates. **The Annals of Statistics,** 1285–1297.
41. Bowman, A. W. (1984). An alternative method of cross-validation for the smoothing of density estimates. *Biometrika, 71*(2), 353–360.
42. Habbema, J. D. F., Hermans, J., & Van den Broek, K. (1974). A stepwise discriminant analysis program using density estimation.
43. Kneip, A., Simar, L., & Van Keilegom, I. (2015). Frontier estimation in the presence of measurement error with unknown variance. *Journal of Econometrics, 184*(2), 379–393.
44. Yagi, D., Chen, Y., Johnson, A. L., & Kuosmanen, T. (2020). Shape-constrained kernel-weighted least squares: Estimating production functions for Chilean manufacturing industries. *Journal of Business & Economic Statistics, 38*(1), 43–54.
45. Bădin, L., Daraio, C., & Simar, L. (2019). A bootstrap approach for bandwidth selection in estimating conditional efficiency measures. *European Journal of Operational Research, 277*(2), 784–797.
46. Zervopoulos, P. D., Triantis, K., Sklavos, S., & Kanas, A. (2023). An alternative Bayesian data envelopment analysis approach for correcting bias of efficiency estimators. *Journal of the Operational Research Society, 74*(4), 1021–1041.

Chapter 16
Can Machine Learning Enhance the Forecasting of Herding Behavior in International Stock Markets?

Panagiotis G. Artikis, Georgios A. Papanastasopoulos, Polyxeni G. Tsitsiri, and Antonios M. Vasilatos

Abstract Behavioral finance uses psychology to make financial decisions. Investor herd behavior—imitating rather than making independent decisions—drives financial crises. Due of the challenging of herding behavior prediction, our machine learning approach is innovative although while COVID-19 epidemic is well-studied. We employ machine learning algorithms to predict stock investors' behavior before and after the COVID-19 pandemic to test rational asset pricing in the Americas, Europe, and Asia–Pacific. Empirical research shows the COVID-19 epidemic increased herding in all regions. In Europe, herding-related variables are more important. These findings demonstrate that market-related dynamics, such as liquidity and transaction procedures, greatly influence investor herding. We also noticed good herding behavior prediction findings using machine learning. Our machine-learning model exceeded traditional regression methods in prediction accuracy due to its advanced algorithms and ability to understand complex robust data patterns. Machine learning captures nonlinear relationships, revealing herding behavior causes Our research has two main consequences. It shows that market conditions impact herding. The liquidity and trading methods of investors may affect crowdsourcing. Machine learning predicts herding behavior, thus advanced algorithms and data-driven approaches may help investors make better decisions.

Keywords Machine learning · Financial forecasting · Herding behavior

P. G. Artikis (✉) · G. A. Papanastasopoulos · P. G. Tsitsiri · A. M. Vasilatos
University of Piraeus, 80, M. Karaoli & A. Dimitriou St, 18534 Piraeus, Greece
e-mail: partikis@unipi.gr

G. A. Papanastasopoulos
e-mail: papanast@unipi.gr

P. G. Tsitsiri
e-mail: polyxenits@unipi.gr

A. M. Vasilatos
e-mail: vasilatos@unipi.gr

© The Author(s), under exclusive license to Springer Nature Switzerland AG 2024
A. Emrouznejad et al. (eds.), *Business Analytics and Decision Making in Practice*, Lecture Notes in Operations Research, https://doi.org/10.1007/978-3-031-61589-4_16

Introduction

In uncertain times, behavioral finance uses psychology to make financial decisions. Among all behavioral biases, academic and research has examined herd behavior, which causes mirroring in international financial markets. Herding may spread financial crisis. Herding behavior—investors following the actions of others rather than making their own decisions—has been researched in finance [1].

Herding has been studied in stock and cryptocurrency markets [2]. Ali et al. [3] examine how financial markets react to volatility and decline as the coronavirus epicenter spreads from China to Europe and the US. Kizys et al. [4] examine if the government can reduce global stock market herding during Covid-19. Espinosa-Méndez and Arias [5] study herding in Europe during COVID-19.

Herding behavior in stock markets may affect asset value, market dynamics, and asset pricing, and the overall efficiency of financial markets. When individuals base their choices on market behavior, not their individual decisions. This trend defies Fama [6] Efficient Market Hypothesis. Chang et al. [1] reported limited herding in Japan, South Korea, and Taiwan. Chiang and Zheng [7] show that herding affects 18 advanced and developing economies' asset markets.

Researchers have tried to understand and predict herding behavior in stock markets to provide valuable insights. Positive investor sentiment may enhance herding [8]. However, if investors lack private experience, they follow well-informed individuals [9].

Herding behavior during the COVID-19 pandemic is little examined in global stock markets simultaneously from July 10, 2019, to July 15, 2020. So, our novelty is the examination of three major regions, America, Europe, and Asia–Pacific during the COVID-19 pandemic. Also, herding behavior prediction is challenging, hence we want to create a machine learning model.

Psychological factors, market sentiment, informational cascades, and market microstructure challenge forecasting models. New econometric methods, machine learning algorithms, and behavioral finance theories offer promising forecasting accuracy [10]. So, if we manage to create a complex herding forecasting model, it will predict herding behavior in a future economic crisis, which is the major implication of our approach.

Our study uses Chang et al. [1] model to study rational asset pricing and herding in 30 global stock indices. The association between cross-sectional absolute deviation (CSAD) and squared market returns ($R^2_{m,t}$) indicates increased herding across all three regions. Herding increased in importance in Europe after the COVID-19 outbreak.

Investors, portfolio managers, regulators, and researchers benefit from the findings. Knowing and accurately predicting herding behavior may help investors making better decisions, reduce risk, and improve investment strategies. Early herding identification may assist market regulators avoid instability and systemic risks [11].

Researchers struggle to predict investors' herding behavior. Using, econometrics, machine learning, and behavioral finance is the aim to predict herding behavior in this study. Also, it improves market dynamics understanding and informs policymakers.

This study organized with the following format. Section "Conceptual Framework" covers conceptual framework and hypothesis. Section "Data and Methodology" describes the data, variables, and econometric approach used to evaluate herding behavior before and after the COVID-19 outbreak. Section "Empirical Results and Discussion" examines our data, while Sect. "Empirical Results and Discussion" provides conclusions, and real-world managerial implications.

Conceptual Framework

Herding in stock markets is affected by pandemics, market dynamics, and behavioral biases. This study proposes a well-researched conceptual framework to understand and predict herding.

Economic theory introduced information cascade. According to Bikhchandani et al. [12], "informational cascades" occur when customers rely their choices on others' behaviors while keeping private information. So, doubt or the belief that the group is smarter may motivate conformity.

According to economics, herding is motivated by regret and fear of divergence. Conformity may avoid regretting a different option and its repercussions. This conduct may increase when market participants are cautious, inexperienced, or feel others know more. Furthermore, conformity and social influence may affect herding. Investors may imitate successful or knowledgeable individuals to get approval or escape criticism. Also, the anxiety of missing out (FOMO) or social need may cause blind herding.

Researchers and practitioners studied financial herding behavior. Investors prefer imitation versus their own analysis. According to Christie and Huang [2], market pressure does not affect daily and monthly asset pricing model results. However, Chang et al. [1] found herding in South Korea and Taiwan, low bias in Japan, and no herd effects in the US and Hong Kong markets after model adjustment. The study of herding is crucial for understanding market dynamics, investor behavior, and asset values.

Financial market herding research has contributed valuable insights. Stock market herding has been extensively studied to understand investor behavior. Chang et al. [1] and Hwang and Salmon [13] studied equites markets herding. They examined how liquidity, trading processes, and market microstructure affect investor herding. Bikhchandani and Sharma [11] studied herding and public information and found that investors assemble after public disclosure. Market volatility affects stock market herding under uncertainty, according to [7] and market turbulence makes stronger tendency for investors to follow others. This and other studies explain stock market herding and provide valuable insights into the implications of herding.

COVID-19 outbreak enables us study financial market herding. Investors followed trends more due to the pandemic's unpredictability, volatility, and economic disruption. Most research have studied how the pandemic affected in individual or up to two markets. Espinosa-Méndez and Arias [5] discovered COVID-19 increased Australian stock market herding.

Academics studied financial market herding lately. Various theories predict herding frequency and intensity. For example, investor and social media sentiment may predict short-term herding [14], these indicators of market players' sentiment provide insight for herding. Also, neural networks and random forests improve forecasting by capturing complex interactions and non-linear trends [15].

Stock market herding behavior research has shown its worth. The COVID-19 epidemic caused uncertainty, information asymmetry, and market stress. However, there is little machine learning research on herding behavior and the COVID-19 pandemic's influence on global financial markets. Hence, this research examines global stock market herding during COVID-19 pandemic and forecast herding behavior in stock markets. These findings may explain stock market herding. We test the following hypothesis.

Assuming significant herding behavior, the hypothesis predicts a negative correlation ($\beta_2 < 0$) between the squared market variable and the cross-sectional absolute deviation (CSAD) in the equation $CSAD_{m,t} = \beta_0 + \beta_1 \left| R_{m,t} \right| + \beta_2 R_{m,t}^2 + e_t$ A strong positive relationship ($\beta_2 > 0$) suggests anti-herding behavior, whereas $\beta_1 > 0$ and $\beta_2 = 0$ imply no herding.

Data and Methodology

Machine Learning (ML) predicts stock market herding in our research. The random forest method is our major main algorithm and benchmark [16]. This algorithm is suitable for both classification and regression. It is one of the most multi-purpose algorithms. Additionally, we use a long short-term memory neural network with support vector regression and test them against the benchmark. This ML technique expands ML-based finance research [17].

From July 10, 2019, until July 15, 2020, it is the examined period. We utilize Thomson Reuters hourly data. We concentrate on America, Europe, and Asia–Pacific. SXA1 for America, STOXX for Europe, and SXP1 for Asia–Pacific are general stock indexes for each region. Our investigation was divided into two periods. The pre-COVID-19 phase runs from July 10 to November 29, 2019. The second was from December 2, 2019, until July 15, 2020, which is the post-COVID-19 period. According to Luu and Luong [18], herding behavior rises during pandemics, and our preliminary data analysis reveals this behavior during the post-COVID-19 period.

We examine herding using the approach of Chang et al. [1] and Chiang and Zheng [7]. More specifically, we calculate market returns as:

$$R_{m,t} = lnP_t - lnP_{t-1} \tag{16.1}$$

We use the cross-sectional absolute deviation (CSAD) of stocks returns around the market portfolio return:

$$CSAD_{m,t} = \frac{\sum_{t=1}^{N} \left| R_{i,t} - R_{m,t} \right|}{N} \tag{16.2}$$

where $R_{i,t}$ is the observed stock return of index i on hour t and it is the first logarithmic difference of closing prices for stock index i at time t, as given below.

$$R_{i,t} = lnP_t - lnP_{t-1} \tag{16.3}$$

Following Chang et al. [1], we regress the CSAD of returns with respect to market returns:

$$CSAD_t = \beta_0 + \beta_1 \left| R_{m,t} \right| + \beta_2 R_{m,t}^2 + e_t \tag{16.4}$$

We use the variables form Chang et al. model as features in machine learning algorithms. CSAD is the target feature we seek to predict.

The Machine Learning Algorithms

K-Nearest Neighbors

K-Nearest Neighbors (KNN) algorithm is a simple, non- parametric algorithm which can be used both for classification and regression problems. KNN can be used for linear and non-linear problems. While it can be easily and fast implemented algorithm (relatively to the sample size), it is considered an effective algorithm with many applications [19].

A KNN algorithm maps every class instance of the training sample to the real n-dimensional space. The number of the dimensions is equal to the number of model's features. The basic assumption of KNN is that similar data points (data points of the same class) are in proximity. This proximity in mathematics is called distance. To measure the distance, we rely on distance metrics. The most widely used metric is the Minkowski distance.[1] The distance is calculated in the following way. For two data points $A = (a_1, a_2, \ldots, a_n) and B = (\beta_1, \beta_2, \ldots, \beta_n) \in \mathbb{R}^n$, the Minkowski distance of order p is defined as:

$$D(A, B) = \left(\sum_{i=1}^{n} (|a_i - \beta_i|)^p \right)^{\frac{1}{p}} \tag{16.5}$$

By changing the p-value to 1 we get the Manhattan distance:

$$D_{MNH}(A, B) = \sum_{i=1}^{n} |a_i - \beta_i| \tag{16.6}$$

[1] Other distance metrics like Cosine, Jaccard or any other custom distance measure can be used.

And when p equals to 2, we get the Euclidean distance:

$$D_{EUC}(A, B) = \sqrt{\sum_{i=1}^{n} (a_i - \beta_i)^2} \quad (16.7)$$

When the distance between data points is calculated and stored in the training set (those are the Neighbors), the algorithm is "fed" with the test set. The algorithm will search for the K-Nearest Neighbors to the data point of the test set in order to classify it. The number of K ranges from 1 to the total number of instances. We have used a grid search to find the number of K that maximizes the AUC score. However, we must note that a K = 1 will cause the algorithm to overfit the data and a K equal to the sample size is computationally expensive.

Random Forest

Breiman [16] suggested Decision Tree-based Random Forest. A versatile, general-purpose, yet complicated algorithm for any data and situation. Random Forest averages Decision Tree predictions. This method may be compared to the Fama–Macbeth method, which averages regression coefficients over years [20]. A unique random Decision Tree is created using bootstrap sample of the training set by using a random sample of features. This creates a forest as many times as the algorithm user chooses trees. After training, forest trees use majority votes to predict test data. The method is complicated since the user must hyper-adjust several parameters. Forest tree count must be chosen initially. Many experts believe that a large number of trees is sufficient and not a tuning parameter. More forest trees increase accuracy, but they need more computational power and may not always do so [21].

Grid search yielded the best Random Forest hyperparameters. We adjusted many hyperparameters for maximum tree depth, minimum node splits, and leaf node samples. Each split condition may be identified. Gini impurity or entropy may be used. However, grid searches that include both criteria are computationally expensive and perform comparably. For this reason, we have used the Gini impurity as the splitting criterion.

Cross Validation Approach

It is widely established in machine learning research in accounting and finance that intertemporal financial data cannot be arbitrarily separated in train and test sets [22]. According to Bao et al. [23], a k-fold cross validation is improper for our data since its random splits may anticipate previous data points using future data points.

In their working paper, [24] recommended time-series cross validation. Data are divided k-folds to account for intertemporal nature. Following their model, we divided our data using scikit-learn's library in Python TimeSeriesSplit [25]. This

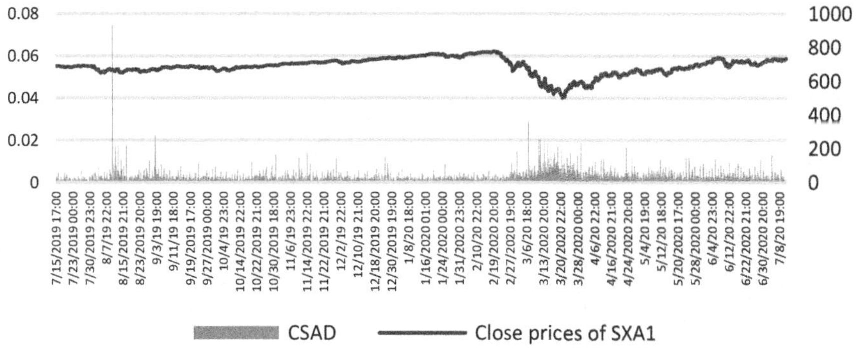

Fig. 16.1 Close prices of SXA1 and American CSAD

form of cross-validation partitions training data by year. First block trains algorithms, second block validates them. The third block validates the first and second blocks after training. This is repeated k-1 times, where k is the folds. Grid search determines our algorithm's optimal hyperparameters during training. After training, we found the best hyperparameters. To calculate our out-of-sample performance, we train the algorithm using the best hyperparameters in the training subsample and forecast the testing fold we left out. We repeat the procedure by adding the test fold we just used to generate our predictions to the training set, retrain the algorithm, and predict the next test fold.

Forecasting Strategy

After we have trained the algorithms in the training set, we have obtained the best performing model. We use this model (and its hyperparameters) to make predictions in the holdout set (test set). We make one step ahead predictions. In each step, the algorithms are fed with the actual observation, and then they predict the next step forward. This approach is called rolling-origin-recalibration evaluation [24] or forward chaining. The model performance is calculated as the average performance of all out-of-sample steps we have predicted.

Empirical Results and Discussion

Figures 16.1, 16.2, and 16.3 show price dynamics in America, Europe, and Asia–Pacific and their cross-sectional absolute deviation (CSAD) values. General stock index values are on the left axis, CSAD values on the right axis.

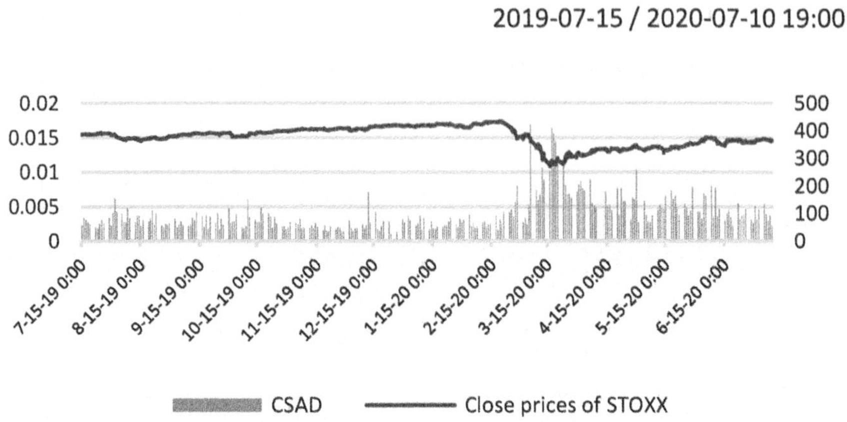

Fig. 16.2 Close prices of STOXX and European CSAD

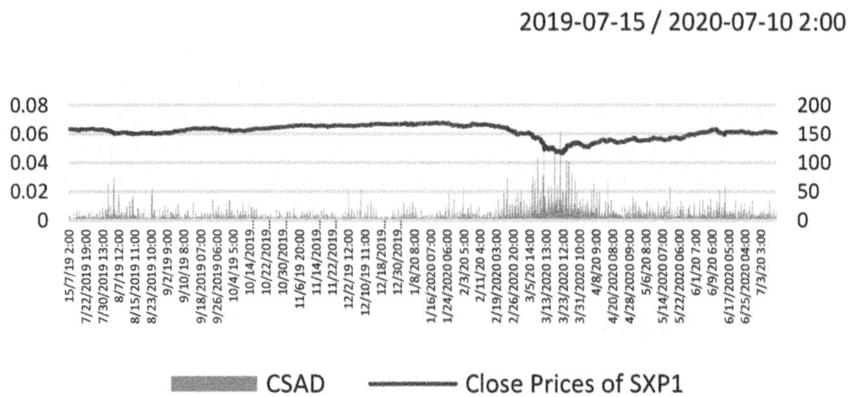

Fig. 16.3 Close prices of SXP1 and Asian-Pacific CSAD

Between February 17, 2020, and April 24, 2020, the closing prices of the general stock indexes fell, but CSAD values rose in all regions. The first coronavirus deaths had been reported, and the epidemic had spread worldwide.

We seek to study the ability of herding prediction. Due to CSAD volatility after the COVID-19 pandemic, we expect our forecasts to be inaccurate using pre-Covid-19 period as the training sample and the post-Covid-19 as a test sample. However, separating into two time periods should improve our forecasts, as after Covid-19, the training sample will be volatile. Herding behavior should also affect our forecasts differently over time. Before Covid-19, it will not have a greater impact on the CSAD forecast compared to squared market returns that is expected to affect the CSAD forecast after the Covid-19.

Table 16.1 Out-of-sample forecasting scores

Algorithm:	KNN		Random forest	
Scoring:	Mean squared error (MSE)			
Region	Before covid	Post covid	Before covid	Post covid
US	0.000004457	0.000009735	0.00000341	0.00000841
Europe	0.000000433	0.0000025	0.000000366	0.00000196
ASIA–PACIFIC	0.000005295	0.000020364	0.000005010	0.000019717

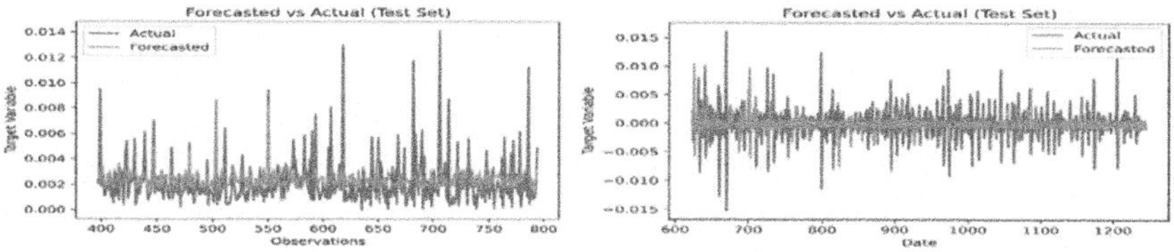

Fig. 16.4 Pre and post covid forecasts using Random Forest US sample

Given that herding behavior is common during the COVID-19 pandemic, we expect squared market returns to affect CSAD prediction. This will prove herding occurred during the examined period and improve the CSAD forecast. We want to find the squared return-CSAD correlation. If we find a strong negative relationship, we will prove herding. Anti-herding behavior is indicated by a positive relationship.

The Out-Of-Sample Forecasting Results

As expected, the Random Forest method beats K-Nearest Neighbors in out-of-sample data (Table 16.1). Moreover, CSAD is less volatile in the pre-Covid era, therefore we predict it more accurately. Mean Squared Error is better since it's closer to zero. However, feature scale must be considered. Based on descriptive statistics, CSAD values are near to zero, however MSE score close to zero does not guarantee perfect prediction accuracy. That is obvious from the actual versus predicted plots. We overlay actual vs predicted observations to visually assess forecasting results (Figs. 16.4, 16.5 and 16.6).

Shedding Light into Machine Learning

One of the criticisms of the ML methods (and especially neural networks) is that they are considered "black box" methods. The user can just pick an algorithm, fit it

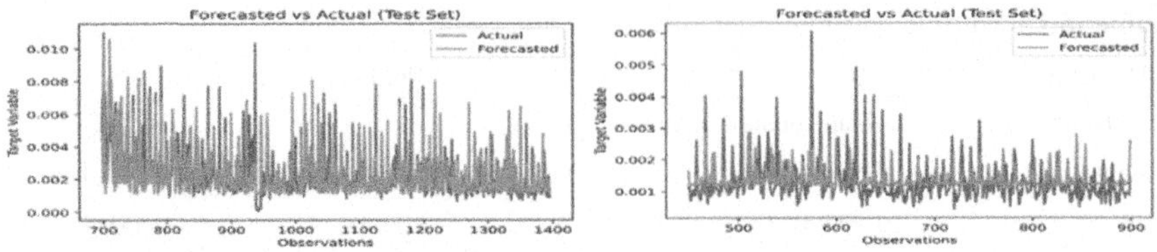

Fig. 16.5 Pre and post covid forecasts using Random Forest European sample

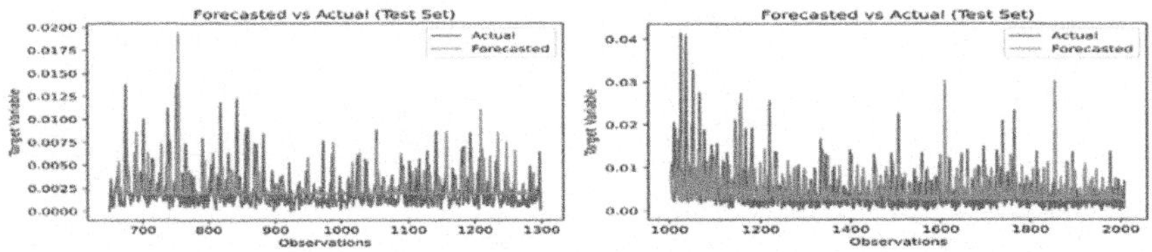

Fig. 16.6 Pre and post covid forecasts using Random Forest Asia–Pacific sample

to the data and obtain some results. This is also the case too with the Random Forest, especially if it is compared to a single Decision Tree. The single Decision Tree is interpretable while the Random Forest contains so many trees that no visual interpretation is possible. Even though visual interpretation is impossible with Random Forest, we can calculate feature importance. We create feature importance plots (Figs. 16.7, 16.9 and 16.11) and PDP plots (Figs. 16.8, 16.10 and 16.12), in the pre- and post-covid period for all the regions.

We notice that in the post covid period in the US, there is an increase in the importance of R_m^2, which indicates an increase in herding activity (Fig. 16.7).

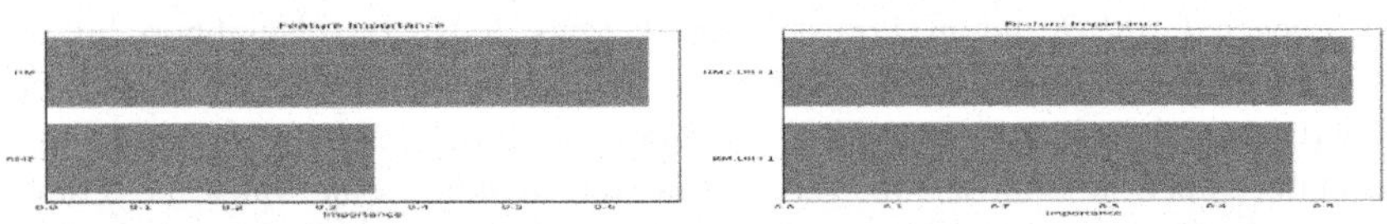

Fig. 16.7 Feature importance plots for US sample

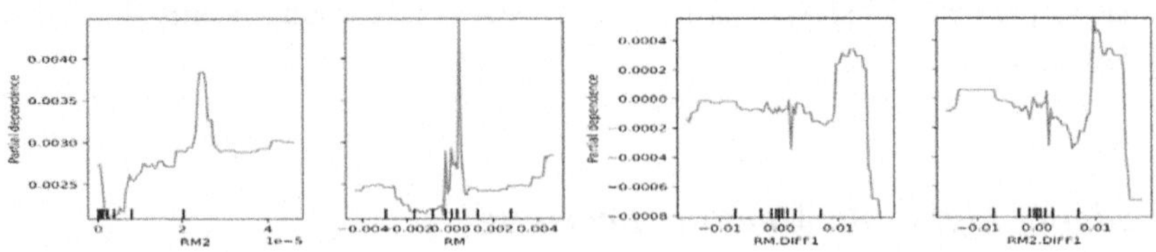

Fig. 16.8 PDP plots for US sample

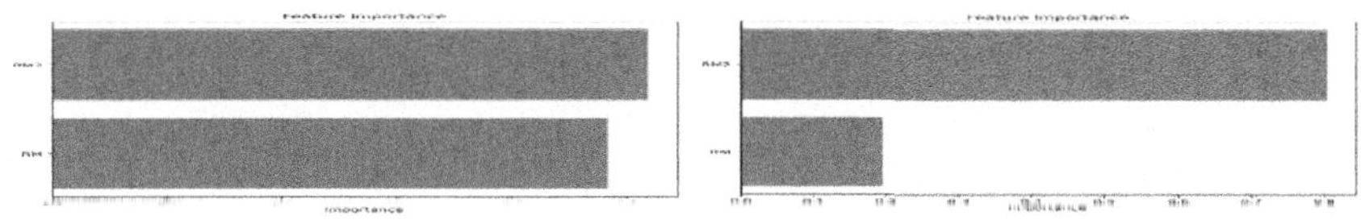

Fig. 16.9 Feature importance plots for European sample

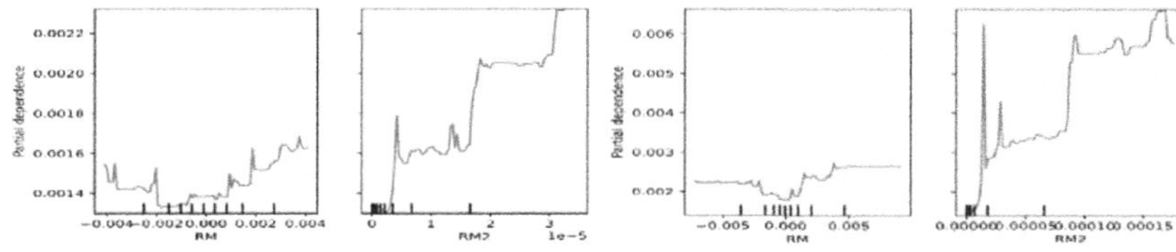

Fig. 16.10 PDP plots for European sample

From the PDP plots we get the relationship of features with the target we are predicting. The negative relationship of CSAD and R_m^2 in the post covid period indicates herding activity in the US (Fig. 16.8). We follow the same approach for the rest of the regions.

We notice the same increase in the importance of R_m^2 in the post-covid period in Europe. We notice though that in Europe, R_m^2 is more important compared to US, which may suggest even higher levels of herding activity (Fig. 16.9).

What is interesting in Europe, is that we can find a negative relationship of R_m^2 with CSAD only for the lowest and highest values of R_m^2. So, if herding behavior is existent, it exists only for very low and high values of R_m^2 (Fig. 16.10). Finally, in the Asia–Pacific region, we notice a small change in the importance of R_m^2; however, in the post-covid period it is the most important variable (Fig. 16.11). We notice that if herding exists in the post-covid period, it exists only for the highest values of R_m^2, where there is a negative relationship between R_m^2 and the target (CSAD) (Fig. 16.12).

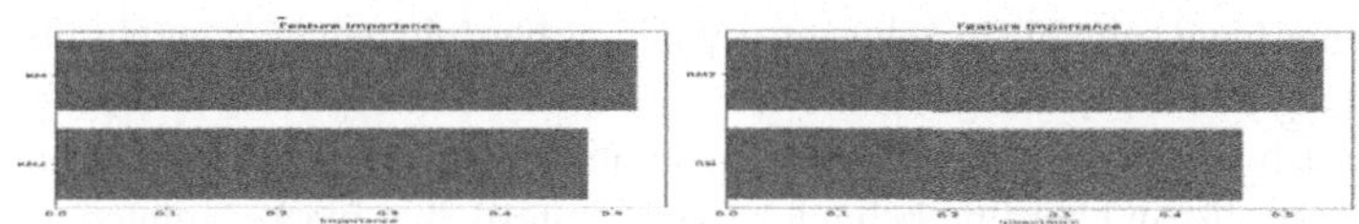

Fig. 16.11 Feature importance plots for Asia–Pacific sample

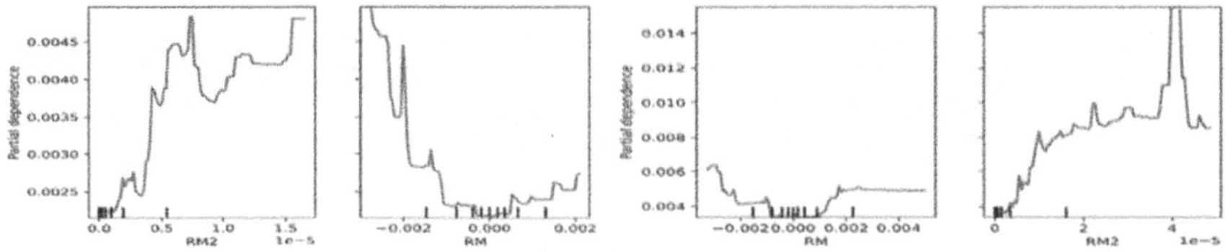

Fig. 16.12 PDP plots for Asia–Pacific sample

Conclusions

Various factors impact investor decision-making, making understanding of forecasting herding behavior in stock markets challenging. This study examines herding behavior and its forecasting implications using a conceptual framework and extensive literature review.

We examined the significant role of investor sentiment in driving herding behavior by incorporating the findings from studies such as using Baker and Wurgler [8], and Bikhchandani et al. [12]. Positive relationship with market returns and CSAD means that investors rely on their own views, whereas negative relationship inspire imitation and herding behavior.

This analysis forecasts stock market herding by adopting Chang et al. [1] conceptual framework, who employing OLS regression. We enhanced our forecasting model using machine learning.

Hourly returns for 30 indexes listed in major global markets from July 2019 to July 2020 are presented. To analyze these indexes globally, we split them into America, Europe, and Asia–Pacific to draw global conclusions and use the cross-sectional dispersion approach and machine learning approaches to forecast potential herding effects.

The COVID-19 outbreak boosted herding in all the examined regions. Herding is increased in the importance of herding related variable, particularly in Europe. Goal variability influenced our forecast accuracy after the COVID-19 pandemic, as predicted.

Findings have valuable insights into dynamics herding behavior and its prediction. Like Chang et al. [1], our machine learning study demonstrated a cross-sectional connection between the cross-sectional absolute deviation of stock returns around the market portfolio and herding behavior. This suggests that market-related dynamics, such as trading strategies and liquidity appear to greatly influence investor herding.

We also saw promising findings for incorporating machine learning in herding behavior forecasting. Our machine-learning model exceeded traditional regression techniques in prediction accuracy due to its advanced algorithms and ability to capture complex data patterns. Using machine learning nonlinear relationships can be captured and we can understand the underlying factors that affect herding behavior.

While our study helps explain and predict herding behavior, but it has limitations. Machine learning models need robust data quality and potential overfitting. Alternative methods and additional variables may improve herding behavior forecasting in future research.

Future research could be focus on the forecasting of herding behavior during Covid-19 pandemic taking into consideration other variables including market sentiment indicators, investor sentiment measures, news sentiment, and social media sentiment, to capture the multifaceted nature of herding behavior [17, 26]. Researchers may examine how macroeconomic data, financial news, and laws impact on herding behavior to enhance the accuracy and reliability of their predictions [27].

Real-World Managerial Implications and Impact

Research has two main effects. Our study shows market-related factors affect herding. Investors should examine liquidity and trade dynamics as may affect investor herding. Moreover, advanced algorithms and data-driven techniques may help investors make better informed decisions than regression methods since machine learning may forecast herding behavior.

Academics and practitioners benefit from this study. Investments, portfolio managers, and politicians may benefit from herding behavior study and prediction. Herding behavior accurate forecasts may aid risk management, decision-making, and investment strategy strategies. Regulators may detect herding early to reduce market volatility and systemic risks.

Building on Chang et al. [1] framework and using machine learning for forecasting, this research sheds light on stock market herding. Financial market participants and regulators learn about investing, risk, and herding from this study.

In conclusion, stock market herding predicting is difficult but essential. Past data are used to create a complete forecasting framework that encompasses herding behavior's complexity. Herding conduct may help practitioners and regulators understand markets.

Acknowledgements This work has been partly supported by the University of Piraeus Research Center.

The research work was supported by the Hellenic Foundation for Research and Innovation (HFRI) under the 4th Call for HFRI PhD Fellowships (Fellowship Number: 9116).

References

1. Chang, E. C., Cheng, J. W., & Khorana, A. (2000). An examination of herd behavior in equity markets: An international perspective. *Journal of Banking & Finance, 24*(10), 1651–1679. https://doi.org/10.1016/S0378-4266(99)00096-5
2. Christie, W. G., & Huang, R. D. (1995). Following the pied piper: do individual returns herd around the market? *Financial Analysts Journal, 51*(4), 31–37. https://doi.org/10.2469/faj.v51.n4.1918
3. Ali, M., Alam, N., & Rizvi, S. A. R. (2020). Coronavirus (COVID-19)—An epidemic or pandemic for financial markets. *Journal of Behavioral and Experimental Finance, 27*, 100341. https://doi.org/10.1016/j.jbef.2020.100341
4. Kizys, R., Tzouvanas, P., & Donadelli, M. (2021). From COVID-19 herd immunity to investor herding in international stock markets: The role of government and regulatory restrictions. *International Review of Financial Analysis, 74*, 101663. https://doi.org/10.1016/j.irfa.2021.101663
5. Espinosa-Méndez, C., & Arias, J. (2021). COVID-19 effect on herding behaviour in European capital markets. *Finance Research Letters, 38*, 101787. https://doi.org/10.1016/j.frl.2020.101787
6. Fama, E. F. (1965). The behavior of stock-market prices. *The Journal of Business, 38*(1), 34–105. https://www.jstor.org/stable/2350752.
7. Chiang, T. C., & Zheng, D. (2010). An empirical analysis of herd behavior in global stock markets. *Journal of Banking & Finance, 34*(8), 1911–1921. https://doi.org/10.1016/j.jbankfin.2009.12.014
8. Baker, M., & Wurgler, J. (2006). Investor sentiment and the cross-section of stock returns. *The Journal of Finance, 61*(4), 1645–1680. https://doi.org/10.1111/j.1540-6261.2006.00885.x
9. Devenow, A., & Welch, I. (1996). Rational herding in financial economics. *European Economic Review, 40*(3–5), 603–615. https://doi.org/10.1016/0014-2921(95)00073-9
10. Feng, L., & Seasholes, M. S. (2004). Correlated trading and location. *The Journal of Finance, 59*(5), 2117–2144. https://doi.org/10.1111/j.1540-6261.2004.00694.x
11. Bikhchandani, S., & Sharma, S. (2000). Herd behavior in financial markets: A review. *IMF Staff Papers, 47*(3), 279–310. https://doi.org/10.2307/3867650
12. Bikhchandani, S., Hirshleifer, D., & Welch, I. (1992). A theory of fads, fashion, custom, and cultural change as informational cascades. *Journal of Political Economy, 100*(5), 992–1026. https://doi.org/10.1086/261849
13. Hwang, S., & Salmon, M. (2004). Market stress and herding. *Journal of Empirical Finance, 11*(4), 585–616. https://doi.org/10.1016/j.jempfin.2004.04.003
14. Da, Z., Engelberg, J., & Gao, P. (2015). The sum of all FEARS investor sentiment and asset prices. *The Review of Financial Studies, 28*(1), 1–32. https://doi.org/10.1093/rfs/hhu072
15. Zhang, G. P. (2003). Time series forecasting using a hybrid ARIMA and neural network model. *Neurocomputing, 50*, 159–175. https://doi.org/10.1016/S0925-2312(01)00702-0
16. Breiman, L. (2001). Random Forests. *Machine Learning, 45*, 5–32. https://doi.org/10.1023/A:1010933404324
17. Chen, X., Cho, Y. H., Dou, Y., & Lev, B. (2022). Predicting future earnings changes using machine learning and detailed financial data. *Journal of Accounting Research, 60*(2), 467–515. https://doi.org/10.1111/1475-679X.12429
18. Luu, Q. T., & Luong, H. T. T. (2020). Herding behavior in emerging and frontier stock markets during pandemic influenza panics. *The Journal of Asian Finance, Economics and Business, 7*(9), 147–158. https://doi.org/10.13106/jafeb.2020.vol7.no9.147.
19. Triguero, I., García-Gil, D., Maillo, J., Luengo, J., García, S., & Herrera, F. (2019). Transforming big data into smart data: An insight on the use of the k-nearest neighbors algorithm to obtain quality data. *WIREs Data Mining and Knowledge Discovery, 9*(2), e1289. https://doi.org/10.1002/widm.1289
20. Varian, H. R. (2014). Big data: New tricks for econometrics. *Journal of Economic Perspectives, 28*(2), 3–28. https://doi.org/10.1257/jep.28.2.3

21. Probst, P., & Boulesteix, A.-L. (2018). To tune or not to tune the number of trees in random forest. *The Journal of Machine Learning Research, 18*(181), 1–18.
22. Bertomeu, J., Cheynel, E., Floyd, E., & Pan, W. (2021). Using machine learning to detect misstatements. *Review of Accounting Studies, 26*, 468–519. https://doi.org/10.1007/s11142-020-09563-8
23. Bao, Y., Ke, B., Li, B., Yu, Y. J., & Zhang, J. (2020). Detecting accounting fraud in publicly traded U.S. firms using a machine learning approach. *Journal of Accounting Research, 58*(1), 199–235. https://doi.org/10.1111/1475-679X.12292.
24. Bergmeir, C., & Benítez, J. M. (2012). On the use of cross-validation for time series predictor evaluation. *Information Sciences, 191*, 192–213. https://doi.org/10.1016/j.ins.2011.12.028
25. Pedregosa, F., Varoquaux, G., Gramfort, A., Michel, V., Thirion, B., Grisel, O., Blondel, M., Prettenhofer, P., Weiss, R., Dubourg, V., Vanderplas, J., Passos, A., & Cournapeau D. (2011). Scikit-learn: Machine learning in Python. *The Journal of Machine Learning Research, 12*, 2825–2830. https://doi.org/10.48550/arXiv.1201.0490.
26. Zhang, D., Hu, M., & Ji, Q. (2020). Financial markets under the global pandemic of COVID-19. *Finance Research Letters, 36*, 101528. https://doi.org/10.1016/j.frl.2020.101528
27. Yao, J., Ma, C., & He, W. P. (2014). Investor herding behaviour of Chinese stock market. *International Review of Economics & Finance, 29*, 12–29. https://doi.org/10.1016/j.iref.2013.03.002

Chapter 17
Comparing the Performance of Classification Algorithms for Predicting Electric Vehicle Adoption

Shamma AlRashdi, Aysha AlHassani, Fatima Haile, Rauda AlNuaimi, Thouraya Labben, and Gurdal Ertek

Abstract In this study, the Electric Vehicle (EV) purchase decisions of European consumers are predicted using supervised machine learning (ML), specifically classification. Following the replacement (imputing) of missing data values through predicted values and continuizing of all predictor features, the predictor features are ranked according to the Information Gain Ratio and the Gini coefficient. The results suggest that suiting daily driving needs (Q17), belief that society must reward electric cars instead of petrol and diesel cars (Q14), and opinion change regarding electric cars during the past year (Q21) ranked the highest with respect to the Gini coefficient metric. The same predictor features rank the highest with respect to the Information Gain Ratio metric, yet in a different rank (Q17, Q21, and Q14). For predictive analytics, a multitude of classification algorithms are applied to predict the decision of EV purchase, and the performance of the applied algorithms is compared. The results suggest that gradient boosting performed best in predicting EV adoption decisions, followed by the logistic regression and random forest algorithms.

Supplementary Information The online version contains supplementary material available at (https://doi.org/10.1007/978-3-031-61589-4_17).

S. AlRashdi · A. AlHassani · F. Haile · R. AlNuaimi · T. Labben · G. Ertek (✉)
College of Business and Economics, United Arab Emirates University, Al Ain, UAE
e-mail: gurdal@uaeu.ac.ae

S. AlRashdi
e-mail: 201908453@uaeu.ac.ae

A. AlHassani
e-mail: 202009415@uaeu.ac.ae

F. Haile
e-mail: 201950204@uaeu.ac.ae

R. AlNuaimi
e-mail: 201909003@uaeu.ac.ae

T. Labben
e-mail: thouraya.labben@uaeu.ac.ae

© The Author(s), under exclusive license to Springer Nature Switzerland AG 2024

A. Emrouznejad et al. (eds.), *Business Analytics and Decision Making in Practice*,
Lecture Notes in Operations Research, https://doi.org/10.1007/978-3-031-61589-4_17

Keywords Electric vehicles · Market adoption · Sustainable development goals (SDG) · Machine learning · Feature ranking · Classification algorithms

Introduction

Electric vehicles (EVs) have emerged as a viable solution to the growing global need for environmentally sustainable transportation [1]. However, a thorough understanding of the many features affecting EV performance and market acceptance is necessary before EVs can be widely adopted across the globe.

The key is to develop an advanced evaluation model that can yield precise and informative predictions regarding the future course of electric vehicle market adoption. This study tackles this problem by analyzing a peer-reviewed and published survey dataset using machine learning, specifically through classification analysis [2].

The main objectives of this study are to identify the classification algorithm(s) that may perform best with respect to prediction and the critical features (factors) that significantly influence EV market acceptance and adoption [3, 4]. These objectives require a thorough investigation using a range of classification algorithms and the application of ranking analysis. The goal is not only to achieve accuracy in EV performance classification; we also aim to produce knowledge that may inform consumers, lead regulatory decision-making, and provide stakeholders in the automotive sector with strategic guidance that can spur further market penetration. Through the integration of data-driven classification algorithms and a comprehensive understanding of the automotive industry, EV industry stakeholders can enhance market forecasts and take data-driven decisions and actions towards the disruptive transformation that EVs are likely to bring. The results of this research are aimed at contributing to initiatives to move transportation toward a more environmentally friendly and sustainable future.

The present research contributes to the understanding of consumers and the features that determine EV adoption using a machine learning technique for classification analysis. Therefore, it is necessary to summarize the literature on this line of research.

Multiple benchmark studies have compared the performance of various supervised machine-learning algorithms in diverse domains, ranging from the classification of textual data [5] to credit risk management [6], cancer detection [7], and wind turbine accidents [8]. However, these benchmark studies did not focus on the specific context of EV purchase intentions.

On the other hand, multiple studies have applied machine learning to predict EV adoption behavior [9–11], however these are studies conducted with earlier and dated data. Furthermore, compared to the plethora of studies that use various mathematical modeling techniques to analyze EV adoption [12], the literature that applies machine learning approaches [9–11] is rather limited. In the rapidly changing EV landscape and world, it is valuable to work with a more recent dataset that passed peer-review and is coming from a data journal (as will be detailed later) and to apply machine learning for analysis. Owing to space limitations, a full literature review is provided in the Supplement [13].

Motivated by the significance of EVs on the economic, environmental, legal, and social dimensions, and by earlier work in the field to understand the influencing factors, the present study applies machine learning, in particular classification and ranking analyses, to uncover more insights and develop an enhanced understanding of EV adoption.

Methodology

Model

Figure 17.1 illustrates the data analytics model developed using Orange data analytics software [14] for conducting machine learning (ML) methods of ranking and classification analysis. Orange, as a visual no-code analytics modeling software, provides a user-friendly visual interface that streamlines the process of loading, processing, and analyzing datasets. The model includes various widgets (nodes with icons) representing different steps in the analysis, such as data preprocessing, model training, and evaluation. This visual representation enhances transparency and allows for a clear understanding of the workflow employed in classification analysis.

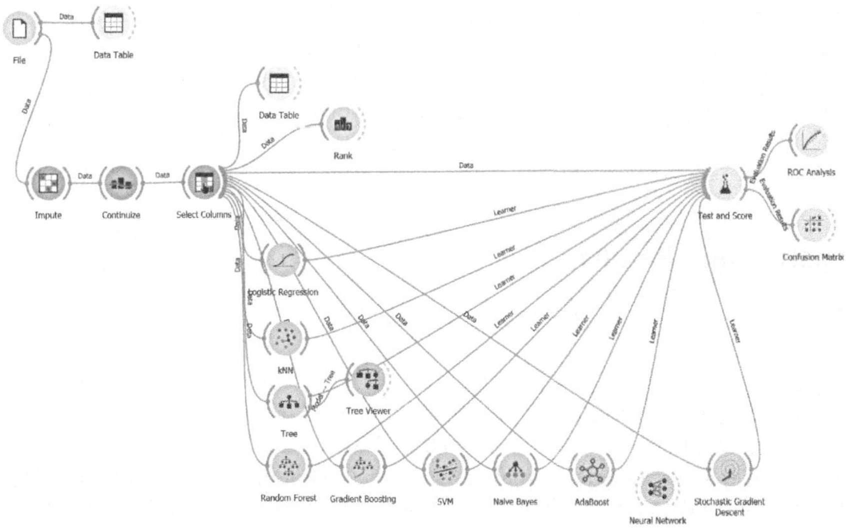

Fig. 17.1 The data analytics model in the Orange software for ranking and classification analysis

Logistic Regression, k-nearest neighbor (kNN), Tree, Random Forest, Gradient Boosting, Support Vector Machine (SVM), Naïve Bayes, AdaBoost, and Stochastic Gradient Descent (SDG) algorithms were included as classification algorithms to predict EV adoption. While the Neural Network (NN) algorithm was also considered, it was later excluded owing to excessive training time. The predictive performance of the models was then compared and benchmarked through stratified fivefold cross-validation experimentation, resulting in a comparison table and ROC. Finally, a confusion matrix was used to better understand and interpret the performance of the best-performing classification algorithm, which was a gradient boost, in this study.

Ranking analysis was conducted by computing the Gain Ratio and Gini Coefficient [15], and sorting in decreasing order of the Gini Coefficient values. In machine learning, ranking refers to ordering a set of items, typically a set of predictor attributes, based on a score or other metric. In this case, higher values of the Gain Ratio and Gini Coefficient indicate a higher influence of a given feature on the predictions. Definitions of these two ranking metrics, as well as others, can be found in the orange documentation [16].

The developed and applied orange model can be found in Supplement [13] as an *. ows file.

Model Evaluation and Comparison

The evaluation and comparison of the employed machine learning algorithms are important components of the analysis. A rigorous scrutiny of each classification algorithm's performance metrics, including AUC, Classification Accuracy (CA), F1 Score, Precision, Recall, and Matthews Correlation Coefficient (MCC), reveals the effectiveness of the predictive algorithms in reflecting the inherent patterns within the dataset. Definitions of these metrics can be found in the orange documentation [17].

Analysis and Results

Data

The data used in the presented research were obtained from [18], published in "Data in Brief" journal, which was accompanied by the full research dataset shared under [19]. This journal is a data journal, meaning that the journal specifically mandates that the source data for any published research in the journal should also be shared as public data. The rise of data journals is part of open data and open research trends [20].

The original dataset, as described in [18], is rich in multiple data tables and features that can potentially be analyzed in multiple research studies. For the research presented in this study, only a subset of the data was selected and used. The selected dataset consists of 69 binary predictor features, available as binary vectors, representing the Yes/No answers to survey question choices, as well as one target feature (response variable), namely Q16, which is the answer to the question "Which of the following statements about electric car suits you the best?". The original responses to Q16 were "If I had an EV, it would be my only car" (A), "If I had an EV, it would be a supplement to a petrol or diesel car" (B), "I would never buy an EV" (C) and "I do not know" (D).

For transparency and a comprehensive understanding of the features used in this analytical study, an exhaustive list of data features, along with their descriptions, is provided in Supplement [13]. However, as mentioned earlier, the source data and metadata for the study are taken directly from [19].

The benefit of using a dataset published in a data journal is that the data has readily passed through peer review and its validity and verification have been confirmed by the journal's reviewers before acceptance and publishing.

Data Preparation

For a focused analysis, for question Q16, which is the target feature, answers A and B have been merged into the single class label AB, which represents purchase intention and adoption, and answers C and D have been merged into the single class label CD, which represents no purchase intention (No Adoption). By reducing the number of values for the target feature Q16 from four to two, a binary classification model with a higher predictive power was obtained, which was subsequently run and analyzed. By designating Question 16 as the target feature, the classification algorithms focused on understanding the factors influencing the merged responses to this specific question. This step is pivotal in framing the analysis towards the business objective of predicting or understanding the factors contributing to electric vehicle adoption, based on the given survey.

There were multiple missing values in the original dataset. Addressing missing data is a critical pre-processing step that ensures the quality of the model, analysis, and results. To impute and fill in the missing values systematically, a model-based imputer that uses a single decision tree [21] was used in the Orange software [14]. As the next step, it was necessary to continue all the predictor features because most machine learning algorithms operate better when the predictor features take continuous numerical values. This step was conducted using the widget in Orange software [22].

Results of Classification Analysis

The classification performance analysis in Table 17.1 reveals notable insights into the effectiveness of various algorithms for predicting electric vehicle adoption based on the selected and prepared datasets. Gradient Boosting emerged as the top-performing algorithm, exhibiting the highest values across all chosen metrics, namely AUC, Classification Accuracy (CA), F1 score, precision, recall, and Matthews Correlation Coefficient (MCC). This signifies its exceptional ability to distinguish between the two classes (AB vs. CD; i.e., adoption vs. No Adoption), leading to an overall accurate prediction of electric vehicle adoption.

Random Forest closely follows gradient boosting by demonstrating robust performance across multiple metrics. Random Forest, Logistic Regression, and Naïve Bayes also showed commendable results with respect to both CA and AUC, demonstrating their reliability in the classification task. kNN and Naïve Bayes performed well with respect to AUC, but not with respect to CA. SDG performed well with respect to CA, but not with respect to AUC.

While logistic regression did not have the top performance, it may be performed due to very fast running time and its ability to generate the parameters of a closed form logistic regression equation that can be used for very fast classification. Similarly, random forest did not have the top performance, yet may be preferred under certain settings, as it generates a multitude of trees, which can be browsed through as a collection through Pythagorean tree visualizations and each tree individually through decision tree visualization.

On the other hand, SVM exhibited the worst performance across *all* metrics, suggesting its limitations in accurately discriminating between classes in this specific context and data. This was despite the fact that SVM was run with the RBF kernel as the default selected kernel, rather than the less robus linear kernel. These findings offer valuable guidance for selecting an appropriate algorithm, with gradient boosting and Random Forest standing out as strong candidates for predicting electric vehicle adoption with high precision and reliability.

Table 17.1 Classification performance analysis based on multiple metrics

Algorithm	AUC	CA	F1	Prec	Recall	MCC
Logistic Regression	0.870	0.826	0.821	0.820	0.826	0.548
kNN	0.827	0.746	0.757	0.790	0.746	0.459
Tree	0.726	0.811	0.810	0.809	0.811	0.525
Random Forest	0.872	0.839	0.831	0.834	0.839	0.575
Gradient Boosting	0.901	0.854	0.849	0.850	0.854	0.621
SVM	0.662	0.707	0.718	0.740	0.707	0.345
Naïve Bayes	0.864	0.799	0.805	0.817	0.799	0.539
AdaBoost	0.732	0.779	0.780	0.782	0.779	0.458
SDG	0.749	0.818	0.813	0.812	0.818	0.572

The Area Under the Curve (AUC) values in Table 17.1 associated with each algorithm served as quantitative measures of their respective Receiver Operating Characteristic (ROC) curves, which are displayed in Fig. 17.2. Notably, Gradient Boosting led to a remarkable AUC of 0.901, indicative of an ROC curve closest to the top-left corner in Fig. 17.2, showcasing its superior discriminatory power. Random Forest closely follows an AUC of 0.872, reinforcing its robustness in distinguishing between the two classes (adoption vs. No Adoption). Logistic Regression and Naïve Bayes exhibited commendable AUC values (0.870 and 0.864, respectively), underlining their competency in predicting electric vehicle adoption. Conversely, the SVM trials had a lower AUC of 0.662, suggesting a comparatively weaker ROC curve. The AUC values provide a concise summary of each algorithm's ability to balance true-positive and false-positive rates, aiding in the selection of the most suitable model for accurate predictions in the context of electric vehicle adoption.

Figure 17.3 displays the confusion matrix for the Gradient Boosting algorithm and offers a detailed perspective on the classification performance of the model. Among the 6,108 instances in the dataset, the algorithm correctly predicted 4,113 cases of adoption (AB) and 1,104 cases of No Adoption (CD). However, 595 No Adoption (CD) instances were incorrectly classified as Adoption (AB) and 296 Adoption (AB) instances were misclassified as no option (CD).

The confusion matrix enabled a nuanced evaluation of the model's strengths and areas for improvement. The high number of true positives and true negatives signifies the model's effectiveness and inaccurate predictions, whereas the occurrences of false positives and false negatives pinpoint specific instances where the model may need refinement.

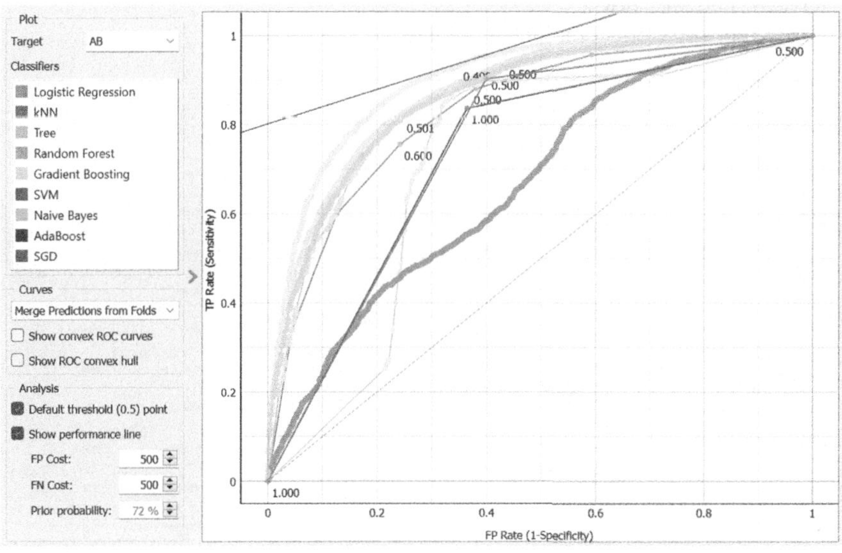

Fig. 17.2 ROC curves for the applied classification algorithms

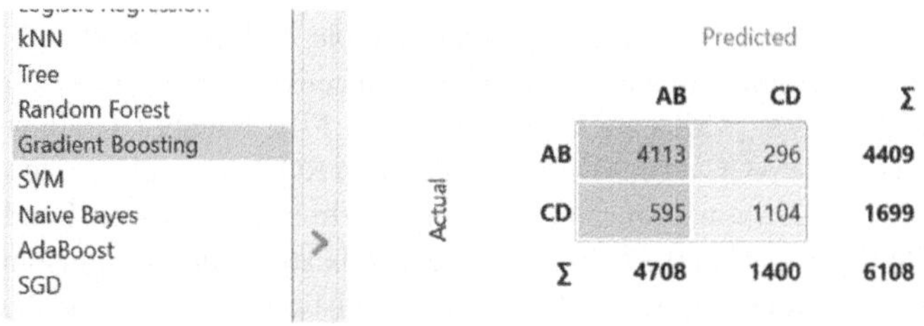

Fig. 17.3 Confusion matrix for Gradient Boosting, which has the highest CA (Classification Accuracy) value

Results of Ranking Analysis

The feature rankings in Fig. 17.4, based on the decreasing values of the Gini coefficients, reveal crucial insights into the factors that influence the prediction of electric vehicle adoption (Q16). The results suggest that suiting daily driving needs (Q17), belief that society must reward electric cars instead of petrol and diesel cars (Q14), and opinion change regarding electric cars during the past year (Q21) ranked the highest with respect to the Gini coefficient metric. The same predictor attributes rank highest with respect to the Information Gain Ratio metric, yet in a different rank (Q17, Q21, and Q14).

Further analysis reveals features such as Q12_4, Q13_2, Q13_3, and Q13_7, which consistently rank high in the Gain Ratio, but not in the Gini coefficient. The features are as follows:

- Q12_4: Purchase price
- Q13_2: Positive effect on global climate

		#	Gain ratio	Gini
1	N Q17		0.112	0.119
2	N Q14		0.075	0.081
3	N Q21		0.082	0.069
4	N Q18_7		0.039	0.045
5	N Q18_3		0.035	0.037
6	N Q18_20		0.029	0.034
7	N Q18_15		0.029	0.033
8	N Q18_17		0.030	0.032
9	N Q18_16		0.028	0.031
10	N Q18_10		0.029	0.031
11	N Q18_18		0.026	0.030
12	N Q18_19		0.028	0.029
13	N Q13_2=1.0		0.059	0.028
14	N Q18_22		0.026	0.028
15	N Q18_23		0.025	0.027
16	N Q18_21		0.024	0.026
17	N Q18_5		0.024	0.025
18	N Q12_4=1.0		0.046	0.025
19	N Q13_3=1.0		0.049	0.021
20	N Q13_7=1.0		0.050	0.021

Fig. 17.4 Feature rankings for EV adoption

- Q13_3: Less noise
- Q13_7: Positive effect on the local environment

Discussion

In this study, an investigation of the factors possibly affecting electric vehicle (EV) adoption was conducted to discern their impact on consumer purchase intention and societal adoption of electric vehicles (EVs) in EU countries. The investigation, through ranking analysis, also aimed to identify the most influential features affecting classification performance. Focusing on a binary categorical target feature, the study analyzed the prediction performance. The values of metrics, such as AUC, CA, F1-score, recall, and precision, were compared across the algorithms, leading to the identification of models exhibiting superior classification performance, thus delivering more accurate predictions of EV adoption. The relevance of features within the models was evaluated to comprehend the unique contribution of each feature to the classification. Receiver Operating Characteristic (ROC) curves were plotted to assess the performance of the classification models.

In the classification performance comparison, gradient boosting was consistently ranked as the best algorithm, followed by random forest and logistic regression. This insight, as well as the results for the other algorithms, has notable ramifications with regard to the selection of classification algorithms for carrying out predictions for the dataset at hand. Most notably, any classification benchmark study on a dataset with a similar structure or scope should consider gradient boosting, logistic regression, random forest, and naïve Bayes in the benchmark.

Despite the actionable insights obtained, there are three potential shortcomings in the conclusions and the presented research. These shortcomings and their discussions are as follows.

First, the dataset may not be representative of all European consumers. In other words, the data may not be representative of the population. However, this risk can be readily reduced by selecting a dataset from a data journal, where peer reviewers evaluate the theoretical and practical validity of the research and source data.

Second, only a few of the numerous classification algorithms in the literature were included in the study, according to their availability in the chosen orange data analytics software [22]. There may be other algorithms, not here by included, that might outperform gradient boosting. This represents an avenue for future research.

The third is the selection of the hyperparameters that govern how the classification models are trained, including control over the model parameters. While parameters learn from data during training, hyperparameters do not learn their values from data; instead, they need to be specified before model training is initiated. In the conducted study, fivefold stratified cross-validation was applied to conduct multiple experiments and obtain multiple results to help alleviate any bias problems with regard to parameter selection and optimization. However, the classification algorithms in the study were always run with their default hyperparameters rather than applying an experimental design, exhaustive evaluation, or hyperparameter optimization [23]. Therefore, different and possibly optimized parameter values for the selected algorithms could potentially change the results and draw conclusions. More specifically, at least in theory, it is possible that there could be certain sets of parameter values for the lagging algorithms at which those lagging algorithms may outperform gradient boosting or other leading algorithms. An ideal future research, therefore, would be to conduct a much more comprehensive study, with proper experimental design and hyperparameter optimization for each algorithm, followed by a posterior comparison, which in itself may be a data analytics study. Ideally, multiple different sets of parameter values should be applied and evaluated systematically. Having mentioned this shortcoming, the present study is valuable as a preliminary study, as most analytics projects in industry also use default parameter settings and values readily set in the software that they are using. Therefore, the presented research in that aspect approximates a typical study that would be conducted in the industry.

Conclusions

In conclusion, our study presents a comprehensive analysis of electric vehicle adoption by utilizing some well-known supervised machine learning algorithms for classification. The identification of key features through rankings, focusing on experiences, satisfaction, and societal attitudes, offers actionable insights for stakeholders in the electric vehicle industry.

As the automotive landscape continues to evolve, our findings contribute to the collective understanding of the factors influencing EV adoption, paving the way for informed decision-making and strategic planning in the pursuit of sustainable and eco-friendly transportation solutions. Building on the insights gained from this analysis, recommendations for stakeholders in the electric vehicle industry could include improving the most significant factors that affect EV purchase intention and adoption, devising targeted marketing strategies, enhancing customer journeys and user experiences, and encouraging and supporting policies and regulations that can address consumer preferences and societal attitudes.

Future work can focus on extending the scope of the analysis to incorporate other classification algorithms, hyperparameter optimization [23], and explainable AI (XAI) [24]. Future work can also focus on exploring temporal trends, incorporating additional data sources such as demographics and charging network infrastructure [25], and refining the analysis to adapt to evolving consumer preferences and industry development.

References

1. Onat, N. C., Abdella, G. M., Kucukvar, M., Kutty, A. A., Al-Nuaimi, M., Kumbaroğlu, G., & Bulu, M. (2021). How eco-efficient are electric vehicles across Europe? A regionalized life cycle assessment-based eco-efficiency analysis. *Sustainable Development, 29*(5), 941–956. https://doi.org/10.1002/sd.2186
2. Caruana, R., Niculescu-Mizil, A. (2006) An empirical comparison of supervised learning algorithms. In *Proceedings of the 23rd International Conference on Machine Learning*, pp. 161–168. https://doi.org/10.1145/1143844.1143865.
3. Anastasiadou, K., & Gavanas, N. (2022). State-of-the-Art review of the key factors affecting Electric Vehicle adoption by consumers. *Energies, 15*(24), 9409. https://doi.org/10.3390/en15249409
4. Bhat, F. A., & Verma, A. (2023). A bibliometric analysis and review of adoption behaviour of electric vehicles. *Transportation in Developing Economies, 9*(1), 5. https://doi.org/10.1007/s40890-022-00175-2
5. Hsu, B.-M. (2020). Comparison of supervised classification models on textual data. *Mathematics, 8*(5), 851. https://doi.org/10.3390/math8050851
6. Coenen, L., Verbeke, W., & Guns, T. (2022). Machine learning methods for short-term probability of default: A comparison of classification, regression and ranking methods. *Journal of the Operational Research Society, 73*(1), 191–206. https://doi.org/10.1080/01605682.2020.1865847
7. Alabi, R. O., Elmusrati, M., Sawazaki-Calone, I., Kowalski, L. P., Haglund, C., Coletta, R. D., ... & Leivo, I. (2020). Comparison of supervised machine learning classification techniques in prediction of locoregional recurrences in early oral tongue cancer. *International Journal of Medical Informatics, 136*, 104068. https://doi.org/10.1016/j.ijmedinf.2019.104068.
8. Asian, S., Ertek, G., Haksoz, C., Pakter, S., & Ulun, S. (2016). Wind turbine accidents: A data mining study. *IEEE Systems Journal, 11*(3), 1567–1578.
9. Bas, J., Cirillo, C., & Cherchi, E. (2021). Classification of potential electric vehicle purchasers: A machine learning approach. *Technological Forecasting and Social Change, 168*, 120759. https://doi.org/10.1016/j.techfore.2021.120759
10. Jia, J., Shi, B., Che, F., & Zhang, H. (2020). Predicting the regional adoption of Electric Vehicle (EV) with comprehensive models. *IEEE Access,* 8, 147275–147285. https://doi.org/10.1109/ACCESS.2020.3014851.
11. Bas, J., Zou, Z., & Cirillo, C. (2023). An interpretable machine learning approach to understanding the impacts of attitudinal and ridesourcing factors on electric vehicle adoption. *Transportation Letters, 15*(1), 30–41. https://doi.org/10.1080/19427867.2021.2009098
12. Maybury, L., Corcoran, P., & Cipcigan, L. (2022). Mathematical modelling of electric vehicle adoption: A systematic literature review. *Transportation Research Part D: Transport and Environment, 107*, 103278. https://doi.org/10.1016/j.trd.2022.103278
13. AlRashdi, S., AlHassani, A., Haile, F., AlNuaimi, R., Labben, T., & Ertek, G. (2024). Supplement to "Comparing the Performance of Classification Algorithms for Predicting Electric Vehicle Adoption". Retrieved from https://ertekprojects.com/ftp/supp/18.zip.
14. Demsar, J., Curk, T., Erjavec, A., Gorup, C., Hocevar, T., Milutinovic, M., Mozina, M., Polajnar, M., Toplak, M., Staric, A., Stajdohar, M., Umek, L., Zagar, L., Zbontar, J., Zitnik, M., & Zupan, B. (2013). Orange: Data Mining Toolbox in Python. *Journal of Machine Learning Research,* 2349–2353. Retrieved from http://jmlr.org/papers/volume14/demsar13a/demsar13a.pdf.
15. Raileanu, L. E., & Stoffel, K. (2004). Theoretical comparison between the Gini Index and information gain criteria. *Annals of Mathematics and Artificial Intelligence, 41*, 77–79. https://doi.org/10.1023/B:AMAI.0000018580.96245.c6
16. Rank. Orange Widget Catalog. Retrieved from https://orangedatamining.com/widget-catalog/data/rank/.
17. Test and Score. Orange Widget Catalog. Retrieved from https://orangedatamining.com/widget-catalog/evaluate/testandscore/.

18. Westerhof, M., García, J. R. R., Haveman, S., & Bonnema, G. M. (2023). Transnational survey data on European consumers' attitude and perceived knowledge about electric vehicles. *Data in Brief, 49*, 109378.
19. García, J. R. R., Bonnema, G. M., Kroon, R., Grinvalds, G., & Troels-Smith, T. (2023). Dataset from "Transnational survey about Electric Vehicles' knowledge and attitude of European consumers". University of Twente, V4. Retrieved from https://data.mendeley.com/datasets/85nz9k5tf5/4.
20. Tedersoo, L., Küngas, R., Oras, E., Köster, K., Eenmaa, H., Leijen, Ä., & Sepp, T. (2021). Data sharing practices and data availability upon request differ across scientific disciplines. *Scientific Data, 8*(1), 192. https://doi.org/10.1038/s41597-021-00981-0
21. Orange Widget Catalog. Impute. Retrieved from https://orangedatamining.com/widget-catalog/transform/impute/.
22. Orange Widget Catalog. Continuize. Retrieved from https://orangedatamining.com/widget-catalog/transform/continuize/.
23. Yang, L., & Shami, A. (2020). On hyperparameter optimization of machine learning algorithms: Theory and practice. *Neurocomputing, 415*, 295–316. https://doi.org/10.1016/j.neucom.2020.07.061
24. Schwalbe, G., & Finzel, B. (2023). A comprehensive taxonomy for explainable artificial intelligence: A systematic survey of surveys on methods and concepts. *Data Mining and Knowledge Discovery*. https://doi.org/10.1007/s10618-022-00867-8
25. Gavranović, H., Barut, A., Ertek, E., Yüzbaşıoğlu, O.B., Pekpostalcı, O., & Tombuş, O. (2014). Optimizing the electric charge station network of EŞARJ. In *Proceedings of 2nd International Conference on Information Technology and Quantitative Management, ITQM 2014*. Procedia Computer Science 31, 15–21.

Chapter 18
Enhancing Inter-Terminal Transport via Early Information

Matteo Brunetti, Eduardo Lalla-Ruiz, and Martijn Mes

Abstract We focus on decoupling decisions of inter-terminal transport (ITT) vehicles at a logistics node, such as a port or business park. We assume the node faces a stochastic flow of trucks delivering and retrieving containers from logistics companies (LCs), i.e., warehouses and terminals. During peak hours, trucks may stop at a parking area, where the ITT fleet, composed of electric and automated vehicles (EAVs), takes over the transport of containers between the parking and the LCs. The decoupling decision (DD) determines whether trucks should proceed to their LC or park. The decision model is based on few parameters, such as the estimated workload of the ITT fleet over a time window. We integrate the decision model into a discrete event simulation of the Port of Moerdijk, the Netherlands, allowing experimentation with various arrival patterns, earliness of information, and DD parameters. The simulation involves a realistic traffic simulation, more than 130 LCs, and up to 100 EAVs. Through parameter calibration, the decision model capitalizes on early information to improve service levels and reduce kilometers driven by conventional trucks.

Keywords Inter-terminal transport · Connected automated transport · Electric vehicles · Decoupling · Logistics nodes

Introduction

The role of freight transport in modern society is expected to expand significantly. It is expected that by 2050, volumes will be more than double compared to 2015 [10]. Due to the growing number of containers transiting intermodal terminals, container trucks can experience considerable delays both inside terminals and in their vicinity [4]. This situation has led to substantial investments in inter-terminal transport (ITT) systems; notable examples are the chassis exchange terminal [5], the extended gate as

M. Brunetti (✉) · E. Lalla-Ruiz · M. Mes
Department of High-Tech Business and Entrepreneurship, University of Twente, Enschede 7500 AE, NL, Netherlands
e-mail: m.brunetti@utwente.nl

© The Author(s), under exclusive license to Springer Nature Switzerland AG 2024

A. Emrouznejad et al. (eds.), *Business Analytics and Decision Making in Practice*, Lecture Notes in Operations Research, https://doi.org/10.1007/978-3-031-61589-4_18

described in [7], and the multi-terminal systems in ports like Shanghai and Rotterdam [8]. Digital platforms that facilitate real-time information exchange have become imperative in port areas to effectively plan, control, and coordinate ITT vehicles [6]. Despite current ITT solutions and decarbonization policies, the growth of freight volumes poses challenges such as increased emissions, shorter delivery times, driver shortages, and disruptions in supply chains.

To address these challenges, the freight sector is turning to Connected Automated Transport (CAT). CAT is an umbrella term for emerging technologies in logistics such as electric automated vehicles (EAVs), drones, and smart infrastructure, which make use of concepts such as the Physical Internet, Internet of Things, Blockchain, and AI through interoperability and connectivity. The European Commission [11] emphasizes that CAT has the potential to revolutionize global trade, making it more efficient, sustainable, and resilient for any mode of transport. However, uncertainties persist in the value and implementation of CAT technologies, which hinders widespread adoption by logistics operators.

In this research, we focus on two aspects of CAT: the transition to electric and automated ITT systems at logistics nodes and the value of early information on arrivals. In the context of ITT, the integration and scaling of EAVs introduce operational challenges in managing dynamic mixed-traffic environments, efficient fleet management, and coordination of decisions and information. Regarding the latter, we argue that proper use of early information on arrivals significantly increases the performance of the EAV fleet by managing the ITT workload. Specifically, we use early information to proactively decide which inbound trucks should park, decouple from their containers, and let the EAVs take care of first- and last-mile transport. We refer to this process as decoupling.

We define a logic for decoupling that utilizes early information and parameters to predict workload, advising operators on job allocation to the ITT fleet. Our decoupling approach relates to truck appointment systems (TASs). TASs are used to schedule truck arrival times at terminals, with the aim of optimizing first- and last-mile operations, reducing congestion, and potentially improving sustainability [1]. In contrast to typical TASs, our decoupling logic uses truck arrival times to determine when a truck should be allowed inside the logistics node or decoupled. We calibrate and test our decoupling approach through a realistic traffic simulation of the Port of Moerdijk (PoM), a major Dutch logistics node with more than 130 logistics companies, i.e., warehouses and terminals, and a total of 18.5 million tons of transshipment per year. After extensive experimentation, we highlight combinations of parameters and ITT fleet setups that improve PoM's logistics efficiency. Additionally, we perform a multi-objective analysis to explore trade-offs between service levels, fleet efficiency, and sustainability.

The remainder of this paper is structured as follows. In section "Enhancing Decoupling Decisions", we define the decoupling logic and its parameters. In section "Simulating a Decoupling Process" we introduce the PoM simulation model. In section "Design of Experiments", we describe our design of experiments. In section "Results", we discuss the results. Finally, we draw conclusions in section "Conclusions".

Enhancing Decoupling Decisions

In this section, we define our Proactive Decoupling Model (PDM). We start by describing the stochastic arrivals at logistics nodes and decoupling process. Next, we describe the decoupling decision (DD) and the model parameters. Finally, we describe the role of early information that leads to three variants of DD.

ITT and Uncertain Arrivals

A logistics node features multiple logistics companies (LCs) that receive, ship, store, and transload freight. Think of a port, such as PoM, consisting of many warehouses, terminals, chemical manufacturers, customs, a port authority, and ITT operators. Due to uncertain arrivals at the LCs' gate, queues may grow and the trucks end up waiting on the roads near the LCs. These queues, coupled with other infrastructural problems, such as narrow roads or busy intersections, lead to unexpected congestion and distressed truck drivers. A port-wide ITT fleet with EAVs could service all LCs, decouple external freight flows from internal logistics at PoM, and prevent said congestion. The results would be a reduced wear on the PoM's infrastructure, reduced emissions inside the port, and the possibility for truck drivers to take official breaks while waiting for containers.

To represent congestion at logistics nodes, we need to model several uncertainties in inbound and outbound freight flows. The arrival of trucks is stochastic, and the trucks can be late or early with respect to an expected arrival time. Containers have a stochastic delivery time window that starts at the expected arrival time of the truck and have a given maximum length. Therefore, trucks and containers can be late or early with respect to the delivery time window. Each truck has one of three job types: dropping off a container at an LC, picking up a container waiting at an LC, or a combined job of dropping off a container and picking up another one at the same LC. The latter is also known as a double move [9]. For the sake of brevity, we refer to these as drop, pick, and drop and pick (D&P), and to the related containers as drop or pick containers. Note that the two containers may have different, but related, time windows for delivery, e.g., a different due date. Furthermore, there are containers to be transported between LCs that do not interact with trucks. These are always transported by the ITT fleet and we refer to them as ITT containers.

The Decoupling Decision and Model Parameters

In the proposed PDM, we redirect trucks and containers to a parking area, the Decoupling Point (DP), where the containers are then separated from the trucks, and the first- and last-mile transport of the containers is carried out by an ITT fleet. We refer

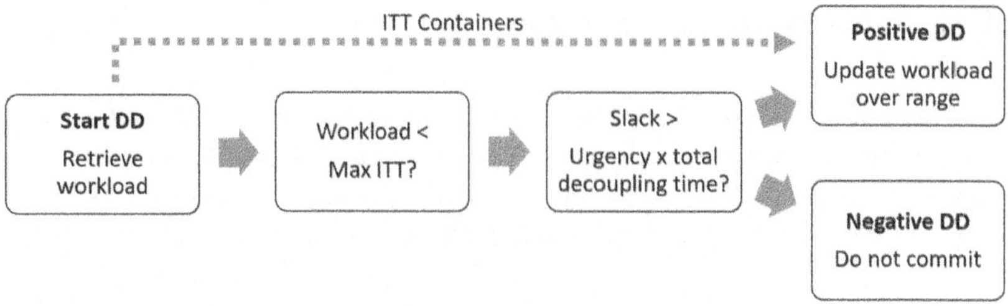

Fig. 18.1 The decoupling decision and separate process for ITT containers

to this process as decoupling, and to its counterpart as direct transport (skipping the DP). Decoupling is done on a positive decoupling decision (DD) where we consider the ITT workload and the urgency of a container. Direct transport is done on a negative DD, e.g., when the time between a container's due date and its truck arrival is too short for decoupling. Figure 18.1 shows the DD procedure. At its core, the DD procedure consists of a series of checks in which we calculate the workload within a time period based on the available arrival information. Then, we compare the workload against given thresholds for the maximum workload and an urgency-adjusted decoupling time. Finally, we update the workload over a certain range of time periods. These time periods, thresholds, and update range are the parameters of our PDM model.

The PDM utilizes four different parameters: the call interval, the urgency, the workload range, and the max ITT. The call interval is used to discretize time into intervals, i.e., time periods, of a certain length. Whenever a DD has to be made, we retrieve the workload of the time period in which first- or last-mile transport will be carried out and compare it against the max ITT threshold. If the workload is above the threshold, we cannot decouple; otherwise, we should decouple. Then, the PDM checks if the container slack time is longer than the total decoupling time. Here, the container slack time is the time between its arrival and its due date, and for D&P truck arrivals the shortest slack time between the two containers is used. The total decoupling time is calculated as the average time it takes to park, decouple, transport, and (un)load the container at the LC. The total decoupling time is then multiplied by the urgency parameter to enforce a longer or shorter slack time. Considering container slack time grants urgent containers direct access to the LCs and prevents urgent containers from being late due to the extra steps of the decoupling process. When the two criteria are met, a positive DD is made. Finally, we update the workload for all time periods included in the workload range around the original arrival time, rounded to the closest time periods. For example, if we were considering an arrival at 13:42, fifteen minute intervals, and a workload range of one hour, the actual range around the arrival would be 13:12–14:12 and the rounded range would be 13:15–14:15.

Early Information and Decision Variants

Early information refers to the availability of truck and container information prior to the arrival of the truck at the logistics node. For example, early information could be released when a truck departs from its origin or some given time before arrival at the logistics node. In our model, it is released before truck arrival, at a given time before the container due date, e.g., twelve hours before. We assume that a pick container can be queued for preparation as soon as the early information becomes available. This leads to pre-emptive loading, release, and transport to the DP by the ITT fleet. Last, we assume that all information related to truck arrivals, containers, and their status with respect to ITT operations is stored and shared on a digital platform at the logistics node.

Early information extends the DD into three variants: DD_1, DD_2 and DD_3, as shown in Fig. 18.2. These DDs are ordered chronologically. DD_1 is the proactive assignment of a truck to decouple upon the release of early information. DD_3 is the reactive decision made upon truck arrival. DD_2 is a dynamic decision between DD_1 and DD_3, triggered when the workload in a time period falls below the threshold level. The workload related to a container's DD is retrieved from the first time period in which that container can be transported by the ITT fleet. For a drop container, the first time period is the truck arrival. For a pick container, it is the release of early information, in DD_1, or the current time period, in DD_2. Finally, for an ITT container, early information is not available and the first time period is its default release time. After decoupling or registering an ITT container, we do not schedule a specific ITT time. Instead, we let an ITT dispatcher prioritize the containers based on the earliest due date.

The DD variants use different workloads: the committed workload (CW), the potential workload (PW), and the estimated workload (EW). The EW is the sum of the CW and the PW. As a general rule, any job that we know of and decide to decouple, i.e., a positive DD, is labeled as CW, while any job that we know of but decide not to decouple at this point in time, i.e., a negative DD, is labeled as PW. However, in case of an urgent job (slack criterion), we do not label the job as PW. The CW and PW are allocated to the first feasible time period for ITT, following the job type logic we described earlier. A job that is PW can later become CW due to DD_2 or DD_3. CW is removed when a drop container reaches its destination or when a pick

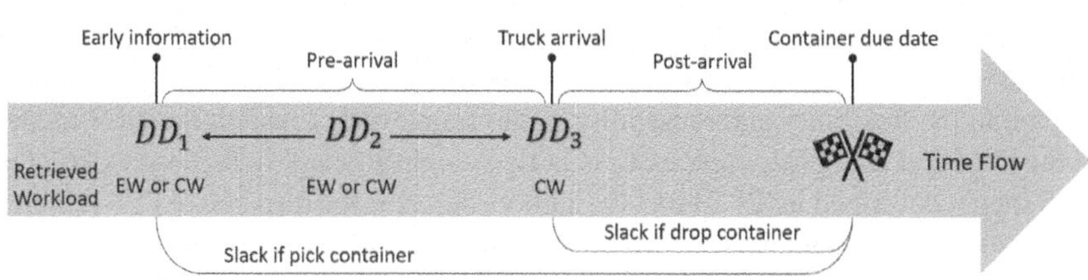

Fig. 18.2 The decoupling decision variants when including early information

container leaves the logistics node. DD_1 and DD_2 use CW when considering the decoupling of pick containers, but use EW for drop containers. Furthermore, if the arrival is a D&P truck, the maximum is chosen between the CW of the pick container and the EW of the drop container. DD_3 uses CW as this decision is made at the last possible moment, upon arrival of the truck, when future PW does not matter for drop containers.

Simulating a Decoupling Process

In this section, we introduce our case study of the Port of Moerdijk (PoM). Then, we describe how we modeled the PoM using a simulation framework and extended it with the PDM.

The PoM is characterized by a compact and enclosed layout, which facilitates the separation of traditional and automated traffic flows within the port. However, the port faces congestion at certain junctions, due to limited maneuvering space, and at the gates of certain LCs. To address this issue, a DP is proposed near the highway. The DP's buffering effect, coupled with efficient ITT dispatching and continuous EAVs operations, can mitigate congestion. The port authority is deliberating whether to adopt the combined DP-EAV system and how to setup the decoupling logic behind it. We aim to support the port authority's decision by studying the effect of a DP-AV system employing the PDM as decoupling logic and considering early information on arrivals.

The discrete event simulation (DES) model of the PoM is based on the simulation framework for CAT at logistics nodes described in [3], with the addition of a network generator as described in [13]. At its core, the framework is a conceptual model, as described by Robinson [12], focusing on freight flows between LCs and the DP, with soft time windows and early information on arrivals. As in most conceptual models, it includes inputs, outputs, assumptions, and high-level process flowcharts. Additionally, it features a multi-agent system to model the interaction between the ITT vehicles, the LCs, and a centralized ITT fleet dispatcher. The network generator transforms OpenStreetmap data to reproduce a multimodal infrastructure at different levels of accuracy, and creates LCs based on user input. The network generator extends the capabilities of our DES model into a realistic traffic simulation, allowing for emerging road congestion. The DES model is coded in Siemens' Tecnomatix Plant Simulation, which has the advantage of reusable libraries and an object-oriented structure.

Due to the flexibility and reusability of the framework, creating the DES model of the PoM and its logistics processes is fairly straightforward. In fact, we extend the DES model described in [2]. There, the authors describe the dispatch of ITT vehicles to transport drop containers or charge their batteries. This decision is made by a centralized agent and is based on heuristics or more advanced approaches, such as reinforcement learning. In [2], the authors assumed that a fixed percentage of arrivals would be decoupled. Instead, we focus on dynamically selecting which arrivals to be decoupled and model drop, pick, and D&P containers. Furthermore, we assume that the charging process takes place whenever an EAV idles in the DP. In more detail, a small amount of battery is recharged every minute an ITT vehicle idles, as long as a charging station is available. The charging stations are assigned to the ITT vehicles with the lowest battery level, and we assume that the time to (un)plug a vehicle (from) to the charging station is negligible.

A major extension to the DES model in [2] is the implementation of the PDM. To transpose the PDM into Tecnomatix Plant Simulation, we created a "workload" table, a "job" table, and functions for the DDs. The workload table represents the partition of the simulation horizon in time periods based on the call interval parameter. The job table acts as the digital platform at the port. The tables and functions are coded as described in section "Enhancing Decoupling Decisions".

Regarding the input of the simulation, we collected a list of all LCs in the PoM, their geographic coordinates, the expected time to service a trailer, and the probability of receiving an arriving truck. We model two types of LC: terminals and warehouses. Terminals operate 24/7, while warehouses operate between 07:00 and 19:00. The capacity of LCs, i.e., the number of docks, is proportional to their surface. Then, from road sensor data, we estimate the hourly arrivals of trucks over a weekday per port entrance, which fluctuate, and the number of ITT containers. We assume that containers' time windows are uniformly distributed between a minimum and a maximum value. Finally, we set the parameters of the DP, e.g., the time to (de)couple a trailer, and the characteristics of the EAVs, e.g., the acceleration and reaction time. We assume that the EAV battery range is 100 km and that the DP has 20 charging stations.

Design of Experiments

The present simulation study evaluates the PDM with respect to its impact on the ITT operations at PoM. We consider three key performance indicators (KPIs): the average share of late containers with respect to their due date, the average turnaround time of trucks at PoM, and the total kilometers driven by inbound and outbound trucks. Our goal is to investigate efficient configurations of the PDM parameters, reduce the size of the ITT fleet, and study the resilience of the decoupling process.

As a preliminary study of the effects of the PDM parameters, we select a default configuration considering all inputs in our simulation, as listed in section "Simulating a Decoupling Process". This default configuration is the center of gravity of our experiments and satisfies two conditions: first, an intensity of arrivals that results in a steady-state simulation for the PoM; second, an ITT fleet size of 100 EAVs not to limit the effect of varying the PDM parameters because of fleet capacity, within reasonable bounds. For example, a small ITT fleet would lead to higher values of the max ITT threshold having the same *(negative) results.

Our Design of Experiments (DoE) covers three stages: parameter calibration, fleet size reduction, and scenario analysis. First, we separately identify the most promising

Table 18.1 Design of experiments

Stage	Experimental factor	Default	Range, increment	Experiments
Parameter calibration	Early info and release	12h	6–24h, 6h	69
	Call interval	15m	5–30m, 5m	
	Workload range	1h	0h–2h, 0.25h	
	Urgency	1	0–3, 0.25	
	Max ITT	25	10–90, 10	
Fleet size reduction	ITT vehicles	100	30–80, 10	74
Scenario analysis	Flow intensity	100%	80–120%, 10%	30
	ITT containers	4.2%	10–20%, 5%	
	D&P %	50%	75%, 90%, and 100%	
	Max due date	24h	6–36h, 6h	

values of each PDM parameter over a sufficiently wide range, while keeping the other parameters at their default values. For parameters that do not have a clear winner, we further test the two best values of each uncertain parameter in combination. Next, we experiment to reduce the ITT fleet size without significantly degrading service levels. When studying the fleet size, we again test promising fleet sizes with those uncertain PDM parameters. Last, we stress-test the newly found configurations of parameters and fleet size by playing with the arrival patterns. Specifically for the type of truck job, we increase the share of D&P jobs and equally reduce the share of separate drop and pick jobs. We compare the response of the newly found configurations with a baseline simulation of the PoM without decoupling. Table 18.1 displays the experimental factors for each stage of the DoE, including the default values, experimental ranges, and increments. Following our DoE, we perform a total of 173 experiments.

When experimenting, a simulation must achieve reliable initial conditions and ensure the statistical significance of its output. We used Welch's graphical procedure to estimate the warm-up period, which is 1 day for our KPIs. Subsequently, we chose a total of 8 d of simulation, consisting of 1 day of warm-up and 7 d of data collection. Taking into account 7 d of data collection, 5 replications are enough to ensure a desired 95% confidence interval for our KPIs.

Results

Following our DoE, we identify both the individual effects of the PDM parameters and specific recommendations for the PDM and fleet size at PoM. We do so by means of a multi-objective analysis on the KPIs mentioned in section "Design of Experiments".

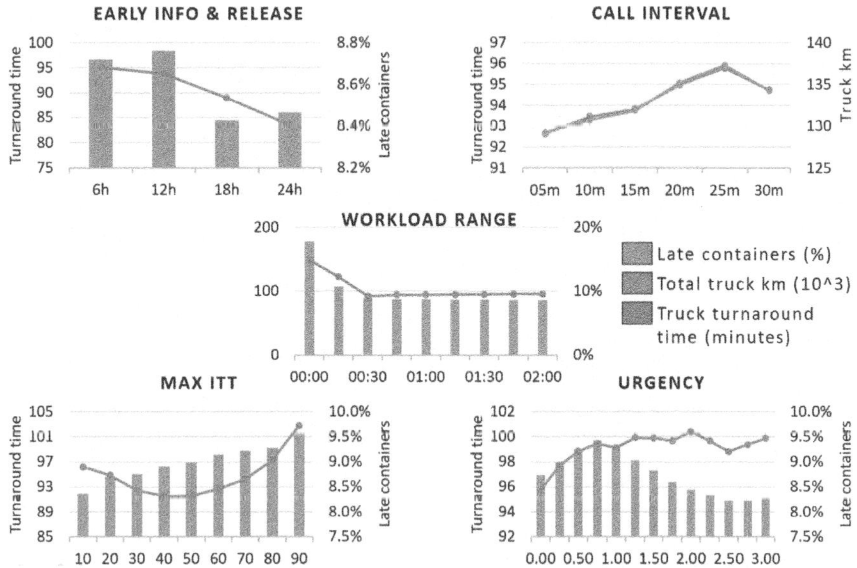

Fig. 18.3 Effects of the PDM parameters when keeping other parameters at default values

Figures 18.3 show the approximate effects of each PDM parameter, focusing on the KPIs that they affect the most. Regarding information availability, we see that earlier information and container preparation (24 h) lead to a shorter truck turnaround time and fewer late containers. Therefore, a longer planning horizon for the ITT workload allows for a more efficient decoupling process. Regarding the call interval, we see that a shorter interval (10 min) slightly reduces the truck turnaround time and the kilometers driven by trucks. Interestingly, it significantly increases fleet utilization and reduces the number of combined trips. This is a direct result of more frequent dispatching. Regarding the workload range, we see that a narrower range (30 min or 1 h) reduces the truck turnaround time and late containers. Additionally, it increases the share of decoupled arrivals, due to a more accurate computation of the ITT workload. However, when the range becomes too narrow, it leads to an underestimation of the ITT workload, with significant negative effects. Regarding urgency, we see a trade-off between truck turnaround time and the share of late containers, where both low and high urgency (0 or 2.5) show promising performance. This is caused by the relationship between the ITT workload and the queues at LCs: when we plan more ITT (lower urgency), we have shorter queues at LCs and thus shorter truck turnaround times. However, too much ITT workload has a negative impact on the delivery of decoupled containers. Finally, regarding the maximum ITT workload, we see a similar trade-off but with a local minimum of truck turnaround time between 40 and 50 maximum workload. We conclude that logistics nodes need to carefully select the urgency and maximum workload.

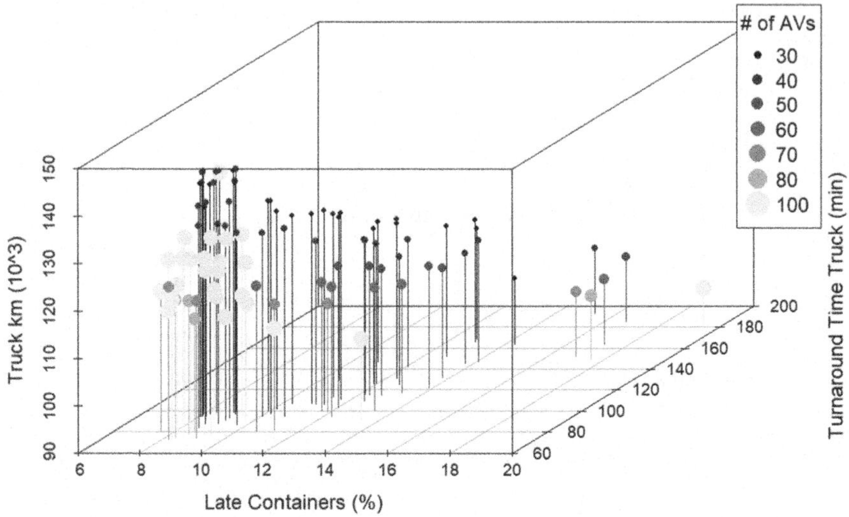

Fig. 18.4 Performance of promising configurations when reducing the fleet size

Table 18.2 Best PDM configuration per fleet size

Fleet size	Best parameters configuration
80	C-Interval: 10m, W-Range: 1 h , Urgency: 0 or 2.5, Max ITT: 50
70	C-Interval: 10m, W-Range: 1 h , Urgency: 2.5, Max ITT: 50
40	C-Interval: 10m, W-Range: 30 m , Urgency: 0, Max ITT: 20

In Fig. 18.4, we present the performance of the ITT fleet when reducing its size and testing the promising values of the PDM parameters. The results are plotted in a 3D scatterplot, and the best performing combination of PDM parameters are closer to the front-left-south corner. In addition, each data point is sized and colored according to the size of the fleet. The best combinations of fleet size and parameters are summarized in Table 18.2.

As expected, a larger fleet performs well for a wider range of configurations. However, with the right PDM calibration, fleets of 80, 70 and 40 EAVs can perform almost as well as 100 EAVs. Still, the 40 EAV fleet falls behind in terms of truck km reduction, as there is simply less capacity to decouple. This gives us two insights. First, logistics nodes can improve service levels with reduced capital expenditure for fleet vehicles. Second, logistics nodes can already reap benefits with a smaller ITT fleet and scale ITT operations over time. Finally, we notice that a greater reduction in truck kilometers leads to an excessive ITT workload and therefore to the degradation of service levels.

When looking at specific recommendations for PoM, the best-performing configuration shows that 80 EAVs achieve an average of 73 min of truck turnaround time, 8.32% late containers a day, and 119 thousand kilometers traveled by trucks during

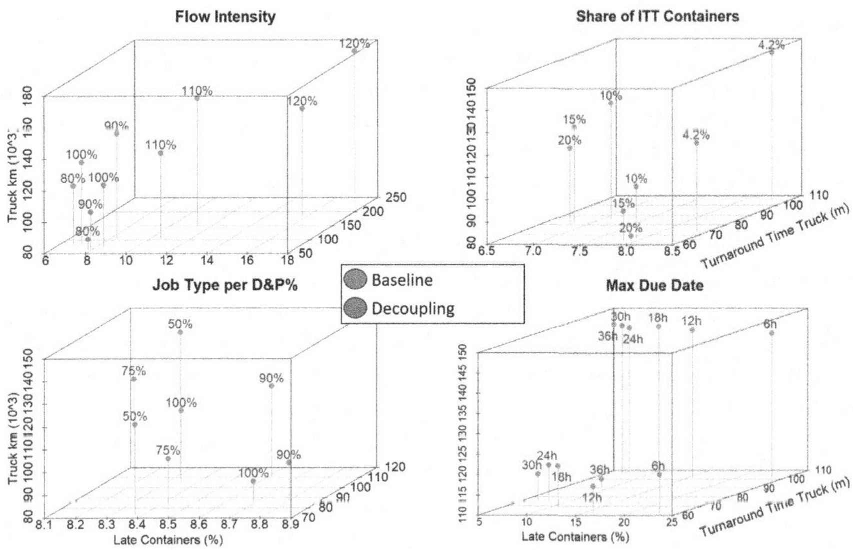

Fig. 18.5 Response of baseline and decoupling systems to changing arrival patterns

seven days of data collection. Compared to the PoM simulation without decoupling and early information, it corresponds to a reduction of 52 min (−41.6%) in the average turnaround time, of 1.6% (−16.6%) in late containers, and of 20 thousand kilometers (−14.5%) over seven days. Specifically, early information is responsible for the reduction in the share of late containers, while decoupling is responsible for the reduction in truck kilometers. Both early information and decoupling lead to a reduction in the average turnaround time, split in 22 and 30 min, respectively. Therefore, assuming sufficient container stacking capacity, early information can immediately contribute to reducing truck turnaround time and the share of late containers.

Figure 18.5 shows the results of our scenario analysis of arrival patterns with the 80 EAV fleet and the previously calibrated PDM. We see that decoupling outperforms the baseline process at PoM with respect to all KPIs and for all changes in arrival patterns. Interestingly, the spread of the results is also reduced, which means a more stable and resilient logistics system. Truck turnaround time is the most resilient KPI when decoupling, while a change in flow intensity is the most impactful factor for all KPIs.

Conclusions

Connected Automated Transport technologies, a promising tool in dealing with growing logistics demands and sustainability goals, are expected to be particularly effective in enhancing inter-terminal transport (ITT). They facilitate smart planning and can optimize the operations of an electric ITT fleet. Our proposed Proactive Decoupling Model (PDM) is central to this, leveraging early truck arrival information to improve ITT efficiency and service levels at logistics nodes. This model guides decisions on when to allocate first- and last-mile transport to an ITT fleet, considering factors such as stochastic arrivals and delivery windows.

We applied this model to the Port of Moerdijk, a major Dutch logistics node, using a realistic microtraffic simulation. Our experiments identified promising combinations of PDM settings and fleet size that markedly improved ITT performance, balancing transport service levels, efficiency, and sustainability. The results show significant improvements in truck turnaround times (−41.6%), reduced truck travel (−14.5%), and fewer late containers (−16.6%), underscoring the benefit of early information and the effectiveness of the PDM in connected and automated ITT operations.

Future research streams involve (i) a thorough mapping of the effects of the PDM parameters in combination with a larger set of CAT interventions and equipment for the ITT fleet, and (ii) the approximation of optimal parameter values, for example, by studying the delayed effects of sequential (decoupling) decisions via reinforcement learning.

Acknowledgements This research has been funded by the Dutch Research Council (NWO) and TKI Dinalog as part of the CATALYST Living Lab (reference 439.18.458 B), of which the Port of Moerdijk is a partner.

References

1. Abdelmagid, A. M., Gheith, M. S., & Eltawil, A. B. (2022). A comprehensive review of the truck appointment scheduling models and directions for future research. *Transport Reviews, 42*(1), 102–126.
2. Brunetti, M., Campuzano, G., & Mes, M. (2022). Simulation of the internal electric fleet dispatching problem at a seaport: A reinforcement learning approach. In *Winter simulation conference (WSC)* (pp. 1–12). IEEE.
3. Brunetti, M., Mes, M., & van Heuveln, J. (2020). A general simulation framework for smart yards. In *2020 Winter simulation conference (WSC)* (pp. 2743–2754). IEEE.
4. Chen, R., Meng, Q., & Jia, P. (2022). Container port drayage operations and management: Past and future. *Transportation Research Part E: Logistics and Transportation Review, 159*, 102633.
5. Dekker, R., Van Der Heide, S., Van Asperen, E., & Ypsilantis, P. (2013). A chassis exchange terminal to reduce truck congestion at container terminals. *Flexible Services and Manufacturing Journal, 25*, 528–542.
6. Evers, J. J. M. (2006). Real-time hiring of vehicles for container transport. *European Journal of Transport and Infrastructure Research, 6*(2), 173–198.
7. Gerrits, B., & Schuur, P. (2022). Innovations in self-organizing maritime logistics. In *Arctic maritime logistics: The potentials and challenges of the Northern Sea Route* (pp. 193–214). Springer.
8. Hu, Q., Wiegmans, B., Corman, F., & Lodewijks, G. (2019). Critical literature review into planning of inter-terminal transport: In port areas and the hinterland. *Journal of Advanced Transportation, 1–15.*

9. Huynh, N., Smith, D., & Harder, F. (2016). Truck appointment systems: Where we are and where to go from here. *Transportation Research Record, 2548*(1), 1–9.
10. International Transport Forum. (2021). ITF transport outlook 2021.
11. Meyer, G., Blervaque, V., & Haikkola, P. (2019). Stria roadmap on connected and automated transport-road, rail and waterborne. European Commission, STRIA-Strategic Transport Research and Innovation Agenda.
12. Robinson, S. (2008). Conceptual modelling for simulation part ii: A framework for conceptual modelling. *Journal of the Operational Research Society, 59*(3), 291–304.
13. van Steenbergen, R., Brunetti, M., & Mes, M. (2021). Network generation for simulation of multimodal logistics systems. In *2021 Winter simulation conference (WSC)* (pp. 1–12). IEEE.

Chapter 19
Insights into Domain Names and Product Categories of e-Commerce Websites: A Case Study of the UAE

Rachidatou Ingrid Traoret, Salama AlDhaheri, Fatima AlAmeemi, and Gurdal Ertek

Abstract Understanding e-commerce markets and online consumer behavior is critical for e-commerce retailers in the United Arab Emirates (UAE) and across the world. This study conducted a comprehensive analysis of unique e-commerce data to reveal patterns in domain names, product categories, and associations with other attributes. Using data analytics techniques, the initial dataset was cleaned and augmented to conduct the analysis. Key findings reveal '.com' and '.ae' as dominant domain extensions, numeric patterns in domain names, and 'apparel' as the most prevalent product category. An in-depth association mining analysis identifies and lists frequent categories and category combinations. This study provides actionable insights into the naming and category planning of e-commerce retailers in the UAE. Multidimensional analysis establishes an analysis and planning framework for evidence-based e-commerce research that can guide the two critical strategic decisions of domain name and category selection, which are among the important strategic decisions when an e-commerce business is first established. Most importantly, while providing insights for UAE e-commerce markets specifically, the methodology can be applied to any country or region.

Keywords E-commerce · UAE · Online shopping · Association mining · Domain names · Product categories

Supplementary Information The online version contains supplementary material available at (https://doi.org/10.1007/978-3-031-61589-4_19).

R. I. Traoret · S. AlDhaheri · F. AlAmeemi · G. Ertek (✉)
College of Business and Economics, United Arab Emirates University, Al Ain, UAE
e-mail: gurdal@uaeu.ac.ae

R. I. Traoret
e-mail: 700042159@uaeu.ac.ae

S. AlDhaheri
e-mail: 201801296@uaeu.ac.ae

F. AlAmeemi
e-mail: 201207945@uaeu.ac.ae

© The Author(s), under exclusive license to Springer Nature Switzerland AG 2024

A. Emrouznejad et al. (eds.), *Business Analytics and Decision Making in Practice*, Lecture Notes in Operations Research, https://doi.org/10.1007/978-3-031-61589-4_19

Introduction

In an era marked by rapid digital transformation and the spread of e-commerce, shopping has overcome the limitations of brick-and-mortar stores. In 2022, the disparity in growth between e-commerce and traditional retail was minimal, with only a marginal difference of 0.2%, highlighting consumers' strong inclination towards online shopping [1]. Online shopping, once a convenience, has evolved to shape the way consumers explore, select, and purchase products and services. This transformation has not only revolutionized the retail landscape, but has also given rise to new consumer behaviors, preferences, and expectations that continue to reshape the virtual marketplace.

The e-commerce industry is rapidly expanding worldwide. By the end of 2023, the global e-commerce market is expected to reach a total value of $6.3 trillion, with a projected growth rate of 11.17% from 2023 to 2027, resulting in a market volume of around $5.56 trillion by 2027 [2].

Across the Middle East and North Africa (MENA) region, 91% of consumers make online purchases, and 70% prefer digital payment methods [3]. The e-commerce sector in the Middle East is expected to achieve a market volume of US$50 billion by 2025. This growth is supported by factors such as the widespread adoption of advanced technologies, government initiatives, and a young, technology-savvy population with increasing disposable income [3].

The United Arab Emirates (UAE), with its rapid technological advancement and high Internet penetration rate, has emerged as a prominent hub for online shopping in the Middle East. Previous studies have highlighted the growth potential and dynamics of e-commerce in the UAE, emphasizing the importance of understanding the underlying factors that contribute to the success of online platforms [4]. UAE is positioned as the 27th largest e-commerce market in the world and is anticipated to generate revenue of approximately US$10,205.5 million by the year 2023 [5]. Recognizing the significance of e-commerce and e-commerce platforms, this research aims to conduct a comprehensive analysis of the top 10,000 e-commerce websites operating in the UAE, examining aspects such as domain-naming patterns and product categories.

To the best of our knowledge, the present study is the first of its kind, by presenting a methodology and sample country-specific analysis that can guide two critical strategic planning problems that every e-commerce start-up across the globe faces:

- What should be our domain name?
- Which product categories should we offer?

The contributions of the current study become more pronounced when related literature is reviewed. However, owing to space limitations, the literature review section has been moved to the supplemental document, available within the Supplement package [6], which complements this paper.

Methodology

The rapid evolution of e-commerce has necessitated a deeper understanding of the ecosystem, particularly concerning domain name and category selection. The unsupervised machine learning technique of association mining [7–9] has emerged as a vital tool to uncover frequent categories and association patterns, offering insights into frequent product categories and category combinations in e-commerce platforms.

The methodology developed and used in this study is presented in Fig. 19.1 and described in this section.

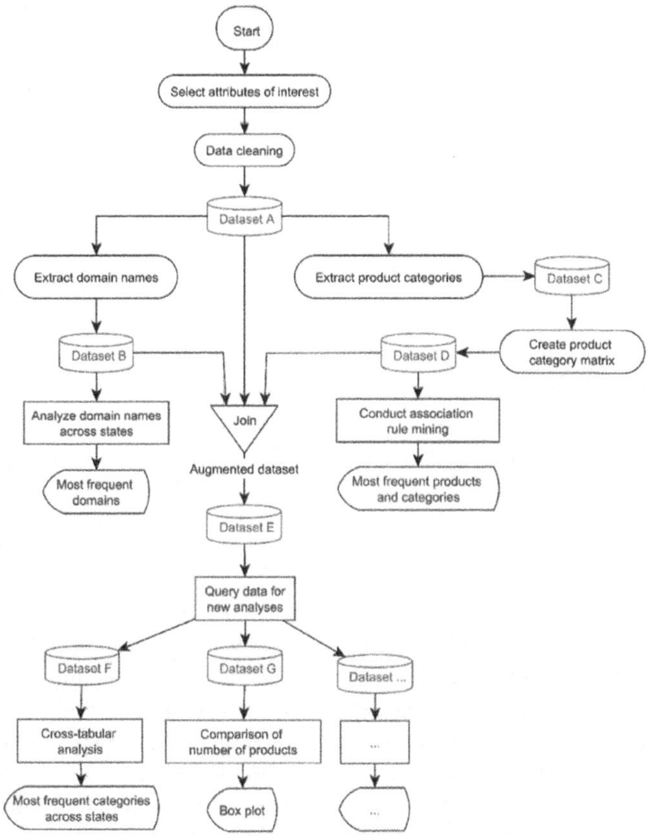

Fig. 19.1 Methodology flowchart showing the steps taken throughout the project

Data Acquisition

In the e-commerce sector, the quality and relevance of data plays a pivotal role in determining the accuracy and depth of analytical insights. To this end, the research sourced its dataset from "My Lead Fox" [10], a data provider Software-as-a-Service (SaaS) platform. The extensive data obtained from this service provides a foundation for exploring consumers' purchasing patterns, ensuring a comprehensive representation of online shopping behaviors. Permission was granted by the platform's founder to publicly share the results of the analysis and source data. Given the vast amount of available data available within My Lead Fox related to all e-commerce platforms worldwide, filtering was applied to include only e-commerce websites within the UAE. The top 10,000 results were downloaded in a tabular format. After this selection, the source data (Dataset A) for the research were obtained.

Data Preparation

For the data analysis in this paper, Microsoft Excel was used for data cleaning and augmentation, as well as tabular analysis [11, 12]. Orange software [13, 14] was used to conduct association rule mining, box plot visualization, and parametric hypothesis testing.

The data augmentation process was conducted on domain names to enhance the dataset, fill any gaps, and add layers of information that could enhance the depth of subsequent analyses. The process, although time-intensive, was instrumental in creating a robust and comprehensive dataset for a focused yet rich analysis of the data. The result was Dataset B.

Additionally, the categories that the website offers were subjected to detailed cleaning, resulting in Dataset C, which was later used as a list of categories in association rule mining.

The categories in each row of the original source Dataset A were then mapped in a matrix format, namely, Dataset D, where rows are websites and columns are individual categories. Dataset D was analyzed using association mining.

Furthermore, the columns in Dataset D corresponding to the top 10 categories were merged with Datasets A and B to obtain the final augmented Dataset E.

Table 1 in the Supplement [6] lists the names of the attributes of interest (that were considered in this particular study) in the augmented Dataset E, along with their data type (strings and integer attributes), sample values for a selected observation, and description.

The enriched and extensive Dataset E can be analyzed in many ways in future research. Within the scope of this study, two sample analyses were conducted, whose source data (derived from Dataset E) were Datasets F and G, respectively, and the results are presented in Figs. 19.2 and 19.3.

Fig. 19.2 Expanded view of the most frequent categories and category combinations

Itemsets	Support	%
Apparel=1	1533	18
Clothing_Accessories=1	374	4.392
Women's_Clothing=1	257	3.018
Footwear=1	128	1.503
Home_Garden=1	1238	14.54
Home_Furnishings=1	298	3.5
Home_Improvement=1	225	2.642
Home_Interior_Decor=1	147	1.726
Home_Appliances=1	145	1.703
Kitchen_Dining=1	101	1.186
Food_Drink=1	1057	12.41
Food=1	642	7.54
Baked_Goods_Desserts=1	110	1.292
Beverages=1	234	2.748
Coffee_Tea=1	155	1.82
Coffee_Tea=1	155	1.82
Baked_Goods_Desserts=1	110	1.292
Beauty_Fitness=1	1051	12.34
Face_Body_Care=1	653	7.669
Perfumes_Fragrances=1	235	2.76
Skin_Nail_Care=1	149	1.75
Make-Up_Cosmetics=1	142	1.668
Perfumes_Fragrances=1	235	2.76
Skin_Nail_Care=1	149	1.75
Make-Up_Cosmetics=1	142	1.668
Hair_Care=1	141	1.656
Fitness=1	126	1.48
Business_Industrial=1	674	7.915
Business_Services=1	284	3.335
Office_Supplies=1	128	1.503
Office_Supplies=1	128	1.503
Face_Body_Care=1	653	7.669
Perfumes_Fragrances=1	235	2.76
Skin_Nail_Care=1	149	1.75
Make-Up_Cosmetics=1	142	1.668
Food=1	642	7.54
Baked_Goods_Desserts=1	110	1.292
Health=1	444	5.214
Nutrition=1	151	1.773
Vitamins_Supplements=1	90	1.057
Vitamins_Supplements=1	90	1.057
Computers=1	428	5.026
Computer_Hardware=1	152	1.785
Clothing_Accessories=1	374	4.392

The cleaning, augmentation, and pinning of domain and category data aimed to lay a solid groundwork for the subsequent analysis, ensuring that the insights derived were both accurate and rich [12].

State (Emirate)	Apparel	Beauty_Fitness	Business_Industrial	Computers	Food_Drink	Health	Home_Garden
Abu Dhabi	32.51	20.49	8.13	4.24	10.60	8.13	15.90
Ajman	23.53	24.71	9.41	4.71	8.24	4.71	24.71
Dubai	31.00	17.81	7.93	6.11	10.85	7.11	19.19
Fujairah	31.03	34.48	3.45	3.45	13.79	6.90	6.90
Ras al Khaimah	25.00	25.00	10.42	2.08	10.42	6.25	20.83
Sharjah	25.45	18.64	14.09	4.09	7.27	4.09	26.36
Umm al Quawain	21.43	28.57	7.14	7.14	7.14	0.00	28.57

Fig. 19.3 Percentage of categories across states

Category Analysis with Association Rule Mining

Association rule mining is used to uncover the relationships between variables in large datasets [7–9]. It works by identifying combinations of items that frequently occur together in transactions, such as products purchased by customers. The goal is to find interesting associations and correlations that can provide insights into sales and marketing strategies [15]. The key metrics in association rule mining are support and confidence [9]. Support indicates how frequently an itemset appears in the dataset, whereas confidence shows the frequency with which a rule is found. As in [15], insignificant associations can be filtered out to focus only on the strongest associations by setting the minimum thresholds for support and confidence.

In e-commerce analytics, association rule mining holds a pivotal position, offering granular insights into how products interact with consumers' shopping carts. In this research, instead of analyzing consumer purchase transactions and items on e-commerce websites, their product categories were analyzed. In other words, each website was treated as a transaction, and each category was treated as an item.

Association rule mining was conducted using the associate extension within the Orange data mining software [13, 14], a tool known in the data science community for its versatility and ease of use, despite challenges with respect to scalability to big data. The software has an intuitive interface and versatile options, which will allow the research team to reveal intricate patterns with clarity, presenting data in an accessible and comprehensible manner.

Analysis of Domain Names

In contemporary research, the depth and breadth of a dataset can significantly impact the comprehensiveness of derived insights. Hence, to overcome the limitations of Dataset A, the data were first augmented to Dataset B, enhancing its scope and details. Augmentation was achieved by appending newly derived attributes related to domain names, ensuring that the resulting Dataset B was seamlessly integrated with the original, while providing more details. This process not only bridges the gaps in the data but also introduces new dimensions that can be leveraged for more in-depth analysis [16].

Additional Analysis

While the primary focus of the study revolved around the analysis of domain names and product categories, an additional analysis was conducted with Datasets F and G to provide a more comprehensive methodology and a wider perspective on the e-commerce landscape in the UAE. This approach, which merges both the core and auxiliary elements of source data and derived data, aims to provide stakeholders with an all-inclusive understanding, catering to both strategic and operational needs.

Results and Discussions

Domain Extension Distribution

Table 19.1 shows the distribution of the most frequent domain extensions (types) across the UAE e-commerce landscape. Table 19.1 reveals the patterns in domain naming with a Generic Top-Level Domain (gTLD) ".com' and Country Code Top-Level Domain (ccTLD) ".ae" emerging as dominant domain extensions. The findings suggest a strong preference for these two domain extensions, possibly due to their global recognition and regional relevance. Furthermore, it can be observed that the rank of the domain type follows the Zipf's Law distribution [17]. The law in probability asserts that the frequencies f of certain events are inversely proportional to their rank r. For the first and second domain type, ".com" appears 6176 times and domain type and ".ae" appears 3118 times (around half times) as often as domain type ".com".

Figure 19.4 plots the percentage of the top three domain types across the different Emirates (states) of the UAE in a bar chart matrix. It offers insights into domain

Table 19.1 Most frequent domain types

Domain type	Frequency	Percentage (%)	Rank
com	6176	61.76	1
ae	3118	31.18	2
net	151	1.51	3
co	114	1.14	4
me	84	0.84	5
shop	53	0.53	6
store	52	0.52	7
org	45	0.45	8
online	20	0.20	9
fr	11	0.11	10
uk	10	0.10	11

Fig. 19.4 Percentage of top three domain types in the seven emirates of UAE

State (Emirate)	com	ae	net
Abu Dhabi	59.84	29.92	2.05
Ajman	72.79	16.91	0.74
Dubai	70.25	21.21	1.48
Fujairah	72.73	13.64	0
Ras al Khaimah	77.33	13.33	2.67
Sharjah	73.12	17.05	2.89
Umm al Quawain	70.83	20.83	8.33

naming preferences at state-level granularity. As can be observed, the ".com" domain type is the most preferred in all seven emirates. In "Ras al Khaimah" Emirate especially, 77.33% of the e-commerce websites prefer ". com" domain whereas in "Abu Dhabi," 29.92% of them prefer ". ae" domain type, possibly related to Abu Dhabi City being the capital of the UAE. These findings are significant for marketers and online businesses as they guide domain selection based on regional profiles [16]. For example, an e-commerce company targeting especially Abu Dhabi market can be more open to the idea of having the ".com" gTLD or ".ae" ccTLD extension compared to the other Emirates.

Numeric Strings in Domain Names

Table 19.2 presents the frequency of numeric strings within domain names, suggesting a trend wherein certain platforms incorporate numbers, which could be justified by the will to be unique or for branding strategies, as highlighted in [15]. Table 19.2 reveals that the most frequent single digits in domain names are "4," "2," "3," "1" and "7" across UAE e-commerce websites. Three-digit number strings "360" and "800," as well as '247,' possibly alluding to holistic offerings or round-the-clock availability, are the most frequent three-digit numbers appearing in domain names. Understanding these numbers can help infer potential marketing branding strategies and the appeal of numeric integrations in domain names [16, 18].

The "Benford's Law" also known as "First Digit Law" introduced in 1938 by Frank Benford and mathematically proved in 1995 by Theodore Hill, stated that the probability of the first digit being 1–9 is unequal and satisfied a logarithmic distribution. This means that the larger the digit, the lower is the probability of its appearance as the first digit [19]. From what has been observed, the digit distribution in domain names does not follow Benford's law, possibly because of semantic branding strategies.

Table 19.2 Most frequent digits in domain names

Number string	Frequency
4	35
2	31
3	24
1	16
7	13
360	10
8	9
800	9
24	6
9	5
11	5
247	5

Frequent Product Categories

Table 19.3 presents the top ten most frequent categories offered on UAE e-commerce websites, together with their frequencies (support). This analysis relates to the "Pareto Principle" which states that a big majority of the consequences are produced by a small minority of most significant causes [20].

The association mining analysis on ecommerce transactions reveals 'Apparel' as the most prevalent category in the dataset, with a support of 1,533, meaning that at least 1,533 out of the 10,000 e-commerce websites in the UAE offer apparel products, indicating a substantial supply and possibly demand for apparel among UAE's online shoppers. 'Home_Garden' follows closely as the second most popular category, occurring in 1,238 websites. This finding can be related to [21], who found

Table 19.3 Top ten most frequent categories

Rank	Itemsets	Support
1	Apparel = 1	1533
2	Home_Garden = 1	1238
3	Food_Drink = 1	1057
4	Beauty_Fitness = 1	1051
5	Business_Industrial = 1	674
6	Face_Body_Care = 1	653
7	Food = 1	642
8	Health = 1	444
9	Computers = 1	428
10	Clothing_Accessories = 1	374

that garden and DIY segment purchases increased during the Covid-19 pandemic due to the lockdown. The high presence of home- and garden-related purchases implies significant interest in products that enhance living spaces and environments. 'Food_Drink' emerges as the top category with 1,057 websites, underlining a robust appetite for culinary items among customers, and confirming the results in [21] on the increase of "food" category purchases online since 2019.

Figure 19.2. presents an expanded view of the most frequent categories *and* combinations. For example, from the first two rows of the columns on the left-hand side, we read the following: Apparel is the most popular category, offered on 1533 websites, whereas the Apparel and Clothing-Accessories categories are jointly offered on 374 websites. The percentages of columns are higher than the (support/10,000) values because of the missing category data for approximately 15% of websites (rows).

In an in-depth analysis, Fig. 19.2 highlights differences within broader categories, distinguishing between sub-types like 'Food,' 'Beverages,' and 'Baked_Goods_Desserts.' This granular view reveals specific inclinations, potentially guiding targeted intervention. For instance, beverage brands may recognize opportunities to increase their market share based on subset frequency. Furthermore, the repetition of certain categories indicates frequent co-occurrence. 'Face_Body_Care' appears multiple times, suggesting it commonly complements other categories during transactions (Fig. 19.2). Uncovering such combinations can empower businesses to devise complementary product recommendations or bundling strategies to boost their sales.

As an additional analysis, the percentage of categories across the Emirates (states) of the UAE was determined by querying Dataset F from Dataset E. Dataset F denotes the state of the e-commerce website for each website that had a category (e.g., Apparel). The results displayed in Fig. 19.3 provide several insights: Abu Dhabi, Dubai, and Fujairah have the highest percentages of apparel e-commerce websites, Fujairah has a notably higher percentage of Beauty_Fitness e-commerce websites, Sharjah has a notably higher percentage of Business_Industrial websites, and Umm al Quawain, Sharjah, and Ajman have notably higher percentages of Home_Garden websites.

The last analysis concerned the number of products sold on e-commerce websites in relation to product categories. Due to space limitations, this analysis is provided in the Supplement [6]. This analysis was conducted by querying Dataset G from Dataset E. Dataset G denotes the number of products sold on the e-commerce website for each website that had a category (e.g., Apparel).

The above findings provide actionable insights into category preferences within the UAE e-commerce landscape. The category distribution analysis identifies popular categories, such as apparel and home products, that businesses can consider for strategic planning and resource allocation. Meanwhile, association mining analysis revealed the frequencies and relationships between the categories. Recognizing these patterns and associations can guide targeted marketing and merchandising strategies and plans for online retailers in the UAE [4].

Conclusion and Future Work

In conclusion, this study offers novel insights and perspectives on the e-commerce landscape of the UAE through a comprehensive analysis of domain names, product categories, and associated metrics. The findings reveal key patterns and relationships that can empower businesses to make data-driven decisions about branding, product offerings, and marketing strategies tailored to consumer preferences. For instance, the prevalence of '. com' and '.ae' domains point to an opportunity for international and localized branding approaches. The high frequency of apparel and home products guides strategic prioritization of offerings and inventory planning. Meanwhile, insights into sub-categories and recurring itemsets provide a granular view of demand patterns to inform targeted promotions.

While this study makes significant contributions, avenues exist to further enrich the understanding of the UAE online shopping ecosystem. First, augmenting the dataset with additional parameters, such as pricing, ratings, and reviews, can reveal crucial factors influencing purchase conversions, as in [22–24]. Second, a comparative analysis across e-commerce platforms and demographics can highlight the preferences of diverse consumer segments [21, 25, 26]. Moreover, integrating temporal variables can uncover seasonal and event-based spikes, guiding inventory and delivery management [27] to better satisfy customers and increase retention and sales. Further data augmentation with emerging data sources, such as delivery performance metrics [28] can enhance the decision-making capabilities further in this continuously evolving landscape.

This study equips businesses with pivotal insights to gain a competitive edge. The study also establishes a methodological blueprint for similar analytical projects in any e-commerce market, inspiring more evidence-based research on e-commerce.

Acknowledgements The authors thank Mr. Sumit Ghugharwal, Founder of the Dubai-based My Lead Fox SaaS platform, for allowing the research team to use the dataset for research and to publish the analysis results of the filtered dataset publicly.

References

1. Keenan, M. (2023). *Global ecommerce statistics: Trends to guide your store in 2024*. Shopify Plus. Retrieved December 25, 2023 from https://www.shopify.com/enterprise/global-ecommerce-statistics.
2. Statista. eCommerce-Worldwide. Retrieved December 25, 2023 from https://www.statista.com/outlook/dmo/ecommerce/worldwide.
3. Lau, L. (2023) *Ecommerce in the Middle East: Statistics and trends in 2023*. Orienteed e-commerce Solutions. Retrieved December 25, 2023 from https://orienteed.com/en/ecommerce-middle-east-statistics-trends-2023/.
4. Alzaabi, K., Hamad, M. O., Omar, R., & Romle, A. R. (2021). Organizational factors and e-commerce adoption in SMEs of United Arab Emirates: Mediating role of perceived strategic value. *International Journal of Entrepreneurship, 25*(1), 1–18.

5. ECBD. (2022). *eCommerce Market UAE - Data, Trends, Top Stores*. ECDB. Retrieved December 25, 2023 from https://ecommercedb.com/markets/ae/all.
6. Traoret, R. I., AlDhaheri, S., AlAmeemi, F., Ertek, G. (2024). *Supplement to "Insights into domain names and product categories of e-commerce websites: a case study of the UAE"*. https://ertekprojects.com/ftp/supp/20.zip.
7. Agrawal, R., Imielinski, T., & Swami, A. N.: Mining association rules be-tween sets of items in large databases. In: P. Buneman, S. Jajodia (Eds.), *SIGMOD Conference* (pp. 207–216). New York: ACM Press. https://doi.org/10.1145/170036.170072.
8. Han, J., Pei, J., & Yin, Y. (2000). Mining frequent patterns without candidate generation. *ACM SIGMOD Record, 29*(2), 1–12. https://doi.org/10.1145/335191.335372.
9. Ertek, G., & Demiriz, A. (2006). A framework for visualizing association mining re-sults. In International Symposium on Computer and Information Sciences (pp. 593–602). Berlin, Heidelberg: Springer Berlin Heidelberg. https://doi.org/10.1007/11902140_63.
10. My Lead Fox. Retrieved December 25, 2023 from https://myleadfox.com.
11. Microsoft Support. Top ten ways to clean your data. (2021). Retrieved October 20, 2023 from https://support.microsoft.com/en-us/office/top-ten-ways-to-clean-your-data-2844b620-677c-47a7-ac3e-c2e157d1db19.
12. Swapna, S., Niranjan, P., Srinivas, B., Swapna, R. (2016). Data cleaning for data quality. In *2016 3rd International Conference on Computing for Sustainable Global Development (INDIACom)*, pp. 344–348 (2016, March). IEEE.
13. Orange Data Mining. Retrieved December 25, 2023 from https://orangedatamining.com.
14. Demsar, J., Curk, T., Erjavec, A., Gorup, C., Hocevar, T., Mi-lutinovic, M., Mozina, M., Polajnar, M., Toplak, M., Staric, A., Stajdohar, M., Umek, L., Zagar, L., Zbontar, J., Zitnik, M., & Zupan, B. (2013). Orange: data mining toolbox in python. *Journal of Machine Learning Research 2349–2353*. http://jmlr.org/papers/volume14/demsar13a/demsar13a.pdf.
15. Diwandari, S., Zaky, U. (2021). Analysis of customer purchase behavior using association rules in e-shop. In 2021 *IEEE 5th international conference on information technology, in-formation systems and electrical engineering (ICITISEE)*. IEEE. https://doi.org/10.1109/icitisee53823.2021.9655892.
16. Gulf Business. Revealed: How to register a domain name in the UAE and how much it costs. Retrieved October 26, 2023 from https://gulfbusiness.com/revealed-register-domain-name-uae-much-costs/.
17. Hosch, W. L. Zipf's Law. Encyclopædia Britannica, Encyclopædia Britannica, inc., Retrieved December 25, 2023 from https://www.britannica.com/topic/Zipfs-law.
18. Hargittai, E. (2000). Open portals or closed gates? Channeling content on the World Wide Web. *Poetics, 27*(4), 233–253. https://doi.org/10.1016/s0304-422x(00)00006-1.
19. Li, F., Han, S., Zhang, H., Ding, J., Zhang, J., & Wu, J. (2019). Application of Benford's law in data analysis. In *Journal of Physics: Conference Series* (Vol. 1168, No. 3, p. 032133). IOP Publishing. https://doi.org/10.1088/1742-6596/1168/3/032133.
20. Dunford, R., Su, Q., & Tamang, E. (2014). *The Pareto principle*. Retrieved December 25, 2023 from http://pearl.plymouth.ac.uk/handle/10026.1/14054.
21. Zumstein, D., Oswald, C., & Brauer, C. (2022). Development of E-commerce in Times of the Corona Crisis. In *International Conferences e-Society 2022* (pp. 115–122). https://doi.org/10.33965/es_ml2022_2022021015.
22. Duan, Y., Liu, T., & Mao, Z. (2022). How online reviews and coupons affect sales and pricing: An empirical study based on e-commerce platform. *Journal of Retailing and Consumer Services, 65*, 102846. https://doi.org/10.1016/j.jretconser.2021.102846.
23. Bhattacharyya, S. S., Chahande, S. V., & Reddy, T. V. K. (2021). Exploring the factors impacting seller performance during online sales events in the E-commerce marketplace in India. *International Journal of Asian Business and Information Management (IJABIM), 12*(3), 1–20. https://doi.org/10.4018/ijabim.294100.
24. Abou-Kassem, T., Alazeezi, F. H. O., & Ertek, G. (2023). A data analytics methodology for benchmarking of sentiment scoring algorithms in the analysis of customer reviews. In: X. S. Yang, R. S. Sherratt, N. Dey, A. Joshi (Eds.), *Proceedings of Eighth International Congress*

on Information and Communication Technology. ICICT 2023. Lecture Notes in Networks and Systems, Vol. 693. Springer, Singapore. https://doi.org/10.1007/978-981-99-3243-6_46.
25. Lucas, G. A., Lunardi, G. L., & Dolci, D. B. (2023). From e-commerce to m-commerce: An analysis of the user's experience with different access platforms. *Electronic Commerce Research and Applications, 58*, 101240. https://doi.org/10.1016/j.elerap.2023.101240.
26. Changchit, C., Cutshall, R., & Pham, A. (2022). Personality and demographic characteristics influence on consumers' social commerce preference. *Journal of Computer Information Systems, 62*(1), 98–108. https://doi.org/10.1080/08874417.2019.1709229.
27. Deshpande, V., & Pendem, P. K. (2023). Logistics performance, ratings, and its impact on customer purchasing behavior and sales in e-commerce platforms. *Manufacturing Service Operations Management, 25*(3), 827–845. https://doi.org/10.1287/msom.2021.1045.
28. Akasheh, M. A., Eleyan, N., & Ertek, G. (2022). A predictive data analytics methodology for online food delivery. In *2022 Ninth International Conference on Social Networks Analysis, Management and Security (SNAMS)*, Milan, Italy (pp. 1–7). https://doi.org/10.1109/SNAMS58071.2022.10062613.

Chapter 20
Data Analytics with Large Language Models (LLM): A Novel Prompting Framework

Shamma Mubarak Aylan Abdulla Almheiri, Mohammad AlAnsari, Jaber AlHashmi, Noha Abdalmajeed, Muhammed Jalil, and Gurdal Ertek

Abstract This study presents a novel framework for conducting data analytics using Large Language Models (LLMs). The proposed framework suggests the construction of prompts and interaction patterns using four fundamental constructs: meta-specifications, specifications, instructions, and prompting patterns. The framework can guide and assist data engineers, analysts, and even non-technical domain experts by providing these four constructs as palettes of options. The LLM can then suggest analytics designs, conduct the analysis, provide posterior interpretations and insights, and produce other outputs, such as code or packaged software. The presented novel framework covers an immense space of possibilities through numerous combinations of selected meta-specifications, specifications, instructions, and prompting patterns. The primary theoretical contribution of this research is that it proposes a theoretical foundation and frame of reference for conducting data analytics using LLM. The primary practical contribution is that LLMs can now be employed much more systematically and extensively than before in designing and conducting data analytics. This opens a new world of applications powered by a countless combination of the four constructs across practically all fields of science, technology, and business, where LLMs can be used to guide, conduct, and interpret the results of data analytics.

S. M. A. A. Almheiri · M. AlAnsari · J. AlHashmi · N. Abdalmajeed · M. Jalil · G. Ertek (✉)
College of Business and Economics, United Arab Emirates University, Al Ain, UAE
e-mail: gurdal@uaeu.ac.ae

S. M. A. A. Almheiri
e-mail: 201003124@uaeu.ac.ae

M. AlAnsari
e-mail: 700040492@uaeu.ac.ae

J. AlHashmi
e-mail: 700040210@uaeu.ac.ae

N. Abdalmajeed
e-mail: 200935171@uaeu.ac.ae

M. Jalil
e-mail: muhammad.jalil@uaeu.ac.ae

© The Author(s), under exclusive license to Springer Nature Switzerland AG 2024
A. Emrouznejad et al. (eds.), *Business Analytics and Decision Making in Practice*,
Lecture Notes in Operations Research, https://doi.org/10.1007/978-3-031-61589-4_20

Keywords Data analytics · Large language models · ChatGPT · Theoretical framework · Bottom-up conceptual analysis

Introduction

The field of Artificial Intelligence (AI) is experiencing a "Cambrian Explosion" [1], with multiple innovations every single day.

Since the introduction of the ChatGPT [2] online service by OpenAI [3] in November 2022, many AI innovations are on or related to generative AI (gen-AI), the field of AI with countless use cases, where AI generates content in response to human prompts, either through interaction or in response to algorithmic instructions [4]. The global generative AI market, which was valued at $8.2 billion in 2021, was projected to reach $126.5 billion by 2031, growing at a CAGR (Compound annual growth rate) of 32% from 2022 to 2031 [5]. However, the growth in the field and related sectors in 2023 is such that the actual growth may exceed the projected market size.

Large Language Models (LLM) are AI models trained with large corpi (collections) of text and can respond to queries/prompts in natural language (e.g., English, Chinese, Arabic, Turkish) [6, 7].

ChatGPT, an LLM tool built on the GPT model, has become the fastest adopted application in human history [8], expanding significantly in terms of technical capability, as well as the user and developer base [9] in a very short time span.

Data Analytics is the modeling and analysis of data, with the goal of uncovering hidden patterns, deriving actionable insights, and enabling better decision making [10, 11]. Data analytics enables the ascension of raw data to information, knowledge, insights, and foresight through the application of a portfolio of mathematical, statistical, and computational techniques and tools.

One of the application areas of LLMs is facilitating various stages and steps of data analytics [8, 12]. In this paper, we first present a case study in which LLM is used to develop designs for analyzing data. Then, as a theoretical contribution to the fields of generative AI and data analytics, we present a novel prompting framework (Fig. 20.1) to enable structured and systematic conduct of data analytics with LLMs.

Preliminary Case Study

In early 2023, after the release of ChatGPT, a preliminary one-month-long data analytics project was conducted by the research team in marketing analytics, where LLM was used to generate designs for pivot tables and charts for a sales transaction dataset. The details of this case study are presented in this paper.

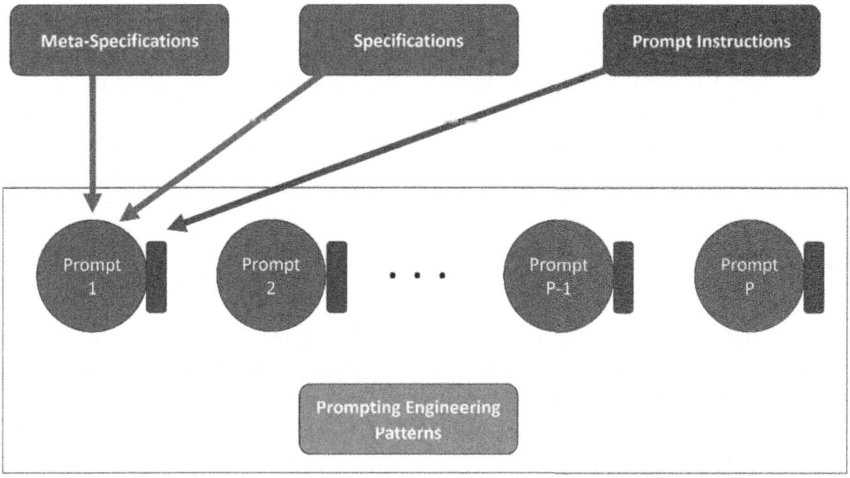

Fig. 20.1 Proposed framework for data analytics with LLM, as the primary contribution of the research study presented in the paper

Recent Developments

Since the preliminary case study was conducted in early 2023, until late 2023, when this research was completed, many developments have taken place in the LLM world. Due to space limitations, these recent developments are summarized in the Supplement [13].

Framework Proposition

The main contribution of this paper is the framework for human-LLM interaction for data analytics. The main idea of the proposed framework is that when interacting with LLM to design and conduct data analytics, each prompt is composed of a combination of four main constructs: (1) meta-specifications, (2) specifications, and (3) prompt instructions as modifiers and performance enhancers, while the interaction takes place through a (4) prompt engineering pattern.

Unique Contributions

The primary theoretical contribution of this research is that it proposes a theoretical foundation and frame of reference for conducting data analytics using LLM. The primary practical contribution is that LLMs can now be employed much more systematically and extensively than before in designing and conducting data analytics. This opens a new world of applications powered by a countless combination of the four constructs across practically all fields of science, technology, and business, where LLMs can be used to guide, conduct, and interpret the results of data analytics.

Literature

Large Language Models (LLM) have emerged as powerful tools in the domain of machine learning and natural language processing, representing a significant advancement in the capabilities of language-based tasks. This literature review aims to delve into the fundamental aspects of LLM, its relevance in data analytics, and the key techniques and parameters that play a crucial role in its configuration.

LLM

Large Language Models refer to a class of advanced machine learning models designed to process and understand natural language at an extensive scale [6, 7, 14]. Unlike traditional language models, LLMs such as GPT-3 (Generative Pre-trained Transformer 3), are pre-trained on vast amounts of diverse data, allowing them to generate coherent and contextually relevant text across a wide range of topics.

Recent developments in LLM as of 2023, including important discoveries and widely used methods, can be found in [7]. The review studies [8, 15] focus particularly on ChatGPT, presenting the background, applications, challenges, ethical issues (including bias), limitations, and future directions for ChatGPT, with ramifications for LLMs at large.

LLM for Data Analytics

How can LLM be used for data analytics?

LLMs offer versatile applications in data analytics, particularly in the analysis of unstructured and structured data [16, 17]. Their ability to comprehend and generate human-like text enables tasks, such as text summarization, categorization, classification, sentiment analysis, and language translation. In data analytics, LLMs can be harnessed for information extraction, content categorization, and the development of sophisticated chatbots, thereby enhancing the efficiency of data processing and interpretation [12].

The research literature has a growing number of studies have reported the application of LLMs in data analytics, as detailed in the Supplement [13].

Generative AI can be used not only to conduct specific tasks and steps of data analytics but also to generate complete data analytics/data science workflows [18].

Constructs of LLM

The effective use of LLMs in data analytics requires careful consideration of several aspects. Due to space limitations, we summarize these aspects in terms of the techniques and LLM parameters that should be carefully configured for optimal LLM performance in the Supplement [13].

Prompt Engineering

Crafting effective prompts or queries is essential when using LLMs in data analytics. The choice of language and framing of questions impact the quality and relevance of the model's responses [19]. The collection of techniques for constructing better prompts is referred to as prompt engineering.

Summary

To summarize, LLMs offer a distinctive opportunity in the landscape of data analytics, offering unprecedented capabilities for understanding and generating natural language. By leveraging transfer learning, attention mechanisms, and careful parameter tuning, LLMs can be tailored to excel in various data analytics applications, from text classification to information retrieval. As these models continue to evolve, their integration into data analytics workflows promises to enhance the efficiency and depth of insight derived from unstructured textual data.

Preliminary Case Study

As a preliminary exploration and to test the applicability of LLM for designing data analytics, a case study was conducted in marketing analytics. Public sales transaction data [20] that accompanies a paper [21] published in the data journal "Data in Brief" [22] were used to ensure the validity and reliability of the source data. Data from data journals are typically superior in quality compared to datasets from Kaggle or public data repositories, as data journals publish papers and accompanying source data for papers only after academic peer review.

Interaction with the LLM

The steps of interaction with the LLM are provided below. The user prompts during the interaction with the LLM are provided in a table in the Supplement [13].

Problem and Data Description

Step 1. Describe to the LLM (e.g., ChatGPT), the dataset to be analyzed.

Step 2. Provide metadata (attribute names, their data types, and descriptions) for the data.

Step 3. Describe the goal for the data analytics project.

Exploratory Data Analysis

Step 4. Request top-N cross-tabular analysis (pivot table) designs for exploratory data analysis.

Step 5. Ask for justification for the design choices.

Step 6. Ask for supporting evidence (academic research, business whitepapers).

Step 7. Apply the suggested designs.

Step 8. Request chart/plot designs for each cross-tabular analysis (pivot table).

Step 9. Ask for justification for the design choices.

Step 10. Ask for supporting evidence (academic research, business whitepapers).

Step 11. Generate charts/plots/visualizations through pivot tables manually (possibly through crowdsourcing) or automatically.

Step 12. (Not executed in the case study) Interpret the plots using a Vision Encoder + LLM (e.g. MiniGPT [23]).

LLM Outputs

The LLM recommended the following pivot table analyses in the given order: Store Performance, Product Performance, Sales by Day of the Month, Product Performance by Store, Monthly Profit Margins, Store Inventory Management, Top Selling Products by Store, Product Profitability by Store, Sales Trend Analysis, Sell Backs Analysis.

As pivot charts, LLM recommended line charts for analyses 3 and 5 and stacked column charts for all remaining analyses.

Given the guidance and design templates from LLM, the analysis in MS Excel was manually conducted by human users (the research team) using MS Excel pivot table and pivot chart features. In some cases, the user did not implement the analysis as is, but used his/her judgement to make minor modifications to the design suggested by the LLM. The details of the full analysis, including pivot tables, charts, and insights, can be found in the Supplement [13].

Observations from the Case Study

Our observations from the case study are as follows:

1. LLM made the exploraty and descriptive data analytics much faster.
2. LLM recommended some of the most popular pivot tables and charts.
3. LLM recommended very basic analysis, which were very similar to each other.
4. LLM did not introduce diversity in the analysis by itself. LLM recommended only two chart types.
5. LLM attempted to combine multiple analyses into a single analysis. For example, instead of breaking the analysis of sales, profit, and quantity sold, it combined all into a single analysis. Although this is efficient when creating the pivot table, plotting the same resulted in a single chart that combined everything and was harder to read and interpret.

From the preliminary case study, we concluded that a more extensive interaction, with further prompting, is needed to introduce non-conventional designs, a higher variety of chart types, and creative designs that may yield surprising insights.

Proposed Framework

Driven by the findings from the preliminary case study, a novel framework is proposed in the current paper, for conducting data analytics with LLM.

The main idea of the proposed framework is that when interacting with LLM to design and conduct data analytics, each prompt is composed of a combination of four main constructs: (1) meta-specifications, (2) specifications, and (3) prompt instructions as modifiers and performance enhancers, while the interaction takes place through a (4) prompt engineering pattern.

In this section, lists are presented for the meta-specifications and specifications. Furthermore, essences and examples of instructions and prompt engineering patterns are discussed.

Meta-Specifications

Meta-Specifications are high-level specifications that are not directly related to the design or analysis to be conducted, but rather to the **inputs** (information and data) provided and actions taken during prompting.

The meta-specifications in the proposed framework are the following:

1. **Upload the data.**
2. **Upload images, audio, video,** or other type of data.
3. Specify the **data structure** (e.g., tabular, graph/network, text, image, video, audio).
4. Mention the **attribute (data field) names.**
5. **Describe the attributes** (e.g., ID: Unique ID that is key, SalesAmount: sales amount in USD in that transaction).
6. Specify the **data type** of the attributes.
7. Specify the **hyperparameter** values.
8. Provide an **overall problem context** (e.g., the company wants to increase sales).
9. Provide the **detailed context** (ex: "the company wants to identify the products that can be bundled for cross-selling").
10. Provide the **metrics** in the problem context (ex: "the company's top KPI is the annual profit").
11. Specify the **reasoning** for the prompt. In other words, specifying the "Why?" data analytics.
12. Specify **temporal information**. In other words, answering the "When?" (e.g., "the study was conducted in 2020"). Such a specification may help the LLM to relate the output to what happened in that specified time frame, such as the Covid pandemic in 2020).
13. Specifying **geographical information** In other words, answering the "Where?" (e.g., "the study was conducted in the United Arab Emirates"). Such a specification may help the LLM relate the output to what happened in that geographical location.
14. Specify **persons/people information**. In other words, answering the "Who?" (e.g., "the study was conducted by a team of bioinformatics researchers"). Such a specification may help the LLM relate the output to who was involved, possibly at a specified stage of data analytics.
15. The **process** steps were specified until the analysis step. In other words, answering the "How?" (e.g., "the DELPHI method was applied to collect the data"). Such a specification may help the LLM relate the output to the process(es) that took place until the analysis.
16. Specify the **persona/role/avatar** (e.g., "you are a data modeler," "you are a data scientist") [24].

Specifications

Specifications are directly related to analytics designs and analyses to be delivered as **output**.

The specifications in the proposed framework are the following:

1. Specify **no information or specifications** yet ask for designs or provide information and specifications.

2. Specify the **data analytics maturity level** and ask for designs accordingly (descriptive, diagnostic, predictive, prescriptive) [25].
3. Specify the **meta-objective** and ask for designs (e.g., "explore the data," "make predictions").
4. Specify the **data analytics objective (goal)** and ask for designs (ex: Deviation, Correlation, Ranking, Distribution, Change over Time, Part-to-Whole, Magnitude, Spatial, Flow) [26].
5. Specify the **class of the data analytics method** and ask for designs (e.g., summary statistics, visual analytics, machine learning).
6. Specify the **data analytics method** to be applied and ask for designs (e.g., cross-tabular analysis, pie chart, regression, classification).
7. Specify the **data analytics workflow** (data science pipeline) and ask for designs (e.g., text mining workflow and classification workflow).
8. Specify the **data analytics software tool, library, or package** and ask for designs (ex: "using Pytorch library in Python," "using TensorFlow library in Python," "using ggplot2 package for R").
9. Specify the **desired type of patterns and insights** and ask for designs (e.g., gaps, similarities, outliers, historical patterns, predicted future patterns).
10. Specify the **desired type of output** and ask for designs (e.g., analytics dashboard, list of association rules, classification comparison table).
11. Ask to generate a (possibly optimized) **prompt** [27, 28].
12. Ask to generate only **designs, or also analysis** (ex: "generate designs and conduct the analysis").
13. Ask to generate **visualizations** (ex: "generate charts") [29].
14. Ask to **create a set of designs** (i.e., a single design, 10 designs).
15. Ask to explain the **reasoning** behind the design(s) and/or analysis In other words, expecting LLM to answer the "Why?" question.
16. Ask for the **citation of sources** for the designs and/or any aspect of the data analytics process.
17. Ask to **score and rank the designs**.
18. Mention **how many designs** to select.
19. Ask to **generate synthetic data** [30].
20. Ask to **clean the data**.
21. Ask to **convert data types** [30].
22. Ask to **break attributes to their atomic elements** [30] (e.g., based on the date time format, create new attributes year month day hour minute second).
23. Ask to **augment the data** (e.g., "add additional derived attributes based on the existing attributes, providing your method of obtaining those derived attributes"').
24. Ask to **ensure data integrity** [30].
25. Ask to **ensure data quality** [30].
26. Ask to **anonymize the data**.
27. Ask to **handle missing values** [30].
28. Ask to **manage outliers** [30].
29. Ask to **optimize the hyperparameter** values.

30. Ask to design/generate **data analytics workflow** [18].
31. Specify the **experimental design settings** (ex: "apply stratified fivefold cross validation").
32. Ask to generate the **code**.
33. Ask to generate **packaged/running software** (ex: "create the docker file," "create executable jar file that runs the generated Java code").
34. Ask to use **image-to-text to interpret** generated or uploaded image files [23].

Instructions

These instructions are used to modify the prompts to enhance performance. More formally stated, Prompt Instructions are natural language text snippets used to modify analytic-oriented prompts to enhance performance in terms of accuracy, speed, and quality. Such instructions could be structuring the process expected from the LLM, such as "Let's think step by step," "Take a deep breath and work on this problem step-by-step." Other instructions could use motivational techniques, such as in the case of instructions such as "I'm going to tip $xxx for a better solution!" and "You will be penalized," which use positive and negative motivation, respectively.

A list of prompt instructions and their evaluation applicable to any context, including the context of data analytics, can be found in Bsharat et al. [31]. Yang et al. [28] is a notable study that discusses the evaluation of instructions in the context of prompt optimization. The instructions in [28, 31] and other similar studies, which result in empirically verified performance enhancements, can be used to construct data analytics prompts by inserting/appending the selected instruction(s) into the prompt(s).

Prompt Engineering Patterns

These are prompt engineering techniques or patterns used during the overall human-LLM interaction process, including the input of the prompts and the way the interaction is carried out. A list of such patterns, brilliantly represented as infographic, can be found in Benveniste [32]. The prompting patterns included in Benveniste [32] are Zero-shot, MemPrompt, Chain of thoughts, Self-ask, Prompt chaining, Act, ReAct, Few shots, Inception, Memetic proxy, Self-consistency, and Action Plan Generation and Execute.

Many generative AI innovations on a daily basis are related to generative AI and/or prompt engineering. Therefore, rather than providing a static list for this construct of the framework, we recommend that readers refer to an up-to-date list of prompting patterns, possibly generated through LLM tools, as they design their prompts.

One notable study that touches on the issue of systematic prompt construction and interaction within the context of data analytics is Vargas-Solar et al. [33]. The authors list six "conversational meta-patterns" during interaction with their LLM-empowered CHATIN "conversational data exploration" platform. The patterns in Vargas-Solar et al. [33] were similar in nature to the prompt engineering patterns mentioned in the current research.

Conclusion

A novel theoretical framework is proposed in this study to systematically conduct data analytics with LLMs.

The study can be enhanced in the future by generating sample prompts for different combinations of constructs and executing them under various LLM models, and then analyzing their performance through experimentation, usability testing, and formal evaluation [34, 35]. In addition to automated benchmarking, interactive visual analytics can be employed to compare and analyze the outputs obtained from different prompt combinations and LLMs [36].

Besides benchmarking, two critically needed future research areas are the conduct of data analytics with LLM ethically by humans [37] and the ethical conduct/behavior of LLM [38] in the context of data analytics.

Other future research areas can be ideated and planned by brainstorming on the research lines outlined in [7, 8, 15], possibly with the help of LLMs.

Acknowledgements This research was funded by the "CBE Annual Research Program (CARP)" for the academic year 2023-2024. The funding is provided by the College of Business and Economics (CBE) of the United Arab Emirates University (UAEU).

References

1. Freund, K. (2019). *A Cambrian explosion in deep learning*, Part 1. Forbes. Retrieved January 7, 2024 from http://tinyurl.com/2sjbswvf.
2. ChatGPT. Retrieved January 7, 2024 from https://chat.openai.com/.
3. OpenAI. Retrieved January 7, 2024 from https://openai.com/.
4. United Arab Emirates, Minister of State for Artificial Intelligence, Digital Economy & Remote Work Applications Office. (2023). 100 practical applications and use cases of generative AI. Retrieved January 7, 2024 from http://tinyurl.com/2r2rxv2u.
5. Allied Market Research. Retrieved January 14, 2024 from http://tinyurl.com/3674f6nk.
6. Oxford Learner's Dictionaries. Retrieved January 7, 2024 from http://tinyurl.com/3x4j6rew.
7. Zhao, W. X., Zhou, K., Li, J., Tang, T., Wang, X., Hou, Y., Min, Y., & Wen, J. R. (2023). A survey of large language models. arXiv:2303.18223.
8. Wu, T. Y., He, S. Z., Liu, J. P., Sun, S. Q., Liu, K., Han, Q. -L., Tang, Y. (2023) A brief overview of ChatGPT: The history, status quo and potential future development. *IEEE/CAA Journal of Automatica Sinica, 10*(5), 1122–1136. https://doi.org/10.1109/JAS.2023.123618.
9. Fui-Hoon Nah, F., Zheng, R., Cai, J., Siau, K., & Chen, L. (2023). Generative AI and ChatGPT: Applications, challenges, and AI-human collaboration. *Journal of Information Technology*

Case and Application Research, 25(3), 277–304. https://doi.org/10.1080/15228053.2023.2233814.
10. Tsai, C. W., Lai, C. F., Chao, H. C., & Vasilikos, A. V. (2015) Big data analytics: a survey. *Journal of Big Data, 2*(21). https://doi.org/10.1186/s40537-015-0030-3.
11. Runkler, T. A. (2020). *Data analytics*. Springer Fachmedien Wiesbaden.
12. Nasseri, M., Brandtner, P., Zimmermann, R., Falatouri, T., Darbanian, F., & Obinwanne, T. (2023) Applications of large language models (LLMs) in business analytics–exemplary use cases in data preparation tasks. In H. Degen, S. Ntoa, A. Moallem (Eds.), *International conference on human-computer interaction. HCI international 2023–late breaking papers. HCII 2023. Lecture notes in computer science* (vol 14059, pp 182–198). Springer, Cham. https://doi.org/10.1007/978-3-031-48057-7_12.
13. Almheiri, S. A. A., AlAnsari, M., AlHashmi, J., Abdalmajeed, N., Jalil, M., Ertek, G. Supplement to 'data analytics with large language models (LLM): A novel prompting framework.' Retrieved January 15, 2024 from https://ertekprojects.com/ftp/supp/23.pdf.
14. Floridi, L., & Chiriatti, M. (2020). GPT-3: Its nature, scope, limits, and consequences. *Minds and Machines, 30*, 681–694. https://doi.org/10.1007/s11023-020-09548-1.
15. Ray, P. P. (2023). ChatGPT: A comprehensive review on background, applications, key challenges, bias, ethics, limitations and future scope. *Internet of Things and Cyber-Physical Systems, 3*, 121–154. https://doi.org/10.1016/j.iotcps.2023.04.003.
16. Gharehchopogh, F. S., Khalifelu, Z. A. (2011). Analysis and evaluation of unstructured data: text mining versus natural language processing. In 2011 5th international conference on application of information and communication technologies (AICT), pp. 1–4. IEEE. https://doi.org/10.1109/icaict.2011.6111017.
17. Stroganov, O., Schedlbauer, A., Lorenzen, E., Kadhim, A., Lobanova, A., Lewis, D. A., & Glausier, J. R. (2023). Unpacking unstructured data: A pilot study on extracting insights from neuropathological reports of Parkinson's disease patients using large language models. bioRxiv, pp. 2023–09. https://doi.org/10.1101/2023.09.12.557252.
18. Duque, A., Syed, A., Day, K. V., Berry, M. J., Katz, D. S., & Kindratenko, V. V. (2023). Leveraging large language models to build and execute computational workflows. arXiv:2312.07711. https://arxiv.org/pdf/2312.07711.pdf.
19. Brown, T., Mann, B., Ryder, N., Subbiah, M., Kaplan, J. D., Dhariwal, P., & Amodei, D. (2020). Language models are few-shot learners. In H. Larochelle, M. Ranzato, R. Hadsell, M. F. Balcan, H. Lin (Eds.), *Advances in neural information processing systems* 33 (Curran Associates, Inc., 2020), 1877–1901. arXiv:2005.14165.
20. Spaulding, T., Hassler, E., Edwards, C., & Cazier, J. (2020). Sandwich analytics: a dataset comprising one year's weekly sales data correlated with crime, demographics, and weather, mendeley data, V2. Retrieved January 14, 2024 from https://doi.org/10.17632/6htjnfs78b.2.
21. Spaulding, T. J., Hassler, E. E., Edwards, C. H., & Cazier, J. A. (2019). Sandwich analytics: A dataset comprising one year's weekly sales data correlated with crime, demographics, and weather. *Data in Brief, 25*, 104252.
22. Data in Brief. Retrieved January 14, 2024 from https://www.data-in-brief.com/.
23. Zhu, D., Chen, J., Shen, X., Li, X., Elhoseiny, M. (2023). Minigpt-4: Enhancing vision-language understanding with advanced large language models. arXiv:2304.10592. https://arxiv.org/pdf/2304.10592.pdf.
24. Wolfram Prompt Repository. Retrieved January 14, 2024 from https://resources.wolframcloud.com/PromptRepository.
25. Gartner. What Is Data and Analytics? Retrieved January 12, 2024 from https://www.gartner.com/en/topics/data-and-analytics.
26. Visual Vocabulary. Retrieved January 12, 2024 from https://public.tableau.com/views/VisualVocabulary/VisualVocabulary.
27. Pryzant, R., Iter, D., Li, J., Lee, Y. T., Zhu, C., & Zeng, M. (2023). Automatic prompt optimization with" gradient descent" and beam search. arXiv:2305.03495. https://arxiv.org/pdf/2305.03495v1.pdf.

28. Yang, C., Wang, X., Lu, Y., Liu, H., Le, Q. V., Zhou, D., & Chen, X. (2023). Large language models as optimizers. arXiv:2309.03409. https://arxiv.org/pdf/2309.03409.pdf.
29. Tian, Y., Cui, W., Deng, D., Yi, X., Yang, Y., Zhang, H., Wu, Y. (2023). ChartGPT: Leveraging LLMs to generate charts from abstract natural language. arXiv:2311.01920. https://arxiv.org/pdf/2311.01920.pdf.
30. Lingo, R. (2023). The role of ChatGPT in democratizing data science: An exploration of AI-facilitated data analysis in telematics. arXiv:2308.02045. https://arxiv.org/pdf/2308.02045.pdf.
31. Bsharat, S. M., Myrzakhan, A., & Shen, Z. (2023). Principled instructions are all you need for questioning LLaMA-1/2, GPT-3.5/4. Retrieved January 12, 2024 from arXiv:2312.16171. https://arxiv.org/pdf/2312.16171v1.pdf.
32. Benveniste, D. (2023). Prompt engineering and LLMOps: Building LLM applications. Retrieved January 12, 2024 from http://tinyurl.com/5f9eezbb.
33. Vargas-Solar, G., Cerquitelli, T., Espinosa-Oviedo, J. A., Cheval, F., Buchaille, A., & Polgar, L. (2023). Conversational data exploration: A game-changer for designing data science pipelines. arXiv-2311. https://arxiv.org/pdf/2311.06695.pdf.
34. Guo, Z., Jin, R., Liu, C., Huang, Y., Shi, D., Yu, L., Xiong, D. (2023). Evaluating large language models: A comprehensive survey. arXiv:2310.19736. https://arxiv.org/pdf/2310.19736.pdf.
35. Chang, Y., Wang, X., Wang, J., Wu, Y., Zhu, K., Chen, H., & Xie, X. (2023). A survey on evaluation of large language models. arXiv:2307.03109. https://arxiv.org/pdf/2307.03109.pdf.
36. Sevastjanova, R., Vogelbacher, S., Spitz, A., Keim, D., El-Assady, M. (2023). Visual comparison of text sequences generated by large language models. In 2023 IEEE visualization in data science (VDS), pp. 11–20. IEEE.
37. Mozes, M., He, X., Kleinberg, B., & Griffin, L. D. (2023). Use of LLMs for illicit purposes: Threats, prevention measures, and vulnerabilities. arXiv:2308.12833. https://arxiv.org/pdf/2308.12833.pdf.
38. Park, P. S., Goldstein, S., O'Gara, A., Chen, M., & Hendrycks, D. (2023). AI deception: A survey of examples, risks, and potential solutions. Retrieved January 14, 2024 from arXiv: 2308.14752 (2023). https://arxiv.org/pdf/2308.14752.pdf.

Chapter 21
Instantaneous and Limiting Behavior of an n-Node Blockchain Under Cyber Attacks from a Single Hacker

Xiufeng Xu and Liang Hong

Abstract We investigate the instantaneous and limiting behavior of an n-node blockchain which is under continuous monitoring of the IT department of a company but faces non-stop cyber attacks from a single hacker. The blockchain is functional as far as no data stored on it has been changed, deleted, or locked. Once the IT department detects the attack from the hacker, it will immediately re-set the blockchain, rendering all previous efforts of the hacker in vain. The hacker will not stop until the blockchain is dysfunctional. For arbitrary distributions of the hacking times and detecting times, we derive the limiting functional probability, instantaneous functional probability, and mean functional time of the blockchain. We also show that all these quantities are increasing functions of the number of nodes, substantiating the intuition that the more nodes a blockchain has, the harder it is for a hacker to succeed in a cyber attack.

Keywords Cyber ransom attacks · Cyber destructive attacks · Instantaneous functional probability · Limiting functional probability · Mean functional times · Operations research

Introduction

In today's society, exchanges of information between companies and customers take various forms via the Internet. This mode of interaction exposes a company to potential cyber attacks. While shutting down the computer network is not a viable option for any business, the loss due to a cyber attack could be astronomical. For example, the Colonial Pipeline ransomware attack in 2021 costed the company 4.4 million dollars in ransom payment, let alone many other indirect costs. Quite a few other

X. Xu · L. Hong (✉)
Department of Mathematical Sciences, The University of Texas at Dallas, 800 West Campbell Road, Richardson, TX 75080, USA
e-mail: liang.hong@utdallas.edu

X. Xu
e-mail: xiufeng.xu@utdallas.edu

© The Author(s), under exclusive license to Springer Nature Switzerland AG 2024

A. Emrouznejad et al. (eds.), *Business Analytics and Decision Making in Practice*, Lecture Notes in Operations Research, https://doi.org/10.1007/978-3-031-61589-4_21

costly cyber attacks are still as fresh as in recent memory. Therefore, a company faces the challenge of safeguarding itself from cyber attacks and maintaining its normal operations at the same time. One plausible approach is to back up a company's data on a continuous basis. However, such a practice is prohibitively expensive to most companies. The blockchain is one efficient alternative that can greatly reduce the likelihood of cyber losses. Generally speaking, a blockchain consists of several identical computers in a network; each of these computers is called a *node* and stores crucial data for a company's daily operation. Different from a traditional computer network where data is housed on a centralized server, the blockchain stores identical data on each of its nodes, and these nodes continuously verify their data according to the majority rule. That is, any piece of data on a node is considered valid only if it is consistent with its counterparts on at least half of the nodes of the blockchain. Thanks to the public key encryption (e.g. [1–4]), it is basically useless for a hacker to steal the data stored on the blockchain. Therefore, most hackers would either make a malicious attack such as changing or deleting the data, or make a ransom attack by locking data on the blockchain and demand a ransom from the company. For the former type of cyber attack, which will be referred to as a *destructive attack*, a hacker must hack into at least half of the nodes before he can execute his plan. For the latter type of cyber attack, which will be called a *ransom attack*, a hacker must hack into all nodes and lock the entire blockchain. Intuitively, a blockchain is much harder to hack than a traditional centralized server; see Section "$P_m(t)$ and $E[T_m]$ as Functions of m" for a rigorous result. No wonder blockchain technology has been used in a wide range of applications such as cryptocurrencies, smart contracts, and P2P insurance; for a comprehensive review of elements of blockchains and their applications, see [5–11], and references therein.

Despite the enhanced cyber security it provides, a blockchain is not immune to cyber attacks. In particular, it is still possible for a hacker to breach the security of a blockchain. Various authors have studied security issues of a blockchain; see, for instance, [12–14]. However, no paper in the literature has investigated the operations-research-theoretic aspect of a blockchain when it is under cyber attacks. The current paper intends to serve as the first step towards filling this lacuna in the literature. In this article, we study an n-node blockchain under the cyber attack from one hacker when it is under continuous monitoring of the IT department of a company. The blockchain is considered to be functional as long as no data stored on it has been modified, deleted, or locked. However, the IT department will re-set the blockchain if it detects the attack from the hacker; if this occurs, all previous efforts of the hacker will be totally forfeited and the hacker will start over again. The hacker will not stop as far as the blockchain remains functional. Assuming the detecting time and the hacking time follow arbitrary distributions, we obtain the limiting functional probability, instantaneous functional probability, and mean functional time of the blockchain. We also show that all these quantities are non-decreasing functions of the number of nodes of the blockchain. To our knowledge, this is the first time that rigorous results have been given to back up the intuition that the more nodes a blockchain has, the harder it is for a hacker to break into.

The remainder of the paper is organized as follows. In Section "Model Setup and Notation", we give a detailed description of our model and establish notational convention. Next, in Section "Stochastic Behavior of the Proposed Blockchain Model", we investigate the stochastic behavior of our proposed blockchain model. In particular, we derive formulas for the limiting functional probability, the instantaneous functional probability, and the mean functional time of the blockchain. We also establish rigorous results to confirm the intuition that the more nodes a blockchain has, the less likely it will be hacked. In Section "Example", we provide an example based on simulation. Finally, we conclude the paper with some remarks in Section "Concluding Remarks". The python code for the example in this paper can be found at the following website: https://github.com/xuxiufeng/Blockchain_Simulations.

Model Setup and Notation

We consider a blockchain consisting of n identical nodes in a given company. Without loss of generality, we assume that $n \geq 2$. From the moment it starts to operate, the blockchain is under 24-hour continuous monitoring by the IT department of the company for any potential cyber attack. There are two major types of cyber attacks for blockchains: (i) *cyber destructive attacks* in which a hacker invades the blockchain and change or/and delete the data stored on the blockchain and (ii) *cyber ransom attacks* where a hacker intrudes into the blockchain, locks the data stored the blockchain, and demands ransoms from the company. In this paper, we are mainly interested in the instantaneous and limiting behavior of the n-node blockchain under these two types of cyber attacks from one single hacker. We assume that blockchain is under immediate cyber attacks from the hacker the moment it starts to operate. For a hacker to change or delete the data on an n-node blockchain, he must hack into at least $m = \lfloor n/2 \rfloor + 1$ nodes. Without loss of generality, we assume that a hacker will start the malicious attack once he succeeds in hacking into m nodes, and that he hacks these m nodes in sequential order. That is, once the hacker penetrates the firewall of one node, he will hold onto it and immediately start hacking the next node. However, if the IT department of the company detects that the blockchain is under attack, they will immediately re-set the whole blockchain so that all previous efforts of the hacker are forfeited. Therefore, the hacker will not be able to change or delete data on the blockchain until he hacks into the m-th node without being detected. We assume that the hacker will perform a cyber destructive attack the moment he succeeds in hacking into m nodes without being detected. Similarly, if the hacker intends to conduct a cyber ransom attack, he will do so immediately after he hacks into the n-th node without being detected. On the other hand, each time the IT department re-sets the blockchain, there will be associated costs due to disruption of business operation. For this reason, the IT department will only re-set the blockchain when they detect that a cyber attack against the blockchain is under way.

Without loss of generality, we label the n nodes of the blockchain as node 1, nodes 2, ..., node n. We assume that it takes a random amount of time X_i for the hacker to hack into node i, $(i = 1, \ldots, n)$, and that $X_1, \ldots, X_n$ are independent and identically distributed (iid) according to a distribution function F_X with a probability density function $f_X(x)$. We also assume that it takes a random amount of time Y for the IT department to detect that the blockchain is under a cyber attack from the moment it starts to operate or has just been re-set, that Y has a distribution function F_Y with a probability density function $f_Y(y)$, and that Y is independent from $X_1, \ldots, X_n$. We call each X_i a *hacking time* and Y a *detecting time*.

The stochastic behavior of the blockchain under a cyber destructive attack can be described as follows. At time $t = 0$, the blockchain is put into operation and immediately under continuous monitoring for cyber attacks. At the same time, a hacker starts to work on bypassing the firewall of a node. Since all n nodes are identical, we may assume that the hacker attacks the nodes of the blockchain according to the ascending order of their labels. More specifically, the hacker attacks node 1 first. If $X_1 < Y$, i.e., the hacker penetrates the firewall of node 1 without being detected, then he will immediately attack node 2. On the other hand, if $X_1 \geq Y$, i.e., the hacker fails to bypass the firewall of node 1 before being detected, the blockchain will be re-set and he has to start over. In general, if $\sum_{i=1}^{m} X_i < Y$, i.e., the hacker is able to hack into m nodes before being detected, he will perform a cyber destructive attack. But, if $\sum_{i=1}^{m} X_i \geq Y$, i.e., he is being detected before he is able to change any data on the blockchain, the blockchain will be re-set, rendering all previous efforts of the hacker in vain. As far as the hacker has not been able to change or delete any data on the blockchain, we say the blockchain is *functional*. The stochastic behavior of the blockchain under a cyber ransom attack can be described in a similar manner by replacing m with n. Therefore, we will focus only on the case of a cyber destructive attack. However, all results extend to the case of a cyber ransom attack if we replace m with n.

We are mainly interested in the limiting behavior and the instantaneous behavior of an n-node blockchain in the aforementioned setup. Recall that $m = \lfloor n/2 \rfloor + 1$. For a given m, we put

$$P_m(\infty) = P\{\text{the blockchain will never be hacked}\}. \tag{1}$$

We say $P_m(\infty)$ is the *limiting functional probability* of an n-node blockchain. We also define T_m to be the total amount of time it takes for the hacker to hack into the blockchain, and call $E[T_m]$ the *mean functional time* of the blockchain. Moreover, for $t > 0$, we define

$$P_m(t) = P\{\text{the blockchain has not been hacked at time t}\}. \tag{2}$$

We call $P_m(t)$ the *instantaneous functional probability* of the blockchain. It follows from the continuity of the probability measure (e.g. [15], Lemma 1.14) that the limiting functional probability and the instantaneous functional probability can be related as follows

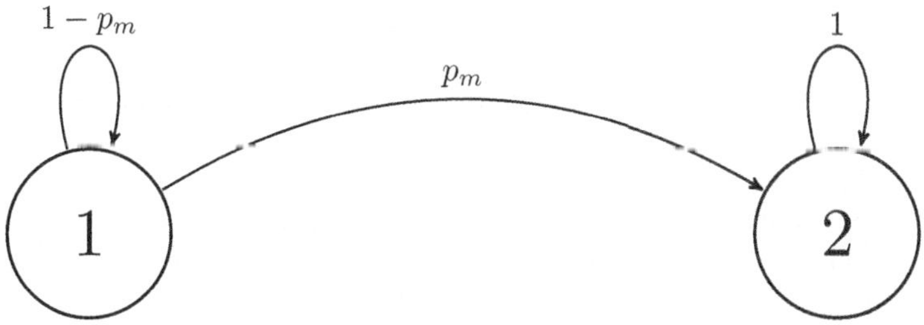

Fig. 21.1 Two-state discrete Markov chain

$$P_m(\infty) = \lim_{t \to \infty} P_m(t).$$

Stochastic Behavior of the Proposed Blockchain Model

Limiting Functional Probability and Mean Functional Time

To investigate the limiting behavior of the blockchain, we define a *cycle* to be the period from the moment the blockchain starts afresh or has just been re-set to the next moment it is being re-set. That is, a cycle is a period from the inception of a detecting time to its end, provided the blockchain has not been hacked. Let N_1 be the number of cycles needed for the hacker to succeed in hacking into the blockchain. Then N_1 is a geometric random variable[1] with parameter p_m given by

$$p_m = P\left\{Y > \sum_{i=1}^{m} X_i\right\} = \int_0^{\infty} F_X^{(m)}(s)dF_Y(s), \tag{3}$$

where $F_X^{(m)}$ is the m-folded convolution of F_X. Note that $p_m > 0$ since we assume both F_X and F_Y admit a probability density function. If we restrict our attention to the epochs the blockchain starts afresh, or has just been re-set, or has just been hacked, then we obtain an embedded discrete-time process. This discrete-time process is a two-state discrete-time Markov chain. In State 1, the blockchain is functional, but in State 2, the blockchain is dysfunctional, i.e., it has been hacked. A diagram of this two-state discrete-time Markov chain is given in Fig. 21.1.

Note that State 2 is an absorbing state. It is clear that the transition probability from State 1 to State 2 is exactly the probability p_m given by (3). Conditioning on the outcome of the first cycle, we have

[1] Here we interpret a geometric random variable with parameter $0 < p < 1$ to be the number of failures (rather than the number of trials) until the first success among a sequence of independent Bernoulli trials with success probability p.

$$P_m(\infty) = p_m \times 0 + (1 - p_m) \times P_m(\infty),$$

implying that $P_m(\infty) = 0$, which is in agreement with our intuition: if we observe the stochastic behavior of the blockchain over an infinite horizon of time, then it will eventually be hacked because $p_m > 0$.

Next, we look at the mean functional time of the blockchain. To this end, we focus only on the moments every possible cycle starts or ends. By doing so, we identify a renewal process $\{N_2(t)\}_{t\geq 0}$ where

$$N_2(t) = \max\left\{k : \sum_{i=1}^{k} Y_i \leq t\right\},$$

and $Y_i \overset{d}{=} Y \mid Y < \sum_{i=1}^{m} X_i$.

$$T_m = \sum_{i=1}^{N_1} Y_i + \left(\sum_{i=1}^{m} X_i | \sum_{i=1}^{m} X_i < Y\right). \tag{4}$$

Also, it is clear that N_1 is independent of $Y_{n+1}, Y_{n+2}, \ldots$, which shows N_1 is a stopping time for the sequence $\{Y_i\}_{i\geq 1}$. Therefore, Wald's Identity implies that

$$\begin{aligned} E[T_m] &= E[N_1]E[Y_1] + E\left[\sum_{i=1}^{m} X_i | \sum_{i=1}^{m} X_i < Y\right] \\ &= E[N_1]E\left[Y|Y < \sum_{i=1}^{m} X_i\right] + E\left[\sum_{i=1}^{m} X_i | \sum_{i=1}^{m} X_i < Y\right]. \end{aligned} \tag{5}$$

Instantaneous Functional Probability

To obtain the instantaneous functional probability of the blockchain, we let S_t be the moment the last time the blockchain is re-set before time t:

$$S_t = \sum_{i=1}^{N_2(t)} Y_i,$$

and F_{S_t} be its distribution function. Then for $t > 0$, we have {the blockchain is functional at t}

$$\begin{aligned} P_m(t) &= P\{\text{the blockchain is functional at t} \mid S_t = 0\}P\{S_t = 0\} \\ &+ \int_0^t P\{\text{the blockchain is functional at t} \mid S_t = s\}dF_{S_t}(s) \end{aligned}$$

$$= P\{t < \sum_{i=1}^{m} X_i \wedge Y\}$$

$$+ \int_0^t P\{\text{the blockchain is functional at t} \mid S_t = s\} dF_{S_t}(s).$$

For $0 < s < t$, we have

$$P\{\text{the blockchain is functional at } t \mid S_t = s\} = P\{Y \wedge \sum_{i=1}^{m} X_i > t - s \mid Y_1 > t - s\},$$

and $dF_{S_t}(s) = P\{Y_1 > t - s\}dG(s)$ where $G(s) = \sum_{i=1}^{\infty} F_{Y_i}(s)$.

Therefore,

$$P_m(t) = P\{t < \sum_{i=1}^{m} X_i \wedge Y\} + \int_0^t P\{Y \wedge \sum_{i=1}^{m} X_i > t - s\} dG(s). \qquad (6)$$

It is clear that the first term in (6) goes to 0 as $t \to \infty$. For the second term, we have

$$\begin{aligned}
0 &\le \lim_{t\to\infty} \int_0^t P\{Y \wedge \sum_{i=1}^{m} X_i > t - s\} dG(s) \\
&= \lim_{t\to\infty} \int_0^\infty P\{Y \wedge \sum_{i=1}^{m} X_i > t - s\} 1_{[0,t]}(s) dG(s) \\
&\le \lim_{t\to\infty} \int_0^\infty P\{Y \wedge \sum_{i=1}^{m} X_i > t - s\} dG(s) \\
&= \int_0^\infty \lim_{t\to\infty} P\{Y \wedge \sum_{i=1}^{m} X_i > t - s\} dG(s) = 0,
\end{aligned}$$

where the last equality follows from the monotone convergence theorem for sequences of decreasing measurable functions. This gives another proof that $P_m(\infty) = 0$.

$P_m(t)$ and $E[T_m]$ as Functions of m

Intuitively, the more nodes a blockchain has, the more secure it is. That is, $P_n(t)$, $P_n(\infty)$, and $E[T_n]$ should all be non-decreasing functions of the number of nodes n, and they should be increasing functions of n if the increment of the number of nodes is at least 2. The next theorem confirms this intuition.

Theorem 1 *$P_m(t)$, $P_m(\infty)$, and $E[T_m]$ are all increasing functions of m, where $m = \lfloor n/2 \rfloor + 1$.*

Proof First, it is obvious that

$$P\{t < Y \wedge \sum_{i=1}^{m+1} X_i\} > P\{t < Y \wedge \sum_{i=1}^{m} X_i\},$$

and

$$P\{t - s < Y \wedge \sum_{i=1}^{m+1} X_i\} > P\{t - s < Y \wedge \sum_{i=1}^{m} X_i\}$$

because each hacking time X_i is positive almost surely. Therefore, (6) implies

$$P_{m+1}(t) > P_m(t), \quad \text{for all positive integer } m. \tag{7}$$

Next, we write (5) as

$$\begin{aligned} E[T_m] &= \frac{1-p_m}{p_m} \times \frac{E[Y 1_{\{Y<\sum_{i=1}^m X_i\}}]}{1-p_m} + \frac{E[(\sum_{i=1}^m X_i) 1_{\{Y>\sum_{i=1}^m X_i\}}]}{p_m} \\ &= \frac{E[Y 1_{\{Y<\sum_{i=1}^m X_i\}}]}{p_m} + \frac{E[(\sum_{i=1}^m X_i) 1_{\{Y>\sum_{i=1}^m X_i\}}]}{p_m}. \end{aligned}$$

Therefore,

$$\begin{aligned} &E[T_{m+1}] - E[T_m] \\ &= (p_m p_{m+1})^{-1} \Bigg\{ p_m E[Y 1_{\{Y<\sum_{i=1}^{m+1} X_i\}}] - p_{m+1} E[Y 1_{\{Y<\sum_{i=1}^{m} X_i\}}] \\ &+ p_m E\left[\left(\sum_{i=1}^{m+1} X_i\right) 1_{\{Y>\sum_{i=1}^{m+1} X_i\}}\right] - p_{m+1} E\left[\left(\sum_{i=1}^{m} X_i\right) 1_{\{Y>\sum_{i=1}^{m} X_i\}}\right] \Bigg\}. \end{aligned}$$

The four terms in the brackets can be written as

$$\begin{aligned} &p_m E[Y 1_{\{Y<\sum_{i=1}^m X_i\}}] - p_{m+1} E[Y 1_{\{Y<\sum_{i=1}^m X_i\}}] \\ &+ p_m E[Y 1_{\{\sum_{i=1}^m X_i < Y < \sum_{i=1}^{m+1} X_i\}}] \\ &+ p_m E[X_{m+1} 1_{\{Y>\sum_{i=1}^{m+1} X_i\}}] + p_m E\left[\left(\sum_{i=1}^{m} X_i\right) 1_{\{Y>\sum_{i=1}^{m+1} X_i\}}\right] \\ &- p_{m+1} E\left[\left(\sum_{i=1}^{m} X_i\right) 1_{\{Y>\sum_{i=1}^{m+1} X_i\}}\right] \end{aligned}$$

$$
\begin{aligned}
&-p_{m+1}E\left[\left(\sum_{i=1}^{m}X_i\right)1_{\{\sum_{i=1}^{m+1}X_i>Y>\sum_{i=1}^{m}X_i\}}\right]\\
&=(p_m-p_{m+1})E[Y1_{\{Y<\sum_{i=1}^{m}X_i\}}]+(p_m-p_{m+1})E\left[\left(\sum_{i=1}^{m}X_i\right)1_{\{Y>\sum_{i=1}^{m+1}X_i\}}\right]\\
&+p_mE[Y1_{\{\sum_{i=1}^{m+1}X_i>Y>\sum_{i=1}^{m}X_i\}}]\\
&-p_{m+1}E\left[\left(\sum_{i=1}^{m}X_i\right)1_{\{\sum_{i=1}^{m+1}X_i>Y>\sum_{i=1}^{m}X_i\}}\right]\\
&+p_mE[X_{m+1}1_{\{Y>\sum_{i=1}^{m+1}X_i\}}]\\
&\geq(p_m-p_{m+1})E[Y1_{\{Y<\sum_{i=1}^{m}X_i\}}]+(p_m-p_{m+1})E\left[\left(\sum_{i=1}^{m}X_i\right)1_{\{Y>\sum_{i=1}^{m+1}X_i\}}\right]\\
&+p_mE\left[\left(\sum_{i=1}^{m}X_i\right)1_{\{\sum_{i=1}^{m+1}X_i>Y>\sum_{i=1}^{m}X_i\}}\right]\\
&-p_{m+1}E\left[\left(\sum_{i=1}^{m}X_i\right)1_{\{\sum_{i=1}^{m+1}X_i>Y>\sum_{i=1}^{m}X_i\}}\right]\\
&+p_mE[X_{m+1}1_{\{Y>\sum_{i=1}^{m+1}X_i\}}]\\
&=(p_m-p_{m+1})E[Y1_{\{Y<\sum_{i=1}^{m}X_i\}}]+(p_m-p_{m+1})E\left[\left(\sum_{i=1}^{m}X_i\right)1_{\{Y>\sum_{i=1}^{m+1}X_i\}}\right]\\
&+(p_m-p_{m+1})E\left[\left(\sum_{i=1}^{m}X_i\right)1_{\{\sum_{i=1}^{m+1}X_i>Y>\sum_{i=1}^{m}X_i\}}\right]\\
&+p_mE[X_{m+1}1_{\{Y>\sum_{i=1}^{m+1}X_i\}}].
\end{aligned}
$$

It is evident from (3) that p_m is decreasing in m. Therefore, $E[T_{m+1}]-E[T_m]>0$.

In view of the above results, one would expect it to be impossible for a blockchain to be hacked when the number of nodes $n\to\infty$. The next theorem confirms this intuition. (Note that $n\to\infty$ if and only if $m=\lfloor n/2\rfloor+1\to\infty$.)

Theorem 2 $\lim_{m\to\infty}P_m(t)=1$ *and* $\lim_{m\to\infty}E[T_m]=\infty$, *where* $m=\lfloor n/2\rfloor+1$.

Proof Since each X_i is positive almost surely, (6) implies

$$\lim_{m\to\infty}P_m(t)=1. \tag{8}$$

Next, (3) implies $p_m\to 0$ as $m\to\infty$. Therefore, $E[N_1]=1/p_m-1\to\infty$ as $m\to\infty$. It follows from (5) that

$$\begin{aligned}\lim_{m\to\infty} E[T_m] &\geq \lim_{m\to\infty} E[N_1]E\left[Y|Y<\sum_{i=1}^{m} X_i\right]\\ &= \lim_{m\to\infty} E[N_1] \lim_{m\to\infty}\left[\frac{E[Y1_{\{Y<\sum_{i=1}^{m} X_i\}}]}{P\left\{Y<\sum_{i=1}^{m} X_i\right\}}\right]\\ &= \lim_{m\to\infty} E[N_1]E[Y]\\ &= \infty.\end{aligned}$$

Cost-Benefit Analysis

Suppose $C_1(m)$ is the cost incurred each time the blockchain is re-set, $C_2(m)$ is the cost per unit time for running the blockchain, and R is the revenue earned per unit time by running the blockchain. Note that $C_1(m)$ is incurred exactly once in each cycle because the blockchain is re-set exactly once in each cycle. By Wald's Identity, the expected number of times $C_1(m)$ is incurred in one unit time equals $1/E[Y]$. Therefore, the expected net revenue (i.e. expected profit) per unit time, denoted as $E_m[NR]$, can be written as

$$E_m[NR] = R - C_2(m) - \frac{C_1(m)}{E[Y]}. \tag{9}$$

In practice, $C_2(m)$ is an increasing function of m, but $1/E[Y]$ is a decreasing function of m. While an optimal m that maximizes the net revenue per unit time does not necessarily exist, one can always use (9) to perform a cost-benefit analysis.

Example

Example 1 Gamma hacking- and Weibull detecting times In this example, we assume that $Y \sim \text{Weibull}(\alpha, \beta)$. That is, Y has the probability density function

$$f(y) = \frac{\beta}{\alpha}\left(\frac{y}{\alpha}\right)^{\beta-1} e^{-(y/\alpha)^{\beta}}, \quad y > 0.$$

We also assume that $X_1, \ldots, X_n \overset{iid}{\sim} Gamma(\eta, \delta)$, and that Y is independent from $X_1, \ldots, X_n$. Then $\sum_{i=1}^{m} X_i \sim Gamma(m\eta, \delta)$.

Based on the formulas derived in the previous section, we perform a simulation study to estimate $E[T_m]$ and $P_m(t)$ as functions of m and t respectively. For $\eta = 0.02$, $\delta = 200$, $\alpha = 110$, and $\beta = 150$, we perform $N_1 = 130,000$ iterations to estimate each $E[T_m]$ for $m = 1, 2, ..., 40$. For each $i = 1, 2, \ldots N_1$, we generate an independent random amount of time T_m^i for a successful cyber destructive attack. Then we estimate $E[T_m]$ as $\sum_{i=1}^{N_1} T_m^i/N_1$. Similarly, we perform $N_2 = 80,000$ iter-

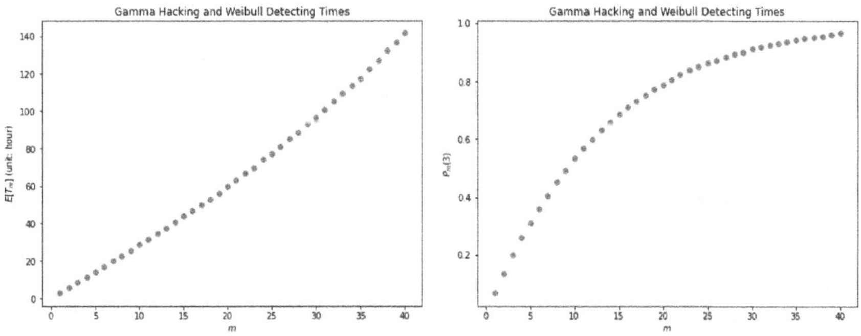

(a) $E(T_m)$ as a function of m (b) $P_m(3)$ as a function of m

Fig. 21.2 Plots of $E[T_m]$ and $P_m(3)$ under gamma hacking- and weibull detecting times

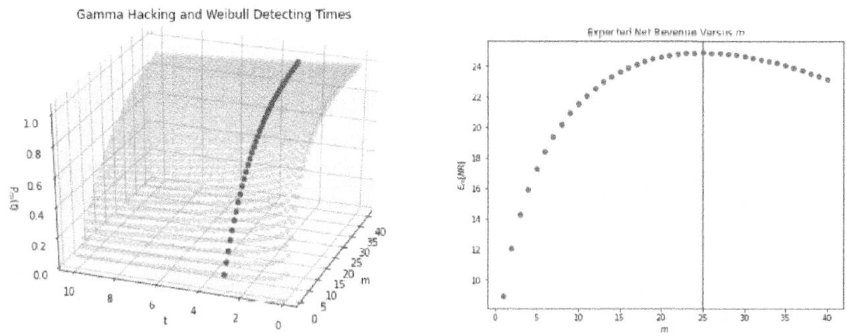

(a) $P_m(t)$ as a bivariate function of m and t (b) $E_m(NR)$ as a function of m

Fig. 21.3 Plots of $P_m(t)$ and $E_m(NR)$ under Gamma hacking- and Weibull detecting times

ations to estimate each $P_m(3)$ for $m = 1, 2, ..., 40$. For each $j = 1, 2, \ldots, N_2$, we set $W_j = 1$ if the blockchain is functional at $t = 5$ and $W_j = 0$ otherwise. Then we take $\sum_{j=1}^{N_2} W_j/N_2$ to be the estimate of $P_m(3)$. Figure 21.2 plots $E[T_m]$ and $P_m(3)$ as functions of m.

Panel (a) of Fig. 21.3 provides a 3D plot of $P_m(t)$ as a bivariate function of m and t, where the blue dot scatter plot is the 3D version of the plot in Panel (b) of Fig. 21.2. Once again, we see that $P_m(t)$ is non-decreasing in m for a fixed $t > 0$ and non-decreasing in t for a fixed m.

To perform the cost-benefit analysis, we choose $R = 10\sqrt{m}$, $C_1 = 6m^{0.2}$, and $C_2 = m$. Then $E_m[NR]$ equals

$$E_m[NR] = 10\sqrt{m} - m - \frac{6m^{0.2}}{E[Y]}.$$

Here an optimal m exists. The red vertical line in Panel (b) of Fig. 21.3 indicates this optimal m equals 25 at which the expected net revenue achieves its maximum value 24.90.

Concluding Remarks

This paper investigates the instantaneous and limiting behavior of an n-node blockchain under cyber attacks from a single hacker. Under arbitrary hacking and detecting times, we have derived the formulas for the limiting functional probability, instantaneous functional probability, and mean functional time. We have also established that these quantities are all non-decreasing functions of the number of nodes. These results substantiate the intuition that increasing the number of nodes can effectively improve the security of a blockchain. To illustrate our results, we also provided an example to illustrate our results.

While blockchain is clearly a "hot topic" that attracts researchers from various fields, it seems that the operations-research-theoretic aspect of blockchains has rarely been explored. This paper aims to stimulate serious interest along this line of research. For future research, one interesting direction is to investigate the operations-research-theoretic aspect of the blockchain when it faces cyber attacks from several hackers.

References

1. Diffie, W., & Hellman, M. E. (1976). New directions in cryptography. *IEEE Transactions on Information Theory, 22*(5), 644–654.
2. Rivest, R. L., Shamir, A., & Adelman, L. (1978). A method for obtaining digital signatures and public-key cryptosystems. *Communications of the ACM, 21*(2), 120–126.
3. Goldwasser, S., & Micali, S. (1982). Probabilistic encryption. *Journal of Computer System and Sciences, 28*(2), 270–299.
4. Goldwasser, S., Micali, S., & Rivest, R. L. (1988). A digital signature scheme secure against adaptive chosen-message attacks. *SIAM Journal on Computing, 18*(1), 283–308.
5. Tama, B. A., Kweka, B. J., Park, Y., & Rhee, K. H. (2017). A critical review of blockchain and its current applications. In *2017 International conference on electrical engineering and computer science IEEE*, (pp. 109–113).
6. Zheng, Z., Xie, S., Dai, H. N., & Wang, H. (2018). Blockchain challenges and opportunities: A survey. *International Journal of Web and Grid Services, 14*(4), 352–375.
7. Casino, F., Dasaklis, T. K., & Patsakis, C. (2019). A systematic literature review of blockchain-based applications: Current status, classification and open issues. *Telematics and Informatics, 36*, 55–81.
8. Kumar, A., Abhishek, K., Ghalib, M. R., Nerurkar, P., Bhirud, S., Alnumay, W., Kumar, S. A., Chatterjee, P., & Ghosh, U. (2020). Securing logistics system and supply chain using Blockchain. *Applied Stochastic Models in Business and Industry, 37*(3), 413–428.
9. Yang, J., Ma, X., Crespo, R. G., & Martínez, O. S. (2020). Blockchain for supply chain performance and logistics management. *Applied Stochastic Models in Business and Industry, 37*(3), 429–441.

10. Lu, J., Wu, S., Cheng, H., Song, B., & Xiang, Z. (2020). Smart contract for electricity transactions and charge settlements using blockchain. *Applied Stochastic Models in Business and Industry, 37*(3), 442–453.
11. Malik, N., Alkhatib, K., Sun, Y., Knight, E., & Jararweh, Y. (2021). A comprehensive review of blockchain applications in industrial Internet of Things. *Applied Stochastic Models in Business and Industry, 37*(3), 391–412.
12. Khan, M. A., & Salah, K. (2017). IoT security: Review, blockchain solutions, and open challenges. *Future Generation Computer Systems, 82*, 395–411.
13. Li, C., & Zhang, L. J. (2017). A blockchain based new secure multi-layer network model for internet of things. In *Proceedings of the 2017 IEEE 2nd international congress on internet of things* (pp. 33–41).
14. Meng, W., Tischhauser, E., Wang, Q., Wang, Y., & Han, J. (2017). When instruction detection meets blockchain technology: A review. *IEEE Access, 6*, 10179–10188.
15. Kallenberg, O. (2002). *Foundations of modern probability* (2nd ed.). Springer.

Chapter 22
A Heuristic Framework for Assessing the Efficiency of Multi-branch Banks Under Big Data Conditions

Vahid Kayvanfar, Hamed Baziyad, Shaya Sheikh, and Frank Werner

Abstract Evaluating the efficiency of organizations and branches within an organization is a challenging issue for managers. Evaluation criteria allow organizations to rank their internal units, identify their position concerning their competitors, and implement strategies for improvement and development purposes. Among the methods that have been applied in the evaluation of bank branches, non-parametric methods have captured the attention of researchers in recent years. One of the most widely used non-parametric methods is the data envelopment analysis (DEA) which leads to promising results. However, the static DEA approaches do not consider the time in the model. Therefore, this paper uses a dynamic DEA (DDEA) method to evaluate the branches of a private bank over three years (2017–2019). The results are then compared with static DEA. After ranking the branches, they are clustered using the K-means method. Finally, a comprehensive sensitivity analysis approach is introduced to help the managers to decide about changing variables to shift a branch from one cluster to a more efficient one.

Keywords Dynamic data envelopment analysis (DDEA) · Branch performance evaluation · Efficiency evaluation · Banking industry · Big data

---

V. Kayvanfar (✉)
Division of Engineering Management and Decision Sciences, College of Science and Engineering, Hamad Bin Khalifa University, Qatar Foundation, Doha 34110, Qatar
e-mail: valikayvanfar@hbku.edu.qa

H. Baziyad
Department of Information Technology, Tarbiat Modares University, Tehran, Iran

S. Sheikh
Department of Supply Chain and Business Analytics, School of Management, New York Institute of Technology, New York, NY 10023, USA

F. Werner
Faculty of Mathematics, Otto-von-Guericke University Magdeburg, P.O. Box 4120, 39016 Magdeburg, Germany

© The Author(s), under exclusive license to Springer Nature Switzerland AG 2024
A. Emrouznejad et al. (eds.), *Business Analytics and Decision Making in Practice*, Lecture Notes in Operations Research, https://doi.org/10.1007/978-3-031-61589-4_22

Introduction

Identifying, evaluating, and improving the operations of bank branches is a necessary structural design. Banks are the most important element of the financial market, which plays a vital role in the economy of countries. Therefore, researchers have been interested in understanding the types of efficiency and measuring the impact of factors and policies on them. Such studies have significant implications for both microeconomics and macroeconomics [2, 13, 7]. From a microeconomic perspective, efficiency enhances competition and institutional and regulatory reforms. From a macroeconomic perspective, efficiency affects the cost and stability of financial intermediation. Improved bank performance indicates better allocation of financial resources and higher investment. In the last two decades, there has been a growing interest in studying the efficiency of the banking system worldwide [12, 27, 35].

DEA is a methodology that measures the relative efficiency of decision-making units (DMUs) and is widely used as a tool for decision-making, especially for analyzing the performance of bank branches [4, 9, 25]. Charnes et al. [8] provided a model that measures the performance with multiple inputs and outputs (a.k.a. CCR model). CCR was first used in assessing the academic achievement of United States school students in 1976. Banker et al. [5] embedded new concepts into DEA and called it the BCC model. The CCR model has a constant return-to-scale that tries to achieve the efficiency of a DMU by selecting the optimal weights for the input and output variables. Unlike the CCR model, the BCC model works based on the variable return-to-scale assumption. Using variable returns to scale makes it possible to provide a very accurate analysis by calculating the technical efficiency in terms of scale and management efficiency values. The same principles of the CCR model are used to build input and output-oriented models in the BCC model. In the input-oriented model, the efficiency increases with decreasing the inputs, in the output model, the efficiency increases with increasing the outputs [18, 30].

DEA has some advantages over other methods of performance evaluation. One advantage is that DEA does not depend on the unit of measurement. It can use inputs and outputs with different units, such as dollars, hours, or kilograms. Another advantage is that DEA compares the units with each other, rather than with a fixed standard. Unlike multi-criteria decision analysis (MCDM) approaches, DEA does not require experts' opinions to assign weights to the criteria. Instead, DEA calculates the optimal weights by itself, which makes it a suitable tool for dealing with uncertainty [10, 38]. However, DEA models suffer from a big data analysis, which may makes the feasibility area smaller and may lead to the infeasibility of the problem under investigation [3, 40, 41]. Nowadays, big data generation has become challenging in different areas, such as text mining [14, 26], healthcare [31], demand forecasting [42], and the Internet of Things [6]. Therefore, utilizing fewer data [24] or developing methods and approaches with the capability of handling a large amount of data [1] was changed into an exciting area. This paper proposes a heuristic framework, enabling DEA models to be run in big data environments. The proposed framework can also enable DEA models to deal with infeasible problems by changing them from infeasible to feasible ones. This process is executed by eliminating some constraints from the model.

A comprehensive overview of the previous studies of DEA-based bank performance analysis is provided in Table 22.1. Different inputs and outputs are provided in different countries as needed. In the case of inputs, variables such as total cost, interest, capital, profit and income, shareholders 'salaries, employees' salaries, number of employees, total assets and fixed assets, as well as deposits, are among the most frequently used items. Among them, a large margin uses items such as the number of employees, staff costs, deposits, fixed assets, and total assets. Some variables such as income and profit, loans, earnings per share, investment, securities and bonds, deposits, and shareholder returns are used more frequently.

Among the many methods that have been registered in the evaluation of bank branches, the interest of researchers in non-parametric methods has increased in recent years, which constitutes a major part of the evaluation of organizations. One of the most widely used non-parametric methods in this field is the use of DEA techniques [22]. DEA can be divided into static and dynamic categories [36]. In the static state, the desired time intervals are not realized, and the research is performed at one point, which may lead to a misleading data analysis. In contrast, dynamic DEA generates more realistic results since it considers the changes that occur over time. In problems where DEA traditionally uses a performance appraisal, the time factor is not considered, despite having time as one of the most important performance factors. Eliminating time intervals could create errors in the output, and as a result, a wrong conclusion by managers becomes more likely. For this purpose, dynamic DEA is used to consider time intervals in evaluating the branches of a private bank in Turkey over three years. The results are then compared with the classical static DEA. After ranking the branches, a clustering of the branches is performed using the K-means method. Finally, to present a situational analysis of the branches and help managers to plan for the future, the amount of change needed in each key indicator (to switch the cluster membership of each branch) is obtained using a comprehensive sensitivity analysis. The research goals can be summarized as follows:

- Investigating the efficiency of the branches of a private bank throughout the country for three years from 2017 to 2019.
- Implementing dynamic DEA in evaluating the bank branches for the first time.
- Ranking of the private bank branches during similar time intervals.
- Clustering of the branches based on the results obtained from their efficiency.
- Performing a sensitivity analysis on the main variables to present a clear picture of each branch's status and upgrade each branch's status.
- Presenting a heuristic framework for dealing with big data and infeasible environments of DEAs.

Based on the provided review table of the literature, previous studies have used DEA for analyzing about 57 DMUs on average. In comparison, we used it for analyzing 531 DMUs which is 9.32 times greater than the existing problems in Table 22.1. The basic DEA models could not find feasible solutions in our case

Table 22.1 Summary of DEA research conducted by country and type of model

Authors/ year	Goal	Country	Model	Input	Output	No. DMUs
Kamarudin et al. [17]	TE, PTE and SE review of 43 banks from 2007 to 201	Gulf Cooperation Council (GCC) countries	CCR–1 and BCC–1	Large assets and deposits	Loans and income	43
Kwon and Lee[19]	Combining two experimental data analysis techniques to evaluate and predict the performance improvement	USA	Two stages: CCR-2 and Back Propagation Neural Network (BPNN)	Step I: Number of employees, equity and costs Stage II: deposits, loans and investments	Step I: Deposits, loans and investments Step II: Profit	13
Yilmaz and Güneş [39]	Comparing the technical and general productivity of ordinary deposit banks and Islamic banks from 2007 to 2013	Turkey	CCR-1 and BCC-1	Deposits and capital	Loans, investments and income	32
Stoica et al. [33]	Analyzing the Internet banking services' performance	Romania	PCA-based DEA	Deposits, total deposits, remittances and total operating costs, number of employees, the value of owned equipment and software programs	Net total revenues, daily "reach" average rate	24
Wanke et al. [37]	A new fuzzy DEA model to evaluate the bank performance in Mozambique from 2003 to 2011	Mozambique	Fuzzy	Total costs (excluding personnel) and personnel costs	Total deposits, pre-tax income and total credit operations	13

(continued)

Table 22.1 (continued)

Authors/ year	Goal	Country	Model	Input	Output	No. DMUs
Kamarudin et al. [16]	Effects of government on the revenue productivity of Islamic and conventional banks in the period 2007–2011	GCC countries	Two-stage - BCC with panel regression	Staff costs and deposits	Income and loans	47
Stewart et al. [32]	The efficiency of the Vietnamese banking system from 1999 to 2009	Vietnam	Two stages: CCR, BCC and bootstrap	Number of employees, deposits from other banks and customer deposits	Loans from customers, other loans and securities	101
Kamarudin et al. [15]	Productivity of Islamic banks by confirming specific factors of banks, industry	Southeast Asian countries	Two-stage - BCC with Malmquist and panel regressions	Staff costs, deposits and fixed assets	Investments and income	29
Fukuyama and Matousek [11]	Evaluating the banks' network revenue performance	Japan		Number of workers, Premises and real estate	Performing loans, securities investment, nonperforming loans	172
Liu et al. [23]	Finding the efficiency gap among different commercial banks	China	Two-stage DEA	Fixed assets, labor operating expenses	Non-performing loan ratio, loans business income	28
Tsolas et al. [35]	Classifying the sampled branches of banks	Greece	Two-stage hybrid DEA-ANN	personnel expenses, rents and depreciation, operational expense	Net interest income, non-interest income,	160

(continued)

Table 22.1 (continued)

Authors/ year	Goal	Country	Model	Input	Output	No. DMUs
Preeti and Roy [27]	Operational performance measurement and prediction	India	DEA + ANN	Operating expenses and deposits	Investments, performing loans and advances, non-interest income	39
Preeti et al. [28]	Comparing the efficiency of private and public banks	India	Frontier-based radial DEA	Operating expenses, assets and deposits	Investment and advances	40
Fukuyama et al. [12]	Analyzing the cost inefficiency levels of Turkish banks	Turkey	two-stage DEA	Total fixed assets, number of employees	Various loans, security investments	26
Li [21]	Analyzing the performance of city commercial banks in China	China	Bootstrapped DEA	Fixed assets, staff expenses, interest expenses	Non-performing loans, deposits, performing loans, non-interest income	101
Lartey et al. [20]	Examining the importance of bank risk exposures through interbank funding on bank efficiency levels	UK	Three-stage DEA with alliance between stage 1 and stage 2, and feedback variables	Net interbank fees, capital measured, total interbank loans	Interbank borrowing, short-term wholesale funding and Long-term wholesale funding	43

(continued)

Table 22.1 (continued)

Authors/ year	Goal	Country	Model	Input	Output	No. DMUs
Omrani et al. [25]	Analyzing the performance of branches of the Agribank in the West Azerbaijan province of Iran	Iran	Two-stage DEA-BWM model	Number of FTE employees, total expenses, number of ATM	Profit	43
Current article	Evaluation of private bank branches	Turkey	Dynamic DEA	Current facilities/ exchange contracts, total costs/ operating expenses;	Total revenue/ unrelated revenues	531

study. Therefore, we present a heuristic framework that can analyze a wide range of DMUs.

The main contributions of this paper are as follows:

- Proposing a new heuristic framework for implementing DEA algorithms under a big data environment comprising many DMUs.
- Employing dynamic DEA (DDEA) on a real case study including many DMUs equipped with a cluster-based hybrid method.
- Presenting an efficient approach for rectifying infeasibility problems in DEA algorithms, so that the exposing probability of infeasible solutions reduces.
- Creating the possibility of employing the proposed framework for every DEA algorithm.

The remainder of this paper is organized as follows. Section 22.2 provides a description of the studied problem, the mathematical model, and the relevant details of the developed framework. Section 22.3 implements the developed framework on a synthetic dataset and shows its capability and applicability. Section 22.4 summarizes the findings and suggests several paths for future research.

Methodology

The statistical population of this research comprises all 531 branches of a private bank in Turkey, and according to the type of method used, there will be no need for sampling. The required data is provided to the researchers by the Studies and

Innovation Center of the bank. However, to prepare the collected data for further analysis, it is necessary to perform the data cleansing process carefully.

A comprehensive list of required indicators is collected based on the availability of data and the goals set by the bank managers. After an initial review of the variables from every aspect, including the review of the subject, consulting with financial experts, and justification of outcomes, the final variables used in the developed dynamic DEA are given in Table 22.2. All data were sorted by year. Each file contains separate sheets such as Resources, Income, Facilities, Deferred, and Expense, in which each data is sorted by the branch code. Please note that the first and second variables are input, the third variable is undesirable link, the fourth variable is desirable link, and the last variable is output.

Although DEA has been previously used in the banking industry, this method examines the system performance at a specific time. It does not provide an accurate understanding of the performance over time. In the present study, the dynamic SBM method is one of the most widely used models in this field (see Tone & Tsutsui, [34]). This paper explains this model in detail. DEA offers many methods for measuring performance changes over time, such as a window analysis and the Malmquist Index. In this regard, it should be noted that these methods do not consider the relationships between different activities at two consecutive times and look at different times independently and separately. However, in a real business under long-term planning, considering the connections between different times helps to evaluate the system and gives a more accurate view. The Dynamic SBM technique considers the logical connections between the variables in consecutive time intervals.

These relationships are divided into four types: Desirable (good), undesirable (bad), free and fixed. The following is an overview of each of these types of communications:

- Desirable links such as net profit transferred to the profit carried forward, and net earned surplus carried to the next term.
- Undesirable links such as loss carried forward, bad debt, idle capital.
- Free and fixed links are optional and forced communications that may vary depending on the problem.

Suppose that we have the number n of DMUs whose information is specified in T periods. Every DMU has a specific input and output at any given time. Also, both consecutive periods (t and $t + 1$) have specific correlations, shown in Fig. 22.1.

Table 22.2 Introduction of branch financial ratios (final list of variables)

The ratio of financial statements and electronic transactions	Variable names
Current facilities/exchange contracts	L1
Total operating costs/expenses	EX1
Total non-operating expenses/expenses	EX2
Total revenue/common revenues	IN1
Total Income/separate revenues	IN2

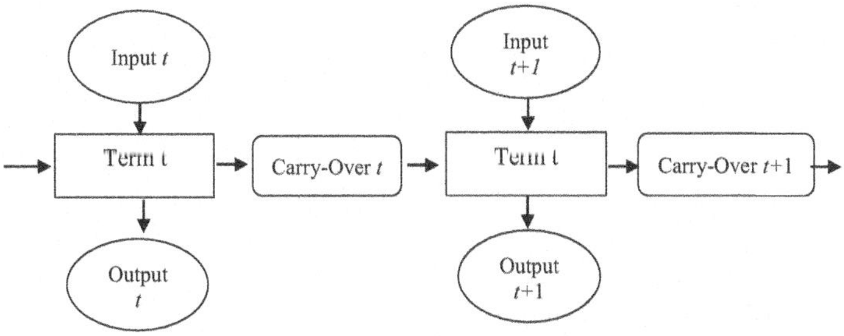

Fig. 22.1 Correlation between the time periods t and t + 1 in a hypothetical DMU

Table 22.3 Inputs and outputs of dynamic data envelopment analysis model

Data	Index	Notes
x_{ijt}	i = 1, 2, 3,…,m; j = 1, 2, 3,…,n; t = 1, 2, 3,…,T	Input
y_{ijt}	i = 1, 2, 3,…,s; j = 1, 2, 3,…,n; t = 1, 2, 3,…,T	Output
z_{ijt}^{good}	i = 1, 2, 3,…,ngood; j = 1, 2, 3,…,n; t = 1, 2, 3,…,T	Desirable link
z_{ijt}^{bad}	i = 1, 2, 3,…,nbad; j = 1, 2, 3,…,n; t = 1, 2, 3,…,T	Undesirable link
λ_j^t	j = 1, 2, 3,…,n; t = 1, 2, 3,…,T	–

Desirable links (optimal links) act as outputs. In this case, the value of the links should not be less than the observed value. Undesirable links act as input, where the value of the links should not exceed the observed value. The comparative lack of links in this section is called inefficiency.

Figure 22.1 is drawn for a specific DMU. As can be seen, there are inputs and outputs at any given time, and different times for a DMU are connected through the links. Sometimes a situation occurs, where the problem's nature requires only one of these items to be presented in the model. We first define the data in the dynamic DEA model, which are explained in Table 22.3. The 'z_{ijt}^{good}' signifies good link values, where 'nbad' and 'ngood' are the number of bad and good links, respectively.

Mathematical Model

The dynamic DEA model under the SBM policy can now be defined as follows, where the base model is taken from Shafiee et al. [29]. In the proposed model, only desirable and undesirable links are considered. Also, to have a better discrimination, we excluded the $\boldsymbol{DMU_o}$ from its reference set.

$$\rho = Min\left[\frac{\frac{1}{T}\sum_{t=1}^{T} W^t\left[1-\frac{1}{m+nbad}\left(\sum_{i=1}^{m}\frac{k.s_{it}^-}{x_{iot}}+\sum_{i=1}^{nbad}\frac{k.s_{it}^{bad}}{z_{iot}^{bad}}\right)\right]}{\frac{1}{T}\sum_{t=1}^{T} W^t\left[1+\frac{1}{s+ngood}\left(\sum_{i=1}^{s}\frac{k.s_{it}^+}{y_{iot}}+\sum_{i=1}^{ngood}\frac{k.s_{it}^{good}}{z_{iot}^{good}}\right)\right]}\right] \tag{22.1}$$

$$s.t.$$

$$\sum_{j=1,j\neq o}^{n} \lambda_j^t x_{ijt} + s_{it}^- = x_{iot};\, i = 1, 2, 3, \ldots, m;\, t = 1,2,3,\ldots, T \tag{22.2}$$

$$\sum_{j=1,j\neq o}^{n} \lambda_j^t y_{ijt} - s_{it}^+ = y_{iot};\, i = 1, 2, 3, \ldots, s;\, t = 1,2,3,\ldots, T \tag{22.3}$$

$$\sum_{j=1,j\neq o}^{n} \lambda_j^t z_{ijt}^{good} - s_{it}^{good} = z_{iot}^{good};\, i = 1, 2, 3, \ldots, n_{good};\, t = 1,2,3,\ldots, T \tag{22.4}$$

$$\sum_{j=1,j\neq o}^{n} \lambda_j^t z_{ijt}^{bad} + s_{it}^{bad} = z_{iot}^{bad};\, i = 1, 2, 3, \ldots, n_{bad};\, t = 1,2,3,\ldots, T \tag{22.5}$$

$$\sum_{j=1,j\neq o}^{n} \lambda_j^t z_{ijt}^{bad} = \sum_{j=1,j\neq o}^{n} \lambda_j^{t+1} z_{ijt}^{bad};\, t = 1, 2, 3\ldots, T-1 \tag{22.6}$$

$$\sum_{j=1,j\neq o}^{n} \lambda_j^t z_{ijt}^{good} = \sum_{j=1,j\neq o}^{n} \lambda_j^{t+1} z_{ijt}^{good};\, t = 1, 2, 3\ldots, T-1 \tag{22.7}$$

$$\sum_{j=1,j\neq o}^{n} \lambda_j^t = 1;\, t = 1, 2, 3\ldots, T \tag{22.8}$$

$$\lambda_j^t, \lambda_j^{t+1}, s_{it}^{bad}, s_{it}^{good}, s_{it}^+, s_{it}^- \geq 0$$

where s_{it}^+, s_{it}^-, s_{it}^{good}, s_{it}^{bad}, and s_{it}^{free} are slack variables which are described in Table 22.4.

Table 22.4 Description of the types of deficiency variables used in the main model

Variable name	Index	Notes
s_{it}^{bad}	i = 1, 2, 3,…,nbad; t = 1, 2, 3,…,T	Undesirable surplus
s_{it}^{good}	i = 1, 2, 3,…,ngood; t = 1, 2, 3,…,T	Lack of optimal source
s_{it}^+	i = 1, 2, 3,…,s; t = 1, 2, 3,…,T	Output shortfall
s_{it}^-	i = 1, 2, 3,…,m; t = 1, 2, 3,…,T	Input excess
λ_j^t	j = 1, 2, 3,…,n; t = 1, 2, 3,…,T	Link coefficient

The primary goal of this method (Eq. 22.1) is to separate efficient and inefficient units and rank them. Although the proposed model is nonlinear, it can be modified to become a linear model. This is achieved by setting the denominator of the fraction in the objective function equal to one and adding it to the constraints. The objective function states that the maximum output limit is obtained for the smallest input values. As can be seen, the two variables of desirable and undesirable links have also been added to the Dynamic SBM model, Eqs. (22.4) and (22.5). Undesirable links are input, where we tend to reduce them. On the other hand, the desirable links are of the output type, and we tend to increase their values. The inputs should not exceed a specific limit. Therefore, we set restrictions such as x_{i0t} (Eq. 22.2) and z_{i0t}^{bad} (Eq. 22.5) on the inputs and undesirable variables, respectively. This is opposite to the output constraints and the constraints on the desirable variables. Since we want these variables not to be less than a specific limit, their constraints are defined to be greater than equal (Eqs. 22.3 and 22.4). We connect consecutive years by the dynamic nature of the model (in which the time factor is involved). Equations (22.6) and (22.7) are used for this purpose. The desirable and undesirable links are linked for two consecutive years. With such an objective function, these constraints must be resolved separately for each unit. Equation (22.8) related to the variable return-to-scale assumption.

DEA Framework

As shown in Table 22.1, most papers have utilized DEA for comparing a low number of DMUs, resulting in a low number of constraints in the DEA model, increasing the feasible solution area. Indeed, as the number of DMUs increases, the feasible solution space becomes smaller, and the probability of reaching a feasible answer increases. However, this paper fed a wide range of DMUs (531 DMUs) to the DEA model, which could not reach feasible solutions. Accordingly, a novel heuristic framework deals with infeasible solutions for implementing DEA algorithms for many DMUs.

According to the proposed heuristic framework, each DMU is considered first, and the rest of the DMUs are randomly categorized into p equal classes. Each time, this opted DMU is added to each cluster, and then the efficiency of this selected DMU is calculated. This process should be conducted for all of the clusters. By doing so, ***p*** numbers, as efficiency values, are obtained. Then, it suffices to take the average of these p numbers, signifying the estimated efficiency of this opted DMU. This process should be carried out for the 531 DUMs considered in this research. The proposed framework can be seen in Fig. 22.2. The goal of taking the average of the obtained numbers is to increase the robustness of the outputs.

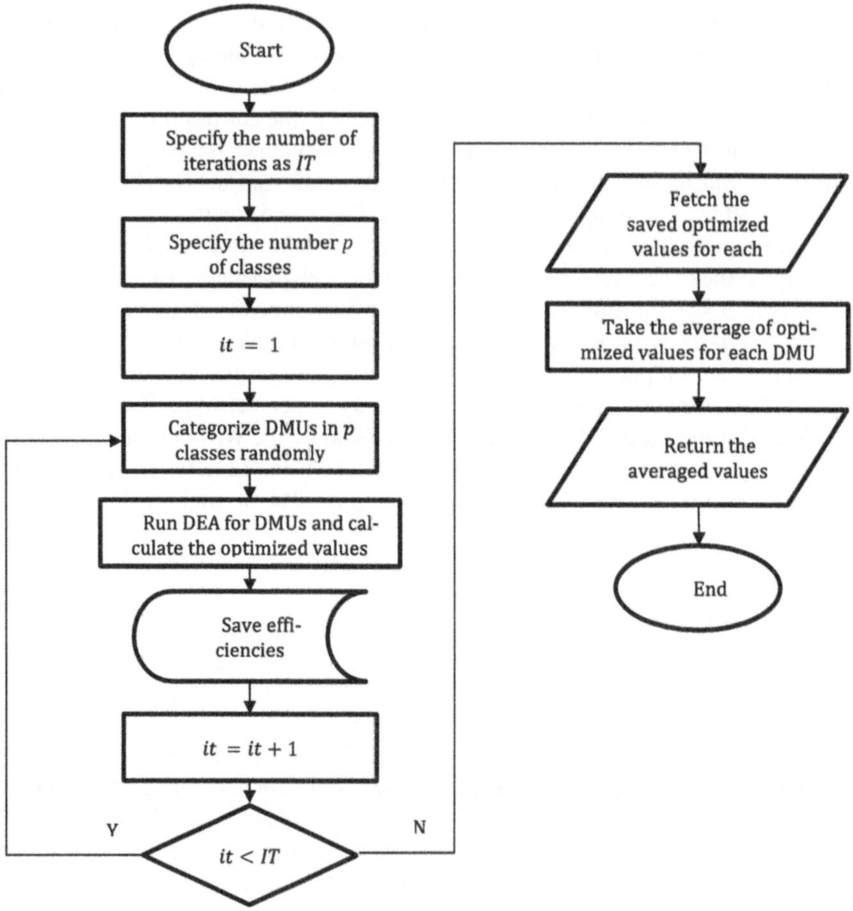

Fig. 22.2 Proposed heuristic framework of DEA

Clustering Process

Clustering is unsupervised learning in which samples are divided into groups whose members are similar to each other. We utilize the k-means algorithm to cluster the branches. Among the essential features of a K-Means clustering, we can refer to.

- the algorithm's performance depends on the primary centers,
- trapping into a local optimum,
- the convexity of the shape of clusters, and
- a practical algorithm for large volumes of data.

The K-means algorithm considers an m × n matrix as m data with n properties; If k is 2, it divides the data into two categories, where the members in the same category have properties that are closer to each other. Consider the set of vectors $a_1, a_2, \ldots, a_n$. The K-means algorithm is used to find the clusters $\pi_1, \pi_2, \ldots, \pi_k$ using the following function (Eq. 22.9):

$$D\left(\{\pi_c\}_{c=1}^{k}\right) = \sum_{c=1}^{k} \sum_{a_i \in \pi_i} \|a_i - m_c\|^2 \tag{22.9}$$

where $mc = \sum_{a_i \in \pi_i} a_i / |\pi_c|$ indicates the average of the cth cluster, where cluster c is represented by π_c.

Experimental Results

In this section, the results of the proposed method are presented. We use Python 3.7 on an Intel Core i7 processor (2.7 GHz) with 12 GB memory to run the experiments. After running the DDEA algorithm as described in the previous section, the top 20 branches of the private bank in Turkey between 2017 and 2019 are listed in Table 22.5 with the current grade of the branch, the efficiency value, and the branch code. As can be seen, most of these branches are first-class, and there are only eight branches of other degrees, of which six are second-class.

The information about the variables is shown in Table 22.6 in the order of the years. The range of variables is almost the same in consecutive years, which indicates that there are not many outliers in the database that may deviate from the final output in a certain direction.

Clustering Implementation

This study uses the silhouette criterion to evaluate different clustering solutions. Figure 22.3 shows different silhouette values for a number of clusters between 2 and 12. As can be seen, the silhouette for 8 clusters has reached its maximum value of 78.21%, and it is equal to 76.48% for 7 clusters.

Based on the Silhouette criterion, seven clusters were disclosed. There are 7 degrees, including privileged, special, 1, 2, 3, 4, and 5, in the classification of the bank branches. The best silhouette values were calculated for clustering algorithms such as hierarchical clustering Ward, K-means, DBSCAN, and OPTICS. The best silhouette values of the K-means method are larger than the amount of the silhouette obtained from the other methods. Therefore, in this research, the K-means method has been used. After ranking the bank branches in the three years, the clustering operation was performed using the K-means method with the number of 7 clusters on the output obtained from the ranking. In this clustering, groups 1 and 2 can be considered as equivalent to special and privileged branches, and the rest of the clusters

Table 22.5 Details of the top 20 branches of the case study during the period from 2017 to 2019 using the DDEA method

Branch type	Efficiency	Branch code	Rank
Special	2.7845	3610	1
Premier	2.5389	1867	2
Premier	2.5379	1887	3
Premier	2.5166	1857	4
1	2.5080	3410	5
2	2.4219	3511	6
2	2.3462	2570	7
2	2.2124	4614	8
4	2.1660	1202	9
2	2.0125	1225	10
2	2.0065	2061	11
Premier	1.9826	1839	12
Premier	1.9746	1818	13
1	1.8744	1110	14
Premier	1.8469	1870	15
3	1.7964	1631	16
Special	1.7895	4410	17
Premier	1.7260	1885	18
2	1.6880	3310	19
1	1.6790	2510	20

Table 22.6 Descriptive statistics related to the variables used for the three years

The ratio of financial statements and transactions	Variable name	Min	Max	Variance	Year
Current facilities/ exchange contracts	L1	0	0.0520	0.0042	2017
		0	0.519	0.00416	2018
		0	0.0517	0.0042	2019
Total operating costs/ expenses	EX1	0	0.0532	0.0037	2107
		0	0.522	0.0035	2018

represent the branches of grades 1, 2, 3, 4, and 5, respectively. Figure 22.4 shows the percentage of dispersion of each of these types.

The clustering results of the top 20 branches of the bank during the three years are given in Table 22.7.

In the clustering, the center values of the clusters are used to compare the clusters. The use of these centers gives a proper representation of the distance between two

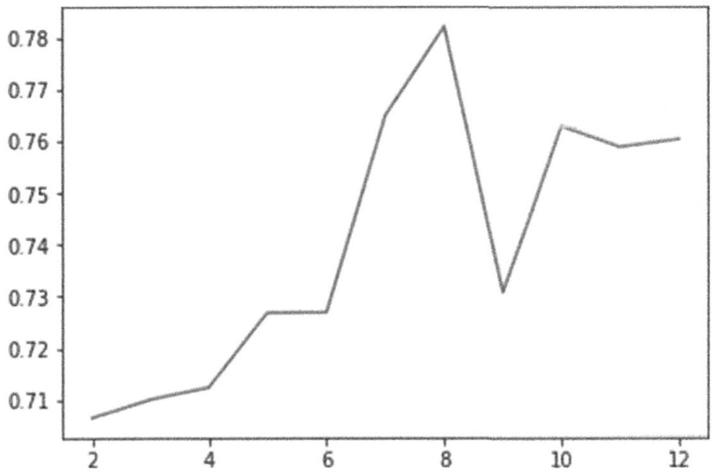

Fig. 22.3 The standard value of silhouette for different clusters

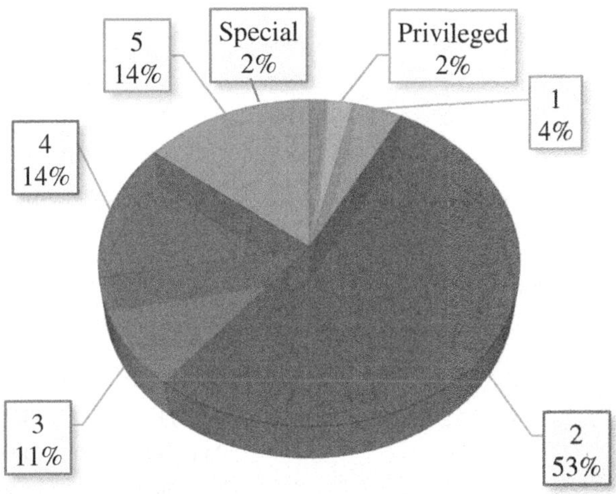

Fig. 22.4 The share of different types of bank branches in the clustering

clusters or the similarity of the two clusters. The efficiency of the cluster centers is shown in Table 22.8.

According to Table 22.8, the efficiency of "special" branches is, on average, equal to 2.48, considered as "super-efficient" branches. It is important to note that in the SBM method, the efficiency of a branch can reach more than one, which indicates that the branch is "super-efficient".

Table 22.7 Clustering results of the top 20 branches of the bank during the years from 2017 to 2019

Cluster #	Performance value	Branch code	Rank
1	2.7845	3610	1
1	2.5389	1867	2
1	2.5379	1887	3
1	2.5166	1857	4
1	2.5080	3410	5
1	2.4219	3511	6
1	2.3462	2570	7
1	2.2124	4614	8
2	2.1660	1202	9
2	2.0125	1225	10
2	2.0065	2061	11
2	1.9826	1839	12
2	1.9746	1818	13
2	1.8744	1110	14
2	1.8469	1870	15
2	1.7964	1631	16
2	1.7895	4410	17
2	1.7260	1885	18
2	1.6880	3310	19
2	1.6790	2510	20

Table 22.8 The efficiency of the cluster centers

Cluster number	Cluster type	Efficiency
1	Special	2.48
2	Privileged	1.878
3	1st rank	1.311
4	2nd rank	1.006
5	3rd rank	0.781
6	4th rank	0.473
7	5th rank	0.033

Sensitivity Analysis

A sensitivity analysis shows the amount of change in each of the five key variables studied, intending to upgrade the level of each branch to a higher level. In other words, conducting such a sensitivity analysis allows the bank managers to assess the current state of the branch over three years and take into account the average performance of each branch in the medium term, which is naturally a more accurate judgement of its status. It also helps to mitigate the negative effect of weaknesses or temporary growth of branches and to increase awareness of branch managers on the amount of the necessary change in each index by each branch. The awareness of the tolerance of these changes gives the bank managers a better view of the allowable change in each index.

Table 22.9 shows the results of the sensitivity analysis of 12 branches located in the second cluster (privileged branches). Based on these results, the change needed in each branch to upgrade that branch to a higher cluster is given. Due to the simultaneous change of the considered variables, the values listed in Table 22.9 are calculated based on the difference in the number of each branch's variables compared to the worst branch in the higher cluster.

For further clarification, we describe the details of one of the branches, branch 1110, located in the second cluster. We want to know what changes can upgrade this branch to a higher cluster, i.e., cluster 1 (special branches). For this purpose, we look up the information related to the same row in the columns related to the variables of cluster 2. According to Table 22.9, this branch should increase its separate income by about 6266 units and reduce its non-operating expenses by about 3332 units simultaneously. While improving the two mentioned variables, the other variables should not be diminished. No discount should occur so that this branch can be upgraded to the first cluster (if the indicators related to the worst branch of cluster 1 remain constant). Table 22.10 shows the results of the sensitivity analysis related to the branches of the third cluster (Grade 1 branches). Based on this sensitivity analysis, it is possible to understand how much change is needed to turn a grade 1 branch into a premium branch. As can be seen, there is no need to change some of these indicators.

Table 22.9 Results of the sensitivity analysis for the second cluster branches (privileged branches)

Non-operating costs	Common income	Separate income	Operating costs	Facilities	Branch code
−3332.33	No change	6265.667	No change	No change	1110
No change	No change	48,904.67	No change	No change	1202
−1961.67	20,978.66	30,688.67	No change	No change	1225
No change	22,063.33	5294	No change	22,974	1631
−19,686	No change	No change	−62,295.7	No change	1818
−60,950.3	No change	No change	−974,662	No change	1839
−5788.67	No change	No change	−75,570.3	No change	1870
−50,661.7	No change	No change	−26,264.3	No change	1885
No change	30,440.66	69,275.67	No change	47,038.33	2061
−12,622	No change	No change	−25,587.7	No change	2510
−2813.33	27,814.66	No change	No change	No change	3310
−155,713	No change	No change	−798,827	No change	4410

Table 22.10 Results of the sensitivity analysis for the third cluster branches (Grade 1 branches)

Non-operating costs	Common income	Separate income	Operating costs	Facilities	Branch code
No change	57,071.67	107,587	No change	1,687,068	1128
No change	21,787.67	No change	No change	1,523,104	1313
No change	53,308	77,444.3	No change	1,687,718	1440
No change	49,338.67	59,115.3	No change	1,670,713	1448
−7416.33	46,334.67	No change	−11,796.3	1,209,070	1861
No change	36,147.67	No change	−39,408.3	1,417,276	1862
No change	No change	78,671.3	No change	1,569,883	1866
−53,387	No change	No change	−270,581	No change	1886
No change	56,365.67	118,305.3	No change	1,698,959	2047
No change	36,198.67	85,478.3	No change	1,669,256	2115
No change	7905	68,022.97	No change	1,522,511	2310
No change	64,203.67	173,368.6	No change	1,738,157	2611
No change	48,579.67	No change	−15,437.7	1,532,845	2734
No change	53,547.33	62,155.3	No change	1,422,109	3110
No change	55,061	159,659.3	No change	1,734,663	3318
No change	56,431	No change	No change	1,619,914	3411
No change	31,804.67	15,928.97	No change	1,625,130	3514
−8047.33	No change	No change	−25,805.7	No change	3813
No change	45,289.67	63,090.63	No change	1,109,168	3822
No change	53,035.67	188,829.3	No change	1,766,943	4020
No change	29,063.33	83,175.63	No change	1,625,689	4617
No change	No change	No change	−10,012	1,188,718	4622

Similarly, Tables A.1-A.4 illustrate the sensitivity analysis of Grade 2 (4th cluster), Grade 3 (5th Cluster), Grade 4 (6th Cluster), and Grade 5 (7th Cluster) branches, respectively. Based on this sensitivity analysis, it is possible to determine how many changes are needed to turn a lower-grade branch into a higher-grade branch. Similar to the previous tables, some sections do not need to be changed. This means that these indexes are good enough, considering the values of the change of the other variables as needed. As mentioned, this main philosophy of the sensitivity analysis in this study is to compare each cluster's worst branch. Thus, if a branch wants to be upgraded from a 4th-degree cluster (worst cluster) to a 3rd-degree cluster (better degree), it must do better than or equal to at least one of the worst members of the 3rd cluster. Let us explain this by an example. Suppose that the worst branch of cluster 3 has a cost of 10 units and the branch we consider, type 4, has a cost of 15. Suppose that this variable is the only criterion for comparing the two branches. In such a situation, if the branch under consideration can at least perform the same as the worst branch of cluster 3 (i.e., reduce its costs to 10 units), it must reduce its costs

Table 22.11 Comparison of DDEA and DEA

Branch code	Efficiency 2017	Efficiency 2018	Efficiency 2019	DDEA efficiency
3610	1	1	1	2.7854
1867	1	1	1	2.5389
1887	1	1	1	2.5378
1857	1	1	0.9916	2.5160
3410	1	1	0.9880	2.5079

by at least 10–5 = 5 units. Finally, it should be noted that the present study used dynamic data envelopment analysis to evaluate the 531 branches. However, no article in the existing literature considers more than 50 branches. One of the reasons for the lack of research could be the high computational volume and complexity of the algorithm used, which prevents a large number of branches from being considered.

Comparison of the Results of the DEA and Dynamic DEA Methods

It is not common to compare the results of the classic SBM and DEA models, as the range of variation between the two methods is different. In the DEA, the performance values are generally between zero and one, while in SBM-based models, the performance values include numbers more significant than one. Nevertheless, we can perform a relative comparison of these two models. It should also be noted that comparing DEA with dynamic DEA (DDEA) would be reasonable and common in the DDEA models with performance values between zero and one. The values obtained in the DEA method for the branches of the studied bank are consistent with the results obtained from the DDEA method. A part of this comparison is illustrated in Table 22.11.

As can be seen, the first five branches in all three time periods have numbers very close to one. This is a testament to the accuracy of the performance obtained from the dynamic data envelopment analysis method.

Conclusions and Future Works

Evaluating the efficiency of organizations and companies or evaluating different branches of an organization is one of the challenging issues that the management has faced. To date, various research has been conducted on this issue. Among the many methods used in the evaluation of bank branches, researchers have become more inclined to non-parametric methods in recent years and constitute a major part of the evaluation of organizational branches. Among these, one of the most widely used non-parametric methods are data envelopment analysis techniques. Data envelopment analysis can be divided into static and dynamic parts, where dynamic DEA leads to results closer to the real world because the real environment changes over time and is not static. Excluding time can increase the possibility of errors in the output of the work and as a result, a wrong analysis of the output increases. After ranking the branches, a clustering of the branches was performed using the K-means method. The silhouette index was used to find the optimal number of clusters, which was the best value for the 8 clusters. However, due to the very small difference in the index value for 7 and 8 clusters and the classification of the bank branches to 7 degrees and to equalize and make the analysis more uniform, we use 7 clusters for clustering the operations. In this clustering, groups 1 and 2 are equivalent with special and privileged branches, while the rest of the clusters show branches of grades 1, 2, 3, 4, and 5, respectively. A comprehensive sensitivity analysis was performed on the results. The sensitivity results show the rank of each branch and the rate of change in each of the five key indicators studied to raise the level of each branch to a higher level. The sensitivity analysis allows the bank managers to recognize the amount of necessary change in each index for each branch. The awareness of the rate of change of each indicator to lower the level of each branch to a lower degree can also be considered in the plans of the bank's general managers. This helps the managers to prevent a violation of the set values and prevent the degradation of the branch.

In the last part of the experiments, the classical DEA method was compared with the DDEA method. The results of this comparison showed how the output obtained from the DDEA method is in full compliance with the three-year analysis of the bank branches through the classical DEA method. The present study used dynamic data envelopment analysis to evaluate 531 branches. However, it can be said that almost no article in the existing literature considers more than 50 branches. Regarding the limitations of the problem under consideration, the "complexity" of the DDEA method can be considered as the most important challenge ahead. In other words, considering a larger number of variables (for example, ten variables) will most likely lead to an infeasible answer. The following future works can be a step towards complete research:

- Providing a smart and user-friendly dashboard for bank experts to take advantage of the method implemented in this research.
- Comparing with other modern methods; this is one of the cases that can always be considered to complete the impact of any research.
- Combining/hybridizing different methods; this should be done considering the strengths and weaknesses of different methods.
- Utilizing big data-based methods; since the data volume is large for performing DEA dynamic analysis, and it takes a lot of time to execute the relevant code, it can be facilitated to implement this method. In this regard, using parallel algorithms and discrete processing seems to be necessary.
- Examining the relationships between variables; Studies show that most articles use domain experts to determine their study variables. Therefore, providing an unsupervised method to identify input and output variables can save time. Also,

measuring the correlation coefficient between these variables and such methods can improve the quality of solving such a problem.

References

1. Anagnostopoulos, I., Zeadally, S., & Exposito, E. (2016). Handling big data: Research challenges and future directions. *The Journal of Supercomputing, 72*(4), 1494–1516. https://doi.org/10.1007/s11227-016-1677-z
2. Ariff, M., & Can, L. (2008). Cost and profit efficiency of Chinese banks: A non-parametric analysis. *China Economic Review, 19*(2), 260–273. https://doi.org/10.1016/j.chieco.2007.04.001
3. Badiezadeh, T., Saen, R. F., & Samavati, T. (2018). Assessing sustainability of supply chains by double frontier network DEA: A big data approach. *Computers & Operations Research, 98*, 284–290. https://doi.org/10.1016/j.cor.2017.06.003
4. Balak, S., Behzadi, M. H., & Nazari, A. (2021). Stochastic copula-DEA model based on the dependence structure of stochastic variables: An application to twenty bank branches. *Economic Analysis and Policy, 72*, 326–341. https://doi.org/10.1016/j.eap.2021.09.002
5. Banker, R. D., Charnes, A., & Cooper, W. W. (1984). Some models for estimating technical and scale inefficiencies in data envelopment analysis. *Management Science, 30*(9), 1078–1092. https://doi.org/10.1287/mnsc.30.9.1078
6. Baziyad, H., Kayvanfar, V., & Kinra, A. (2022). *Chapter 4 - the internet of things—an emerging paradigm to support the digitalization of future supply chains.* In B. L. MacCarthy & D. B. T.-T. D. S. C. Ivanov (Eds.), (pp. 61–76). Elsevier. https://doi.org/10.1016/B978-0-323-91614-1.00004-6
7. Chan, S.-G., Koh, E. H. Y., Zainir, F., & Yong, C.-C. (2015). Market structure, institutional framework and bank efficiency in ASEAN 5. *Journal of Economics and Business, 82*, 84–112. https://doi.org/10.1016/j.jeconbus.2015.07.002
8. Charnes, A., Cooper, W. W., & Rhodes, E. (1978). Measuring the efficiency of decision making units. *European Journal of Operational Research, 2*(6), 429–444. https://doi.org/10.1016/0377-2217(78)90138-8
9. Chu, M., Zhou, G., & Wu, W. (2022). *Data envelopment analysis on relative efficiency assessment and improvement: Evidence from Chinese bank branches BT-Eurasian business and economics perspectives.* In: M. H. Bilgin, H. Danis, E. Demir, & A. Zaremba (Eds.), (pp. 159–178). Springer International Publishing.
10. Filatovas, E., Marcozzi, M., Mostarda, L., & Paulavičius, R. (2022). A MCDM-based framework for blockchain consensus protocol selection. *Expert Systems with Applications, 204*, 117609. https://doi.org/10.1016/j.eswa.2022.117609
11. Fukuyama, H., & Matousek, R. (2017). Modelling bank performance: A network DEA approach. *European Journal of Operational Research, 259*(2), 721–732. https://doi.org/10.1016/j.ejor.2016.10.044
12. Fukuyama, H., Matousek, R., & Tzeremes, N. G. (2020). A Nerlovian cost inefficiency two-stage DEA model for modeling banks' production process: Evidence from the Turkish banking system. *Omega, 95*, 102198. https://doi.org/10.1016/j.omega.2020.102198
13. George Assaf, A., Barros, C. P., & Matousek, R. (2011). Technical efficiency in Saudi banks. *Expert Systems with Applications, 38*(5), 5781–5786. https://doi.org/10.1016/j.eswa.2010.10.054
14. Hosseini, S., Baziyad, H., Norouzi, R., Jabbedari Khiabani, S., Gidófalvi, G., Albadvi, A., Alimohammadi, A., & Seyedabrishami, S. (2021). Mapping the intellectual structure of GIS-T field (2008–2019): A dynamic co-word analysis. *Scientometrics, 126*(4), 2667–2688. https://doi.org/10.1007/s11192-020-03840-8

15. Kamarudin, F., Hue, C. Z., Sufian, F., & Mohamad Anwar, N. A. (2017). Does productivity of Islamic banks endure progress or regress? *Humanomics, 33*(1), 84–118. https://doi.org/10.1108/H-08-2016-0059
16. Kamarudin, F., Sufian, F., & Md. Nassir, A. (2016). Does country governance foster revenue efficiency of Islamic and conventional banks in GCC countries? *EuroMed Journal of Business, 11*(2), 181–211.https://doi.org/10.1108/EMJB-06-2015-0026
17. Kamarudin, F., Sufian, F., Nassir, A. M., & Anwar, N. A. M. (2015). Technical efficiency and returns to scale on banking sector: Empirical evidence from GCC countries. *Pertanika Journal of Social Sciences & Humanities, 23*(9), 219–236.
18. Kazemi Matin, R., Azadi, M., & Saen, R. F. (2021). Measuring the sustainability and resilience of blood supply chains. *Decision Support Systems*, 113629. https://doi.org/10.1016/j.dss.2021.113629
19. Kwon, H.-B., & Lee, J. (2015). Two-stage production modeling of large U.S. banks: A DEA-neural network approach. *Expert Systems with Applications, 42*(19), 6758–6766. https://doi.org/10.1016/j.eswa.2015.04.062
20. Lartey, T., James, G. A., & Danso, A. (2021). Interbank funding, bank risk exposure and performance in the UK: A three-stage network DEA approach. *International Review of Financial Analysis, 75*, 101753. https://doi.org/10.1016/j.irfa.2021.101753
21. Li, Y. (2020). Analyzing efficiencies of city commercial banks in China: An application of the bootstrapped DEA approach. *Pacific-Basin Finance Journal, 62*, 101372. https://doi.org/10.1016/j.pacfin.2020.101372
22. Liu, J. S., Lu, L. Y. Y., & Lu, W.-M. (2016). *Research fronts and prevailing applications in data envelopment analysis BT-data envelopment analysis: A handbook of empirical studies and applications.* In: J. Zhu (Eds.), pp. 543–574. Springer US. https://doi.org/10.1007/978-1-4899-7684-0_18
23. Liu, X., Yang, F., & Wu, J. (2020). DEA considering technological heterogeneity and intermediate output target setting: The performance analysis of Chinese commercial banks. *Annals of Operations Research, 291*(1), 605–626. https://doi.org/10.1007/s10479-019-03413-w
24. Norouzi, R., Baziyad, H., Aknondzadeh Noghabi, E., & Albadvi, A. (2022). Developing tourism users' profiles with data-driven explicit information. *Mathematical Problems in Engineering, 2022*, 6536908. https://doi.org/10.1155/2022/6536908
25. Omrani, H., Alizadeh, A., Emrouznejad, A., & Oveysi, Z. (2022). A novel best-worst-method two-stage data envelopment analysis model considering decision makers' preferences: An application in bank branches evaluation. *International Journal of Finance & Economics.*
26. Pourhatami, A., Kaviyani-Charati, M., Kargar, B., Baziyad, H., Kargar, M., & Olmeda-Gómez, C. (2021). Mapping the intellectual structure of the coronavirus field (2000–2020): A co-word analysis. *Scientometrics.* https://doi.org/10.1007/s11192-021-04038-2
27. Preeti, & Roy, S. (2020). *Performance modelling on banking system: A data envelopment analysis-artificial neural network approach BT - re-imagining diffusion and adoption of information technology and systems: a continuing conversation.* In: S. K. Sharma, Y. K. Dwivedi, B. Metri, & N. P. Rana (Eds.), (pp. 581–597). Springer International Publishing.
28. Preeti, Roy, S., & Kumar, K. (2020). *Radial data envelopment analysis approach to performance measurement: study on indian banking system BT-supply chain intelligence: Application and optimization.* In: K. Kumar & J. P. Davim (Eds.), (pp. 155–171). Springer International Publishing. https://doi.org/10.1007/978-3-030-46425-7_8
29. Shafiee, M., Sangi, M., & Ghaderi, M. (2013). Bank performance evaluation using dynamic DEA: A slacks-based measure approach. *Journal of Data Envelopment Analysis and Decision Science, 2013*(26), 1–12.
30. Shao, B. B. M., Shi, Z. (Michael), Choi, T. Y., & Chae, S. (2018). A data-analytics approach to identifying hidden critical suppliers in supply networks: Development of nexus supplier index. *Decision Support Systems, 114*, 37–48. https://doi.org/10.1016/j.dss.2018.08.008
31. Shirazi, S., Baziyad, H., Ahmadi, N., & Albadvi, A. (2020). A new application of Louvain algorithm for identifying disease fields using big data techniques. *Journal of Biostatistics and Epidemiology, 5*(3 SE-Original Article(s)). https://doi.org/10.18502/jbe.v5i3.3613

32. Stewart, C., Matousek, R., & Nguyen, T. N. (2016). Efficiency in the Vietnamese banking system: A DEA double bootstrap approach. *Research in International Business and Finance, 36*, 96–111. https://doi.org/10.1016/j.ribaf.2015.09.006
33. Stoica, O., Mehdian, S., & Sargu, A. (2015). The impact of internet banking on the performance of Romanian banks: DEA and PCA approach. *Procedia Economics and Finance, 20*, 610–622. https://doi.org/10.1016/S2212-5671(15)00115-X
34. Tone, K., & Tsutsui, M. (2010). Dynamic DEA: A slacks-based measure approach. *Omega, 38*(3), 145–156. https://doi.org/10.1016/j.omega.2009.07.003
35. Tsolas, I. E., Charles, V., & Gherman, T. (2020). Supporting better practice benchmarking: A DEA-ANN approach to bank branch performance assessment. *Expert Systems with Applications, 160*, 113599. https://doi.org/10.1016/j.eswa.2020.113599
36. von Geymueller, P. (2009). Static versus dynamic DEA in electricity regulation: The case of US transmission system operators. *Central European Journal of Operations Research, 17*(4), 397. https://doi.org/10.1007/s10100-009-0099-x
37. Wanke, P., Barros, C. P., & Emrouznejad, A. (2016). Assessing productive efficiency of banks using integrated Fuzzy-DEA and bootstrapping: A case of Mozambican banks. *European Journal of Operational Research, 249*(1), 378–389. https://doi.org/10.1016/j.ejor.2015.10.018
38. Yi-Chia, L., Chompunut, L., Tonapa, P., & Anupong, W. (2022). Measuring the environmental efficiency of the investment for offshore wind power site in Taiwan. *Sustainable Computing: Informatics and Systems, 35*, 100693. https://doi.org/10.1016/j.suscom.2022.100693
39. Yilmaz, A., & Güneş, N. (2015). Efficiency comparison of participation and conventional banking sectors in Turkey BETWEEN 2007–2013. *Procedia - Social and Behavioral Sciences, 195*, 383–392. https://doi.org/10.1016/j.sbspro.2015.06.338
40. Zhou, Z., Gao, M., Xiao, H., Wang, R., & Liu, W. (2021). Big data and portfolio optimization: A novel approach integrating DEA with multiple data sources. *Omega, 104*, 102479. https://doi.org/10.1016/j.omega.2021.102479
41. Zhu, J. (2022). DEA under big data: Data enabled analytics and network data envelopment analysis. *Annals of Operations Research, 309*(2), 761–783. https://doi.org/10.1007/s10479-020-03668-8
42. Zohdi, M., Rafiee, M., Kayvanfar, V., & Salamiraad, A. (2022). Demand forecasting based machine learning algorithms on customer information: An applied approach. *International Journal of Information Technology, 14*(4), 1937–1947. https://doi.org/10.1007/s41870-022-00875-3

Chapter 23
Exploring the Optimal Camera Placement Problem and Its Relationship with the Set Covering Problem

Malek Almousa, Matthias Ehrgott, and Ahmed Kheiri

Abstract Optimal Camera Placement (OCP) is the process of finding a subset of cameras that either maximises the coverage, such that the cost of cameras is reduced, or minimises the total cost of cameras, such that coverage constraints are satisfied. By adopting the latter formulation, the OCP problem can be formulated as a Set Covering Problem (SCP), as the concepts of the two problems are inherently similar. Until recently, the literature has not explicitly discussed this similarity. Hence, this paper examines the OCP problem by leveraging the formulation established in prior research. Our focus lies in the practical application, as we implement the model on all instances to derive meaningful insights. Furthermore, we explore techniques from the SCP literature that can be applied to address the OCP problem in future studies. In this study, we address 69 problem instances, utilising a benchmark set generated by other researchers. These instances were employed as part of the GECCO 2021 competition on the optimal camera placement problem and the unicost set covering problem. We provide detailed results, and we conclude with recommendations for future research.

Keywords Combinatorial optimisation · Set covering problem · Optimal camera placement

Introduction

In recent years, establishing an optimal camera network for surveillance purposes has been the subject of interest in several studies. This increased attention is prompted by the worldwide spread of surveillance systems, which are being employed to address various issues, including analysing crowd movements, monitoring transportation

M. Almousa (✉) · M. Ehrgott · A. Kheiri
Department of Management Science, Lancaster University, Lancaster LA1 4YX, UK
e-mail: m.almousa@lancaster.ac.uk

M. Ehrgott
e-mail: m.ehrgott@lancaster.ac.uk

A. Kheiri
e-mail: a.kheiri@lancaster.ac.uk

© The Author(s), under exclusive license to Springer Nature Switzerland AG 2024
A. Emrouznejad et al. (eds.), *Business Analytics and Decision Making in Practice*, Lecture Notes in Operations Research, https://doi.org/10.1007/978-3-031-61589-4_23

systems, or simply observing certain places for general purposes [13]. The idea of camera placement was first discussed in computational geometry in the 1970s by Chvátal [6]. The author's widely known Art Gallery Problem (AGP) has inspired numerous camera placement studies since its introduction. AGP is an approach for placing guards in an art gallery, where the goal is to minimise the number of guards, ensuring that every point in the art gallery is covered by at least one guard. Transforming this idea to camera placement, guards become cameras, and the general goal is to select the smallest subset of cameras that achieves full coverage.

Generally and more formally, *Optimal Camera Placement* (OCP) is the process of finding a subset of cameras that either maximises the coverage, such that the cost (or number) of cameras is reduced; or minimises the total cost (or number) of cameras, such that coverage constraints are satisfied. When working with the minimisation objective, the problem can be viewed as a *Set Covering Problem* (SCP). However, the similarity between the two problems has not been explicitly discussed in the OCP literature until recently [13]. Therefore, the aim of our study is to explore the connection between the two problems and discuss techniques from the SCP literature that can be applied to the OCP problem.

The rest of the paper is structured as follows: section "OCP Literature" provides a brief summary of the OCP literature, discussing different techniques that have been used to deal with the problem. This is followed by a description of SCP in section "Set Covering Problem (SCP)". Section "Problem Description and Formulation" provides a detailed problem description and formulations of our OCP problem. Then, a summary of the results of the problem is given in section "Computational Results". Finally, section "Conclusions" concludes our study and gives a hint of what can be done in future work.

OCP Literature

Inspired by the AGP, the use of optimal camera networks for surveillance has increased in the past few decades in order to fully monitor different areas, including public places, warehouses, buildings, and so on. The general goal is to maximise coverage or to minimise the cost (or number) of cameras, given a set of constraints [14]. When it comes to solving OCP, researchers have employed various tools. Some used exact methods to deal with the problem and find optimal solutions, while others used heuristic methods to find near-optimal solutions within a reasonable time. To elaborate on the latter, since the OCP is an $\mathcal{NP}$-hard problem [13, 14], finding optimal solutions can be time-consuming for sufficiently large instances. Moreover, if a client's priority is time rather than solution quality, then it might be sensible to use heuristic methods instead of exact methods to find a satisfactory solution in a reasonable amount of time.

The study in [11] employed three heuristic algorithms to tackle an OCP. The authors focused on finding the best algorithm capable of solving OCP problems for the surveillance of bridges. Their model aimed to minimise the total cost of cameras, while ensuring that a minimum coverage level is satisfied. Moreover, this study included different types of cameras to cover specific target points in a three-dimensional space, resulting, for example, in an increase in the number of camera locations, which can lead to the expansion of the size of the problem instances. Because of that, the authors started by examining their OCP instance using two popular heuristic methods, greedy and genetic algorithms, as well as a novel heuristic method that is called the Uniqueness Score with Local Search Algorithm (ULA). For the first method, the greedy algorithm allocates cameras starting from the cheapest one and going up until it reaches the highest possible number of cameras or achieves the minimum coverage level. At this point, the algorithm halts and provides the proposed camera network. For the second method, the genetic algorithms begins by randomly generating a population of chromosomes, where each chromosome represents a camera placement. This process continues until the minimum coverage level is reached by each chromosome. Subsequently, a subset of chromosomes that satisfy the minimum coverage is selected. Mutation and crossover operators are then applied to this subset, creating new generations and ensuring that the solutions do not become stuck in a local optimum. As demonstrated in [11], the last method, proven to be more effective than the other two approaches, begins with the first solution of ULA, inspired by the uniqueness score. It emphasises specific areas that are left uncovered by other cameras. Subsequently, a local search is used to enhance the solution by changing the selected cameras, aiming to find a new solution with a lower cost.

Another study in [22] also examined three different approaches for three OCP cases, one of which utilised an exact method, while the other two employed heuristic methods. In the first case, the objective was to maximise the coverage within a limited budget. The authors formulated the optimisation problem as a set covering problem and then applied dynamic programming to address it. For the second case, the goal was to minimise the total cost while ensuring full coverage constraints. Similar to the first case, the optimisation problem was formulated as an SCP, and branch-and-bound algorithms were employed. This involved relaxing the binary constraints, solving the new optimisation problem using primal and dual simplex methods. A solution to the original problem would be optimal if it satisfied the binary constraints. If not, further steps were taken until a satisfactory solution was found. Lastly, the third case integrated the previous two cases to form a multi-objective problem, aiming to both maximise coverage and minimise cost. Instead of looking for one optimal solution, the goal was to find a set of optimal trade-offs (Pareto optimal solutions). In this context, 'Pareto optimal' refers to a set of selected candidate cameras, where no feasible candidate cameras could improve coverage without worsening the cost simultaneously. To address this problem, the authors suggested using a heuristic method, specifically the multi-objective genetic algorithm NSGA-II [7].

Another method that was used to address an OCP problem is Differential Evolution (DE). In [23], this heuristic method was utilised to improve the performance of greedy algorithms in order to achieve full coverage. Their DE starts by using an array containing a number of cameras, where each array represents an individual within the population. The algorithm then uses the vector differential of two individuals from the previous generation to create a new individual. Consequently, the algorithm continues to generate improved individuals compared to those in the previous generations.

The work in [4] used a variant of DE, called set-based DE, inspired by the study in [15], to address their OCP problem. In contrast to the study in [23], the authors focused on minimising the number of cameras (i.e., cost reduction) while ensuring complete coverage. The difference between set-based DE and the original DE lies in the fact that the former is primarily utilised to solve permutation-based problems, whereas the latter can be applied to solve general problems (including set-based problems).

Other methods employed in recent literature include dynamic algorithms [2], simulated annealing [17], greedy algorithms [19] and hill climbing [1]. While some OCP studies have formulated their problem as an SCP, the study in [12] argues that almost none of these studies have fully exploited this similarity or employed techniques from the SCP literature to address the OCP problem. As a result, many techniques employed in the SCP literature have not yet been utilised to address the OCP problem. This suggests potential opportunities for contributions that can make a difference in the OCP field.

Set Covering Problem (SCP)

The Set Covering Problem (SCP) is a popular combinatorial optimisation problem classified as $\mathcal{NP}$-hard [10]. Throughout the years, numerous studies have explored the SCP to tackle a diverse range of real-world applications. These include solving transit crew scheduling problems [8], optimising transit crew scheduling design with multiple objectives [16], finding optimal quantity and location of gas detectors [20], assigning fire stations with ladder trucks [21], and scheduling wireless sensor networks [24], among others. For more information regarding the SCP and its applications, readers are advised to refer to [5].

A brief definition of SCP would be: given a zero-one matrix, the goal is to obtain a subset of columns that minimises the total cost associated with the selected columns, ensuring that all the rows of the matrix are covered by these columns [18]. To elaborate on that, consider matrix A with three rows and three columns:

$$A = \begin{bmatrix} 1 & 0 & 0 \\ 0 & 0 & 1 \\ 1 & 1 & 0 \end{bmatrix}$$

In this matrix, the value 1 indicates that a given row is covered by a given column, and 0 otherwise. For instance, element a_{11} shows that row 1 is covered by column 1. While element a_{33} shows that row 3 is not covered by column 3. If each column is associated with a specific cost, the objective of SCP would be to find a subset of columns that minimises the total cost while ensuring that all the rows of the matrix are covered by this subset. For example, if the cost is the same for all columns, an optimal solution could be selecting both columns 1 and 3, as this combination covers all given rows.

Formally, given a finite universal set $U = \{1, 2, \ldots, n\}$ and a family of subsets $S = \{S_1, S_2, \ldots, S_m\}$ where each S_i is a subset of U, and each subset S_i has an associated cost c_i, the set covering problem is to find a minimum-cost subset $C \subseteq S$ such that the union of the selected subsets covers the entire universal set, i.e., $\bigcup_{S_j \in C} S_j = U$. The goal is to minimise $\sum_{S_i \in C} c_i$, the total cost of the selected subsets.

From an optimisation perspective, the study in [13] asserts that the Optimal Camera Placement (OCP) can be reformulated as a set covering problem after certain pre-processing steps. In this context, the pre-processing phase involves the transformation of the OCP problem into a visibility matrix. This matrix matches each location in the surveillance area with every possible configuration (position and orientation) of the given cameras, resulting in a 0–1 matrix similar to matrix A. To elaborate, columns in this matrix represent the cameras, and rows represent the locations that need to be covered by the cameras. The objective of the transformed OCP problem is to find a subset of cameras that minimises the cost, ensuring that all the specified locations are covered by this subset. In essence, this conversion enables the application of SCP methodologies to address the OCP challenge effectively.

In the study conducted in [13], an in-depth illustration is provided regarding the possibility of using methods from the SCP literature and applying them to OCP problems. The study introduces different heuristic and metaheuristic methods that were employed to address SCP, and consequently, these methods could be adapted for solving OCP problems. For instance, greedy algorithm is one of the heuristic methods utilised in the SCP literature to address the problem's $\mathcal{NP}$-hard nature. As will be pointed out later, the greedy algorithm was also used in the OCP literature in recent years. Despite being tailored specifically for OCP problems, the main conceptual framework remains the same for both domains. Another approach discussed in this study involves the work of [9], who utilised the Row-Weighting Local Search (RWLS) algorithm for a special type of SCP known as the Unicost SCP. According to [13], this approach has not been explored in the OCP literature, and questions arise regarding its potential efficiency in solving OCP problems. In a subsequent study [12], different approaches from both OCP and SCP literature were employed on real OCP cases. Notably, they utilised the RWLS algorithm and demonstrated its efficacy in solving OCP problems. This illustrates the possibility of leveraging techniques from the SCP literature to effectively address challenges in the OCP domain.

Problem Description and Formulation

In this study, we address 69 three-dimensional problem instances, comprising 32 academic problem instances and 37 real-world instances. These instances were utilised as part of the GECCO 2021 competition on the optimal camera placement problem and the unicost set covering problem [3]. The academic problem instances represent different sizes of rectangularly modelled rooms, where ceilings are used for camera placement. On the other hand, the real-world instances represent diverse sizes of actual urban spaces, with the walls of multiple buildings serving as potential camera locations.

For each problem instance, a set of camera configurations (locations and orientations coordinates), referred to as *candidates*, is provided. Additionally, a grid containing a set of points in three-dimensional space, denoted as *samples*, must be covered by a subset of candidates. For each sample, there exists multiple candidates capable of overseeing it.

Consider the example with two samples (a and b) and four candidates (1, 2, 3, and 4). Suppose sample a can be covered by candidates 1, 2, and 4, while sample b can be covered by candidates 2 and 3. We can create a visibility matrix A for this example, where the first row represents sample a, the second row represents sample b, and columns represent candidates 1, 2, 3, and 4. Matrix A is presented below.

$$A = \begin{bmatrix} 1 & 1 & 0 & 1 \\ 0 & 1 & 1 & 0 \end{bmatrix}$$

The objective of the OCP problem is to cover all the given samples by identifying a subset of candidates that achieves this goal at the minimum cost. In the presented OCP example, assuming equal costs for each candidate, the optimal solution would be to select candidate 2. This choice is optimal because candidate 2 is the only one overseeing both samples, enabling the coverage of both samples with a single candidate. Other solutions, such as candidates 1 and 3 or candidates 3 and 4, would require paying for two candidates to cover the two samples. However, the optimal solution, in this case, is achieved by selecting just one candidate (i.e., candidate 2).

Mathematical Formulation

As previously mentioned, the OCP problem shares similarities with the set covering problem. Consequently, we model the OCP problem as an SCP. The formulation involves a binary matrix A ($a_{ij} = 1$ if row i can be covered by column j, and 0 otherwise) with m rows and n columns, where each column j is assigned a specific cost c_j. The mathematical model is expressed as follows:

$$\begin{aligned} &\text{minimise} \sum_{j=1}^{n} c_j x_j \\ &\text{subject to} \sum_{j=1}^{n} a_{ij} x_j \geq 1, \forall i \in \{1, \ldots, m\} \\ &x_j \in \{0, 1\}, \quad \forall j \in \{1, \ldots, n\} \end{aligned}$$

Here, $x_j = 1$ if column j is selected, and $x_j = 0$ otherwise. The objective function minimises the total cost, and the first constraints ensure that each row i is covered by at least one selected column j.

Now, for our specific OCP problem, all cameras have the same cost [3]. This simplification transform our SCP model into a Unicost SCP (USCP), where the cost parameter c_j is omitted:

$$\begin{aligned} \text{minimse} \quad & \sum_{j=1}^{n} x_j \\ \text{subject to} \quad & \sum_{j=1}^{n} a_{ij} x_j \geq 1, \forall i \in \{1, \ldots, m\} \\ & x_j \in \{0, 1\}, \quad \forall j \in \{1, \ldots, n\} \end{aligned}$$

In this formulation, $x_j = 1$ if candidate j is included in the solution, and $x_j = 0$ otherwise. The objective remains to minimise the total number of selected candidates, ensuring that each sample is covered by at least one candidate. This transition to a USCP model simplifies the objective by focusing on minimising the number of candidates, each having an equal cost, while maintaining the essential constraints for effective camera placement in the OCP context.

Computational Results

For this study, we utilised the PuLP[1] package in Python to build the model, which was then solved using Gurobi, a commercial Linear Programming solver. Experiments were conducted on an Intel(R) Core(TM) i5-8500T CPU @ 2.10GHz 2.11GHz with 8.00GB RAM. A time limit of 3 h was imposed on each problem instance, encompassing both model building and problem solving. The results for both the academic and real-world problem instances are presented in Tables 23.1 and 23.2, respectively. For instances where obtaining the optimal solution was unattainable, we provide the upper and lower bounds along with the calculated optimality gap using the formula:

$$\text{Gap} = \frac{|UB - LB|}{|UB|}$$

where UB is the upper bound, and LB is the lower bound. In cases where the optimal solution was found, both the upper and lower bounds will have the same value. Instances where no results were produced due to insufficient memory or time limitations are denoted by a dash '–' in the respective table entries.

As seen in both tables, there are many cells filled with a dash '-', mostly because the three-hour time limit was reached before completing the model building stage,

[1] https://pypi.org/project/PuLP/.

Table 23.1 Results for academic problem instances. m is the number of samples, n is the number of candidates, UB is the upper bound, LB is the lower bound, Gap is the optimality gap, and Time is the time taken in seconds

Instance	m	n	UB	LB	Gap	Time
AC_01	605	2904	7.00	7.00	0.00%	9.33
AC_02	2205	10584	4.00	4.00	0.00%	140.30
AC_03	4805	23064	3.00	3.00	0.00%	1411.43
AC_04	8405	40344	5.00	5.00	0.00%	7542.09
AC_05	13005	62424	–	–	–	–
AC_06	18605	89304	–	–	–	–
AC_07	32805	157464	–	–	–	–
AC_08	51005	244824	–	–	–	–
AC_09	73205	351384	–	–	–	–
AC_10	605	2904	20.00	17.37	10.00%	10800.00
AC_11	2205	10584	72.00	52.00	26.39%	10800.00
AC_12	4805	23064	168.00	109.81	34.52%	10800.00
AC_13	8405	40344	344.00	187.18	45.35%	10800.00
AC_14	13005	62424	723.00	0.00	100.00%	10800.00
AC_15	18605	89304	–	–	–	–
AC_16	32805	157464	–	–	–	–
AC_17	51005	244824	–	–	–	–
AC_18	73205	351384	–	–	–	–
AC_19	99405	477144	–	–	–	–
AC_20	129605	622104	–	–	–	–
AC_21	163805	786264	–	–	–	–
AC_22	202005	969624	–	–	–	–
AC_23	244205	1172184	–	–	–	–
AC_24	290405	1393944	–	–	–	–
AC_25	340605	1634904	–	–	–	–
AC_26	394805	1895064	–	–	–	–
AC_27	453005	2174424	–	–	–	–
AC_28	515205	2472984	–	–	–	–
AC_29	581405	2790744	–	–	–	–
AC_30	651605	3127704	–	–	–	–
AC_31	725805	3483864	–	–	–	–
AC_32	804005	3859224	–	–	–	–

Table 23.2 Results for real-world problem instances. m is the number of samples, n is the number of candidates, UB is the upper bound, LB is the lower bound, Gap is the optimality gap, and Time is the time taken in seconds

Instance	m	n	UB	LB	Gap	Time
RW_01	153368	32430	–	–	–	–
RW_02	285698	56132	–	–	–	–
RW_03	161099	32040	–	–	–	–
RW_04	304655	59137	–	–	–	–
RW_05	206900	34568	–	–	–	–
RW_06	380420	65691	–	–	–	–
RW_07	214889	42046	–	–	–	–
RW_08	382651	77986	–	–	–	–
RW_09	206816	39003	–	–	–	–
RW_10	368114	71323	–	–	–	–
RW_11	82437	15632	316.00	309.88	1.90%	10800.00
RW_12	136555	28109	–	–	–	–
RW_13	293138	61741	–	–	–	–
RW_14	81062	14916	337.00	334.77	0.59%	10800.00
RW_15	141309	27008	–	–	–	–
RW_16	105829	21063	–	–	–	–
RW_17	180453	35635	–	–	–	–
RW_18	79947	14423	338.00	336.65	0.30%	10800.00
RW_19	141114	26483	–	–	–	–
RW_20	332300	50284	–	–	–	–
RW_21	654068	90050	–	–	–	–
RW_22	83835	17203	399.00	392.93	1.50%	10800.00
RW_23	142326	31038	–	–	–	–
RW_24	201967	33880	–	–	–	–
RW_25	375680	59851	–	–	–	–
RW_26	105566	18043	–	–	–	–
RW_27	181090	32669	–	–	–	–
RW_28	136755	27838	–	–	–	–
RW_29	273964	49267	–	–	–	–
RW_30	263518	49354	–	–	–	–
RW_31	472660	87248	–	–	–	–
RW_32	124289	30189	–	–	–	–
RW_33	229231	55000	–	–	–	–
RW_34	134479	27329	–	–	–	–
RW_35	238546	47590	–	–	–	–
RW_36	135043	28162	–	–	–	–
RW_37	238492	50702	–	–	–	–

rendering it impossible to solve the problem. This issue arises mainly due to the substantial size of these problem instances, requiring hours or even days solely for model building. Another observation is that, even when the model is built for some instances, an optimal solution is not always achieved. In such cases, having the upper and lower bounds is particularly valuable, offering useful information about how close we are to reaching the optimal solution. For some instances, like RW_18, it appears that we are quiet close to the optimal solution. For these, extending the runtime may yield the optimal solutions. However, there are instances, such as AC_13, where the gap is notably large, indicating the need for an extended runtime to solve them.

If we take a look at the first problem instance in Table 23.1, AC_01 contains 605 samples and 2904 candidates. The solution for this problem instance involved using 7 candidates to cover all 605 samples, and the computation took 9.33 s. Figure 23.1 visualises this problem instance, with yellow points representing all candidates and blue points representing samples that must be covered by a subset of those candidates. The solution of AC_01 is also depicted in Fig. 23.1, where 7 yellow points on the graph represent the optimal locations of the candidates that cover all 605 samples.

Similar to AC_01, we visualise another problem instance in Fig. 23.2, namely RW_22. This real-world instance contains 17,203 candidates and 83,835 samples. A feasible solution was obtained with an upper bound of 399, a lower bound of 392.93, and a gap of 1.5%. The visual representation of this problem's solution can be observed in the same figure.

Addressing the time issue is paramount for achieving improved results. One approach involves tackling the size of the problem; for instance, reducing it by eliminating unnecessary sets from the visibility matrix. Additionally, exploring different optimisation methods could be beneficial, such as employing multi-objective optimisation. In this approach, the main objectives would involve maximising area coverage while minimising the number of cameras. Another viable approach is the

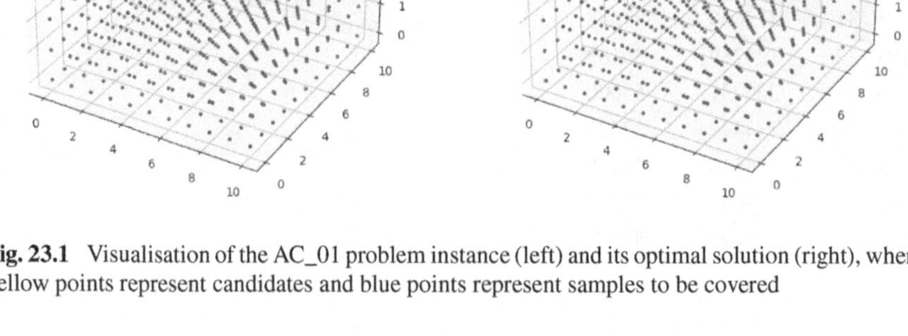

Fig. 23.1 Visualisation of the AC_01 problem instance (left) and its optimal solution (right), where yellow points represent candidates and blue points represent samples to be covered

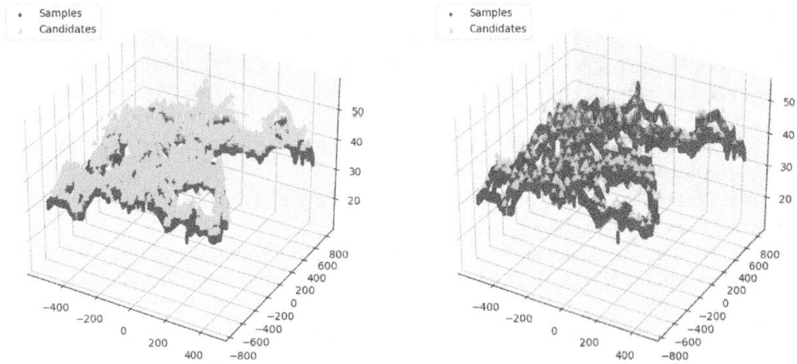

Fig. 23.2 Visualisation of the RW_22 problem instance (left) and its feasible solution (right), where yellow points represent candidates and blue points represent samples to be covered

utilisation of heuristic techniques. By studying various heuristic methods employed in the SCP literature and applying them to our OCP problem instances, we can potentially obtain near-optimal results within a reasonable amount of time.

Conclusions

OCP problem involves determining the optimal locations and orientations for a set of cameras, with the objective either being to maximise the coverage of a given surveillance area or to minimise the total cost or number of selected cameras. Although this problem has been formulated in a manner resembling the SCP, this similarity has only recently been explicitly discussed in the literature.

This paper studied the OCP problem by exploring its literature and understanding its relationship with SCP. Given that OCP is an $\mathcal{NP}$-hard problem, and the majority of problem instances are large, relying solely on exact methods did not provide solutions for all instances. Therefore, addressing the $\mathcal{NP}$-hard nature of the problem becomes crucial for future work. Potential strategies include reducing the size of problem instances, and/or resorting to heuristic techniques.

References

1. Ahn, J. W., Chang, T. W., Lee, S. H., & Seo, Y. W. (2016). Two-phase algorithm for optimal camera placement. *Scientific Programming*
2. Altahir, A. A., Asirvadam, V. S., Hamid, N. H. B., Sebastian, P., Saad, N. B., Ibrahim, R. B., & Dass, S. C. (2017). Optimizing visual surveillance sensor coverage using dynamic programming. *IEEE Sensors Journal, 17*(11), 3398–3405.

3. Brévilliers, M., Lepagnot, J., & Idoumghar, L. (2021). GECCO 2021 competition on the Optimal Camera Placement Problem (OCP) and the Unicost Set Covering Problem (USCP)
4. Brévilliers, M., Lepagnot, J., Kritter, J., & Idoumghar, L. (2018). Parallel preprocessing for the optimal camera placement problem. *International Journal of Modeling and Optimization, 8*(1), 33–40.
5. Caprara, A., Toth, P., & Fischetti, M. (2000). Algorithms for the set covering problem. *Annals of Operations Research, 98*(1), 353–371.
6. Chvátal, V. (1975). A combinatorial theorem in plane geometry. *Journal of Combinatorial Theory, Series B, 18*(1), 39–41.
7. Deb, K., Pratap, A., Agarwal, S., & Meyarivan, T. (2002). A fast and elitist multiobjective genetic algorithm: NSGA-II. *IEEE Transactions on Evolutionary Computation, 6*(2), 182–197.
8. Desrochers, M., & Soumis, F. (1989). A column generation approach to the urban transit crew scheduling problem. *Transportation Science, 23*(1), 1–13.
9. Gao, C., Yao, X., Weise, T., & Li, J. (2015). An efficient local search heuristic with row weighting for the unicost set covering problem. *European Journal of Operational Research, 246*(3), 750–761.
10. Garey, M. R., & Johnson, D. S. (1979). *Computers and intractability*. A Guide
11. Jun, S., Chang, T. W., & Yoon, H. J. (2018). Placing visual sensors using heuristic algorithms for bridge surveillance. *Applied Sciences,8*(1), 70.
12. Kritter, J., Brévilliers, M., Lepagnot, J., & Idoumghar, L. (2019). On the real-world applicability of state-of-the-art algorithms for the optimal camera placement problem. In *6th International conference on control, decision and information technologies (CoDIT)* (pp. 1103–1108). IEEE
13. Kritter, J., Brévilliers, M., Lepagnot, J., & Idoumghar, L. (2019). On the optimal placement of cameras for surveillance and the underlying set cover problem. *Applied Soft Computing, 74*, 133–153.
14. Liu, J., Sridharan, S., & Fookes, C. (2016). Recent advances in camera planning for large area surveillance: A comprehensive review. *ACM Computing Surveys (CSUR), 49*(1), 1–37.
15. Maravilha, A. L., Ramirez, J. A., & Campelo, F.: A new algorithm based on differential evolution for combinatorial optimization. In *BRICS congress on computational intelligence and 11th Brazilian congress on computational intelligence*, (pp. 60–66). IEEE.
16. Owais, M., Osman, M. K., & Moussa, G. (2015). Multi-objective transit route network design as set covering problem. *IEEE Transactions on Intelligent Transportation Systems, 17*(3), 670–679.
17. Rahimian, P., & Kearney, J. K. (2016). Optimal camera placement for motion capture systems. *IEEE Transactions on Visualization and Computer Graphics, 23*(3), 1209–1221.
18. Ren, Z. G., Feng, Z. R., Ke, L. J., & Zhang, Z. J. (2010). New ideas for applying ant colony optimization to the set covering problem. *Computers & Industrial Engineering, 58*(4), 774–784.
19. Suresh, M. S., Narayanan, A., & Menon, V. (2020). Maximizing camera coverage in multi-camera surveillance networks. *IEEE Sensors Journal, 20*(17), 10170–10178.
20. Vianna, S. S. (2019). The set covering problem applied to optimisation of gas detectors in chemical process plants. *Computers & Chemical Engineering, 121*, 388–395.
21. Walker, W. (1974). Application of the set covering problem to the assignment of ladder trucks to fire houses. *Operations Research, 22*, 275–277.
22. Yang, X., Li, H., Huang, T., Zhai, X., Wang, F., & Wang, C. (2018). Computer-aided optimization of surveillance cameras placement on construction sites. *Computer-Aided Civil and Infrastructure Engineering, 33*(12), 1110–1126.
23. Zhang, B., Zhang, X., Chen, X., & Fang, Y. (2016). A differential evolution approach for coverage optimization of visual sensor networks with parallel occlusion detection. In *IEEE international conference on advanced intelligent mechatronics (AIM)*, (pp. 1246–1251). IEEE
24. Zhang, X. Y., Zhang, J., Gong, Y. J., Zhan, Z. H., Chen, W. N., & Li, Y. (2015). Kuhn-Munkres parallel genetic algorithm for the set cover problem and its application to large-scale wireless sensor networks. *IEEE Transactions on Evolutionary Computation, 20*(5), 695–710.

Chapter 24
Human Resources Analytics and Talent Management

Rashmi Bezalwar and Deji Sotunde

Abstract This study examines the impact of AI-enabled HR analytics on organizational performance, using a quantitative design with a sample of 165 h managers and employees from two MNC Indian IT companies as research subjects. The research utilizes a self-completion questionnaire to assess different dimensions of HR practices and organizational performance, with a focus on AI-enabled HR analytics. AI-driven HR analytics are not related to organizational performance; factor analysis reveals that workforce planning and workforce forecasting, components of AI-enhanced HR practices play an important role in determining organizational performance. These results demonstrate the subtlety of AI contribution to the HR processes by highlighting its interrelation, rather than a linear causality in the performance outcomes. The research highlights the role of AI-powered HR with analytics on the enhancement of organizational efficiency working with other HR processes, however, the low regression coefficient shows other unexplored factors are contributing to the organizational performance. This work extends the debate on the contribution of AI-based HR analytics in enhancing organizational outcomes by proposing a more intensive examination of HR subsystem data.

Keywords AI · AI-based HR analytics · AI-powered HR · HR analytics · Organizational performance · Workforce planning · Succession planning · Talent management

Introduction

In the current competitive and dynamic business atmosphere firms must be equipped with new innovative technologies for them to get a strategic advantage. Such modern tech to have of late become very common is AI, particularly for HR analytics [1]. AI-Based HR analytics have massive potential to revolutionizes the traditional role

---

R. Bezalwar (✉) · D. Sotunde
Surrey Business School, University of Surrey, GU 27YW, Guildford, United Kingdom
e-mail: rb01682@surrey.ac.uk; rashmibezalwar@gmail.com

© The Author(s), under exclusive license to Springer Nature Switzerland AG 2024
A. Emrouznejad et al. (eds.), *Business Analytics and Decision Making in Practice*, Lecture Notes in Operations Research, https://doi.org/10.1007/978-3-031-61589-4_24

of human resources understanding of its workforce dynamics and help organizations access information that boosts performance [2]. This paper intends to research HR analytics assisted by AI in making judgments concerning strategic workforce planning, succession and talent management that can be focused on upgrading organizational effectiveness.

Organizations need to engage in strategic workforce planning to ensure that the skills and capabilities of their current and future human resource(s) align with the key objectives of the business its level of competitiveness. AI-driven HR analytics will allow organizations to analyze data related not only to the employee's skills and performance but also demographics, these combined factors can provide valuable insights into both current talent requirements as well as future needs [3]. On the same note, succession planning can incorporate this approach in its process of identifying future leaders and developing them within an organisation; this practice could be improved through the use of workforce modeling that are analytics-driven AI to identify high-potential employees who need specific development opportunities [4]. Moreover, the recruitment retention and development strategy under talent management can be applied using insights generated by AI to identify the best talents available for employment in any organization; predict employee turnover at a given point in time as well designing career paths aimed specially towards such employees [5].

In order to transform the traditional practices of strategic workforce planning, succession planning and talent management; HR analytics based on artificial intelligence would be utilized. With the help of AI algorithms and machine learning techniques, organizations can identify future talent needs which they will adapt accordingly [6]. In addition to the human resource process performance, this also influences organizational effectiveness positivity. Yet on the implementation and uptake of AI-supported HR analytics within organizations there are constraints surrounding issues related to data privacy, data ethics and resistance from organizational entities [7]. Thus, this paper is dedicated to the current state of AI-based HR analytics in such compounds where strategic functions dominate, and their strengths and weaknesses are acutely observed.

Research Objectives

- To investigate the underlying factor structure of AI-driven HR analytics tools and their relationship to organizational performance.
- To assess the reliability and validity of a new AI-driven HR analytics measurement instrument in various organizational settings.
- To determine the impact of specific components of AI-driven HR analytics on the efficiency and effectiveness of HR processes.

Literature Review

The development of AI in the field of HR symbolizes an escalatory change directed towards a data driven decision making approach to strategic workforce planning, succession planning, and talent management [8]. This review of the literature on AI driven analytics in human resources performance, examines whether or not this information can enhance the performance in these fields. When comprehending the advent of profound complexity and change in business landscapes, the introduction of AI applications in HR functions has created innovative solutions to ensure competitive positioning through faster, precise, and farsighted decisions [9]. However, another author noted that AI may be employed by HR managers in strategic workforce planning.

Strategic workforce planning, means connecting an organization's human capital desires with the plans and its long-term goals. It requires comprehension of current and future needs of the workforce [10]. AI driven HR analytics provide a pivotal manifestation of workforce modelling by forecasting future manpower requirements, determining yield dearth as well as offering solutions on ways to meet the training needs of its employees. Nambisan et al. [11] recently emphasised that with artificial intelligence tools, large datasets can be analyzed to predict trends, eliminate biases in planning practices, and provide recruitment practices according to future needs. Also, according to Mohammad and Darwish [12] human drivers behind scenario planning and workforce modelling could be supplemented by using AI that improves the speed and efficiency of change adaptation. Notwithstanding these developments, the incorporation of AI into strategic workforce planning poses challenges such as data privacy and the need to update algorithms constantly to fit the changing dynamics of the business strategies and labor markets as part of the efforts to sustain a desired level of brand equity among organizations [13].

In terms of leadership succession and organizational adaptability, AI-enabled analytics lend a significant advantage to succession planning due to the importance of this process. Through Artificial Intelligence, one of the challenges which the solution will overcome is the ability to identify potential leadership gaps, forecast future leadership needs, and recommend internal candidates for development based on skill, performance, and potential. Kreiser et al. [14] discussed the application of AI algorithms in matching candidates to leadership profiles for both objective and quantified decision-making in the selection process, helping HR professionals reduce biases in their succession decision-making and increase the quality of their selection. However, Khajeh et al. [15] points out that although AI plays a vital role in the study and analysis of leadership traits, significant caution is necessary to avoid over-dependence on algorithms, which may pressure single-mindedness, rigidity and linearity to the point of missing delicate social, cultural, or moral attributes [16].

Talent management involves the recruitment of employees, development of skills and abilities, retention of the employees and their efficient application for the realization of an the organization's strategic goal. Intelligent HR analytics are found to be remarkable as they help personalize employee experience, predict turnover, and optimize the deployment of talent. As Girisha and Nagendrababu [17] posits, AI can make it easier to understand patterns in behaviours, preferences, strengths, and weaknesses of the employees, thus allowing managers to develop a targeted programme of intervention with a framework of competenciesto encourage talent retention, and career development. Furthermore, Farrukh et al. [18] shows how AI can improve talent retention strategies as it identifies at-risk employees while providing insights that would enable management to offer potential employees a solution before, they consider quitting. In this regard, even though there are benefits from AI applications in forecasting outcomes from trends like in the betting industry, there are still major ethical considerations and increased risk of the AI reinforcing any existing bias due to errors in data or algorithms [19].

Achievement of the best performance for an organization remains the main objective behind the surge to incorporate AI into the analytics of HR. There is empirical evidence that the use of AI in HR practices is associated with better managerial results, for instance, higher commitment of employees, better alignment of talent, and flexibility regarding the reaction to environmental changes. As such, a study by Epley et al. [20] indicates companies that apply AI for HR analysis tend to have better performance in terms of innovation, employee satisfaction, and better financial results than their competition. This means that AI is recognized as having the capability to identify the strategic importance of transforming HR into a significant factor driving the success of a business [21].

Although using AI in HR strategy offers clear benefits, its use still raises several challenges including ethical issues, data privacy concerns, and explainability of AI-powered decisions making [22]. The inclusion of AI in the decision-making process in HR also requires following certain ethical principles and putting in place strict standards for the protection of data integrity with continuous monitoring to avoid bias resulting in unfairness within AI-based decisions [23].

AI-directed HR analytics provide several unique options in boosting strategic workforce planning, succession planning, and talent management that supports the organizational results. By using Artificial Intelligence in advancing human resource management, experts can use the predictive powers and analytical skills that the AI has to analyze issues to make better, objective, and strategic decisions [24]. But the use of an artificial mind in HR operations, therefore, entices the right ethical implication to be determined, the privacy of data considered and the human intelligence complementarity with artificial intelligence. Given the dynamic nature of this field, as it continues to grow, such research and practical application will remain necessary to seize as much value as possible from AI-based HR analytics as well as working through the issues it will present [25].

Research Gaps

Integration Challenges: Although the research draws attention to the advantages of AI in the HRM domain, there is little if any literature on the practical challenges organizations face in utilizing or integrating AI into existing HR systems and processes. More detailed studies into the above integration challenges especially in technical, cultural, and operational aspects are required.

Ethical and Bias Considerations: While the possible ethical concerns and risks of bias in AI algorithms are addressed in contemporary research there are no empirical studies on organizations that can be assisted in systematically identifying, reducing, and monitoring ethics and bias in the context of HR analytics. A significant gap literature is research aimed at creating and applying frameworks or regulatory guidelines in the adoption of ethical AI use in HR practices.

Long-term Impact on Workforce Dynamics: While it is not the main goal of this paper, the long-term phenomena of employee morale, job satisfaction, and organizational culture that last after the introduction of intelligent human resource analytics remain underexplored. However, longitudinal studies that investigate these effects can help to understand whether AI makes changes in engagement and retention rates in a period.

Comparative Effectiveness Across Industries and Geographies: A large portion of the research has been confined to the analysis of certain industries or the regions of the fellow countries. Comparative studies conducted across diverse industries and cultural contexts can be used to ascertain the effectiveness of employing AI in HR analytics, evaluating its adaptability and its limitations within specific cultural settings based on its utility.

AI in Small and Medium Enterprises (SMEs): Accordingly, there is a significant dearth in the research literature that revolves around the adoption and implementation of AI-led HR analytics in SMEs. Given the finite resources they have and the individual peculiarities inherent in SMEs, tunderstanding how to effectively leverage artificial intelligence for HR functions becomes a crucial subject. It has the potential to address significant challenges within this vital significant sphere of any economy.

Research Methodology

Research Design

The quantitative research design to investigate the effects of AI-based human resource analytics on organization performance is employed. A questionnaire, which assessed different HR practices and outcomes, with special emphasis on AI enabled HR analytics was designed to collect the necessary data.

Participants

Human resources managers and employees totaling to 165 people constituted the sample for this study. They were selected from two prominent IT companies in India which had fast integration, early adopters, of technology in HR practices. The demographic representation of the study sample featured a collection of roles in the HR function from beginner HR associates to chief HR executive.

Sampling Method

A non-random sampling method was used to identify the study respondents who were integral factors in the practice of HR analytics in their organizations. This sampling technique was selected to ensure that the data gathered were relevant to the research questions as it enabled the recruitment of the individuals who met specific criteria pertinent to the study.

Data Collection Instrument

Information was gathered with aid of a self-completion questionnaire. The questionnaire included items on different dimensions of HR practices and organisational performance, there was a section on AI-driven HR analytics. Questionnaires were designed according to a five-point Likert scale, from 'Strongly Disagree' to 'Strongly Agree' to determine the perceptions of respondents on the capabilities of AI driven HR analytics and its various aspects within the company.

Measures

The questionnaire items were generated from an in-depth literatures review about HR analytics and organisational performance. Constructs that were measured encompassed workforce planning accuracy, efficiency of HR processes, employee engagement and satisfaction, tenure, and innovation, among others. Items, however, proved to measure every construct, which ensured a complete coverage of the study and helped to increase the reliability of the data.

Data Analysis

Statistical analysis of the data was done using SPSS. Descriptive statistics was used to generate a summary of the data. Factor analysis was conducted as a means of investigating the underlying variables or components that contributed to the patterns of correlations among the variables. A Varimax rotation method with the Kaiser Normalization option was applied to facilitate a simpler, clearer and more distinct interpretable factor structure. Next, a regression analysis was undertaken to study the association between AI-related HR analytics and organisational efficiency. To determine the importance of the model and the influences on the predicted variables, an ANOVA test was performed, and a standardized coefficients approach was used to estimate the strength of relationships.

Data Analysis and Results

All 165 cases are valid,, the consistency and stability of the measurements show a positive indication, and the data set is complete with no missing covariate values under this reliability analysis. The high level of reliability justifies the scale for further research study.

Reliability

The SPSS output reveals that the survey or measuring instrument comprising 30 items has the Cronbach's Alpha value of 0.858, meaning high consensus between the items. A Cronbach Alpha of above 0.7 implies acceptable internal consistency, showing that the items measure the construct reliably (Tables 24.1 and 24.2).

Table 24.1 Total cases

		N	%
Cases	Valid	165	100.0
	Excluded a	0	0.0
	Total	165	100.0

a. Listwise deletion based on all variables in the procedure

Table 24.2 Alpha test

Reliability statistics	
Cronbach's Alpha	N of items
0.858	32

Factor Analysis

The factor analysis shown in Table 24.3 indicates the communalities after extraction for a set of items, which represent how much of the variance in each variable is accounted for by the extracted factors. The initial communalities are set at 1.00 for all items, as this represents the total variance before extraction. After extraction, the communalities range from 0.277 to 0.988, suggesting that the factors extracted explain between 27.7 to 98.8% of the variance in each variable.

Variables such as 'talent acquisition strategies' and 'strategic talent management decisions' have very high communalities (0.988), indicating that almost all of their variance is captured by the extracted factors. In contrast, 'HR analytics and enhanced organizational performance' has a relatively low communality (0.277), meaning that the factors extracted do not represent the variance in this item (Tables 24.4 and 24.5).

Regression Analysis

R (Correlation Coefficient): R = 0.058, which states a very weak positive relationship between the predictor and the dependent variable.

R Square: This is the R square, and it is 0.003, on the variance of the dependent variable, indicating that the model explains that 0.3% of the variance that is a negligible part.

Adjusted R Square: Adjusted for the number of predictors it is −0.003 indicating that the model does not have a good predictive power and may yield worse prediction than a simple mean model.

Std. Error of the Estimate: The standard error is 0.99459, quite high, considering the R Square value in light medium insecurity, providing high variance of points of the data around the regression fitting line (Table 24.6).

ANOVA (Analysis of Variance)

Sum of Squares (Regression): 0 0.542 points to the amount of variance captured by the model.

Sum of Squares (Residual): 161.240 reflects the unaccounted variation.

Degrees of Freedom (df): 1 for the regression plus 163 for residuals equal 164 overall hence the number of observations minus the estimated parameters.

Mean Square: It's the square of the sums divided by the respective freedom. For regression, it's 0.542, and for residuals, it's 0.998.

F value: The F statistic is 0.548 and it tests about the explained variance in the model is different and more significant than the uncovered variance.

Table 24.3 Communalities

	Initial	Extraction
Accuracy of our workforce planning	1.000	0.867
Qualiefficiency of our workforce planning process ty	1.000	0.800
Decisions on workforce scaling	1.000	0.920
Identifying skill gaps	1.000	0.954
Better forecasting of workforce needs	1.000	0.907
Adequate skills and resources	1.000	0.917
Succession planning processes ommunication	1.000	0.828
Potential successors	1.000	0.958
Clear information	1.000	0.880
Diverse and inclusive succession	1.000	0.905
Retention of high-potential employees	1.000	0.805
Succession planning outcomes entail sustainability	1.000	0.678
Transparency of our succession planning process	1.000	0.887
Talent acquisition strategies	1.000	0.988
Employee development outcomes	1.000	0.925
Employee wellbeing	1.000	0.582
Talent retention rates	1.000	0.887
Strategic talent management decisions	1.000	0.988
Employee engagement and satisfaction	1.000	0.925
Performance management and feedback	1.000	0.490
HR analytics and enhanced organizational performance	1.000	0.277
Productivity has improved	1.000	0.867
Employee turnover rates	1.000	0.800
Workforce capabilities and business goals	1.000	0.920
Workforce capabilities and business goals	1.000	0.954
Innovation and creativity	1.000	0.907
Challenges in integrating AI-driven HR analytics	1.000	0.917
Training programs	1.000	0.828
Implementation challenges	1.000	0.958
Data infrastructure is well-equipped	1.000	0.880
Change our HR practices	1.000	0.905
Ethical considerations	1.000	0.805
Extraction method: principal component analysis		

Table 24.4 Rotated component matrix

	Component							
	Workforce planning and productivity	Workforce scaling and skills	Talent development and succession planning	Retention and engagement	Diversity and innovation	Employee development outcomes	Strategic HR management	Organizational effectiveness
Accuracy of our workforce planning	0.876	0.107	−0.146	0.194	−0.112	0.005	−0.044	0.122
Qualiefficiency of our workforce planning process ty	0.354	0.554	−0.028	0.395	−0.264	−0.301	0.118	0.191
Decisions on workforce scaling	0.336	0.874	−0.011	0.041	0.08	0.183	−0.043	−0.003
Identifying skill gaps	−0.194	−0.021	0.946	−0.034	−0.052	−0.051	0.101	0.063
Better forecasting of workforce needs	0.77	0.455	−0.169	0.094	0.224	0.083	−0.075	−0.09
Adequate skills and resources	0.35	0.883	−0.026	0.044	0.068	0.07	−0.021	−0.047
Succession planning processes ommunication	0.808	0.311	−0.188	−0.012	0.134	0.103	−0.108	0.044
Potential successors	−0.173	−0.048	0.941	−0.03	−0.141	−0.084	0.116	0.003

(continued)

Table 24.4 (continued)

	Component							
	Workforce planning and productivity	Workforce scaling and skills	Talent development and succession planning	Retention and engagement	Diversity and innovation	Employee development outcomes	Strategic HR management	Organizational effectiveness
Clear information	0.788	0.409	−0.121	0.144	0.216	0.072	−0.034	−0.055
Diverse and inclusive succession	0.185	0.097	−0.171	0.075	0.9	0.053	0.015	0.117
Retention of high-potential employees	0.027	0.181	−0.152	0.792	−0.114	−0.069	−0.067	0.314
Succession planning outcomes entail sustainability	0.132	0.477	−0.164	0.499	0.339	−0.084	0.012	−0.185
Transparency of our succession planning process	0.215	−0.046	0.096	0.859	0.195	0.159	−0.013	−0.169
Talent acquisition strategies	−0.13	−0.022	0.187	−0.049	0.009	0	0.965	0.034
Employee development outcomes	0.092	0.121	−0.137	0.03	0.077	0.931	−0.005	0.102
Employee wellbeing	−0.104	0.068	0.025	0.008	0.289	0.211	−0.078	0.557

(continued)

Table 24.4 (continued)

	Component							
	Workforce planning and productivity	Workforce scaling and skills	Talent development and succession planning	Retention and engagement	Diversity and innovation	Employee development outcomes	Strategic HR management	Organizational effectiveness
Talent retention rates	0.215	−0.046	0.096	0.859	0.195	0.159	−0.013	−0.169
Strategic talent management decisions	−0.13	−0.022	0.187	−0.049	0.009	0	0.965	0.034
Employee engagement and satisfaction	0.092	0.121	−0.137	0.03	0.077	0.931	−0.005	0.102
Performance management and feedback	0.099	−0.127	0.073	0.057	0.028	0.036	0.101	0.666
HR analytics and enhanced organizational performance	0.211	0.051	0.058	0.089	−0.179	0.319	0.042	0.289
Productivity has improved	0.876	0.107	−0.146	0.194	−0.112	0.005	−0.044	0.122
Employee turnover rates	0.354	0.554	−0.028	0.395	−0.264	−0.301	0.118	0.191
Workforce capabilities and business goals	0.336	0.874	−0.011	0.041	0.08	0.183	−0.043	−0.003

(continued)

Table 24.4 (continued)

	Component							
	Workforce planning and productivity	Workforce scaling and skills	Talent development and succession planning	Retention and engagement	Diversity and innovation	Employee development outcomes	Strategic HR management	Organizational effectiveness
Workforce capabilities and business goals	−0.194	−0.021	0.946	−0.034	−0.052	−0.051	0.101	0.063
Innovation and creativity	0.77	0.455	−0.169	0.094	0.224	0.083	−0.075	−0.09
Challenges in integrating AI-driven HR analytics	0.35	0.883	−0.026	0.044	0.068	0.07	−0.021	−0.047
Training programs	0.808	0.311	−0.188	−0.012	0.134	0.103	−0.108	0.044
Implementation challenges	−0.173	−0.048	0.941	−0.03	−0.141	−0.084	0.116	0.003
Data infrastructure is well-equipped	0.788	0.409	−0.121	0.144	0.216	0.072	−0.034	−0.055
Change our HR practices	0.185	0.097	−0.171	0.075	0.9	0.053	0.015	0.117
Ethical considerations	0.027	0.181	−0.152	0.792	−0.114	−0.069	−0.067	0.314
Extraction method: principal component analysis								
Rotation method: varimax with kaiser normalization								
a. Rotation converged in 8 iterations								

Table 24.5 Components and variables

Component name	Variables with high loadings
1. Workforce planning and productivity	Accuracy of our workforce planning, better forecasting of workforce needs, productivity has improved
2. Workforce scaling and skills	Decisions on workforce scaling, adequate skills and resources, challenges in integrating AI-driven HR analytics
3. Talent development and succession planning	Identifying skill gaps, potential successors, succession planning processes
4. Retention and engagement	Retention of high-potential employees, employee engagement and satisfaction, transparency of our succession planning process
5. Diversity and innovation	Diverse and inclusive succession, change our HR practices, innovation and creativity
6. Employee development outcomes	Employee development outcomes, talent retention rates, employee wellbeing
7. Strategic HR management	Talent acquisition strategies, strategic talent management decisions, HR analytics
8. Organizational effectiveness	Performance management and feedback, HR analytics and enhanced organizational performance, ethical considerations

Table 24.6 Model summary

Model summary									
Model	R	R square	Adjusted R square	Std. error of the estimate	Change statistics				
					R square change	F change	df1	df2	Sig. F change
1	0.058a	0.003	−0.003	0.99459	0.003	0.548	1	163	0.460

a. Predictors: (Constant), AI Driven HR Analytics

Sig.: The p-value for the F statistic is 0.460 which is greater than the 0.05 alpha level meaning the model is not statistically significant (Table 24.7).

Coefficients

Constant: The intercept is 3,88 with a very low p-value (<0.001) that implies that the intercept is significant.

AI Driven HR Analytics: According to the coefficient, 0.046, it means that for each unit increase in AI Driven HR Analytics, there is 0.046 unit increase in Organisational

Table 24.7 ANOVA analysis

ANOVA						
Model		Sum of squares	df	Mean square	F	Sig.
1	Regression	0.542	1	0.542	0.548	0.460b
	Residual	161.240	163	0.989		
	Total	161.782	164			
a. Dependent Variable: Organisational Performance						
b. Predictors: (Constant), AI Driven HR Analytics						

Table 24.8 Coeffecients

Coefficients a						
Model		Unstandardized coefficients		Standardized coefficients	t	Sig.
		B	Std. error	Beta		
1	(Constant)	3.880	0.225		17.227	<0.001
	AI driven HR analytics	0.046	0.063	0.058	0.740	0.460
a. Dependent Variable: Organisational Performance						

Performance. With p-value for this predictor being 0.460; however, it does not have statistical significance in explaining Organisational Performance (Table 24.8).

There is no significant association between AI Driven HR Analytics and Organisational Performance in the regression model. Summary of the R Square value the model does not explain much of the variance of the dependent variable is very low. More, the ANOVA table hints at a non-chance significance ($p > 0.05$). Technically, the coefficients table corroborates this, indicating that AI Driven HR Analytics is insignificant as a predictor of Organisational Performance due to a high p-value.

Discussion

The quantitative analysis based on regression shows a negligible positive connection, which shows an R square value of 0.003, between AI-driven HR analytics and organisational performance. AI implementation in HR analytics does not appear to add significantly to the variance in organisational performance outcomes.

But, factor analysis shows a more multilayered picture of the organisational factors involved. It reveals that the workforce planning and productivity rely heavily on the accuracy of workforce planning and the projection of workforce requirements. Similarly, the link between workforce scaling and skills is very high with the results of decisions regarding workforce scaling, along with the integration of AI-driven HR analytics, this is an interesting reversal from the weak results of regression for workforce scaling and skills, as AI-driven HR analytics was very negatively associated with managers' self-assessments.

As such, results support literature that shows AI integration in HR and its use by managers as a tool for improving organisational efficiency [26]. Yet, they differ from previous studies like Al-[27], which revealed a significantly stronger AI impact on HR outcomes. Organizational performance's multidimensional nature is also shown by the factor analysis, which seems to be in line with [28, 29] and his findings that 'measuring such a construct is more than complicated.

These results suggest that from a theoretical standpoint, AI-human resources analytics' contribution to the organizational performance is less about the direct cause-and-effect relationship that the methodology implies but concerns its integration with other HR processes. At a more practical level, human resource professionals should consider how analytics from AI-based tools could be utilized in combination with workforce planning and scaling strategies to achieve maximum effectiveness.

A limitation of the study lies in the low regression coefficient, indicating that unexamined variables exert a substantial influence on organizational performance. Moreover, the cross-sectional design of the data prevents us from drawing causal inferences or conclusions.

Future studies should use panel data to evaluate AI-promoted HR analytics over a long period. The study could benefit in another way from studying a wider diversity of organisational performance indicators and adding some qualitative data to show the subtleties of using AI in HR procedures.

This study contributes to the current discussion about AI-based HR Analytics' effectiveness in improving organisational outcomes. While the apparent effect seems negligible, the interaction effects with other HR processes cannot be ignored. The exploration must continue into the full potential of AI as a tool in the ecosystem that is HR management in organisational interaction.

References

1. Thite, M. (2019). Electronic/Digital HRM: A Primer. *E-HRM: Digital Approaches, Directions & Applications*, 1–21.
2. Shrouf, H., Al-Qudah, S., Al Khawaldeh, K., Obeidat, A. M., & Al Rawashdeh, A. (2020). A study on relationship between human resources and strategic performance: The mediating role of productivity. *Management Science Letters, 10*(13), 3189–3196. https://doi.org/10.5267/j.msl.2020.5.002.
3. Walumbwa, F. O., Christensen-Salem, A., Perrmann-Graham, J., & Kasimu, P. (2020). An identification based framework examining how and when salient social exchange resources facilitate and shape thriving at work. *Human Resource Development Review, 19*(4), 339–361. https://doi.org/10.1177/1534484320946208.
4. Waheed, A., Xiaoming, M., Waheed, S., Ahmad, N., & Tian-Tian, S. (2020). E-HRM implementation, adoption and its predictors: A case of small and medium enterprises of Pakistan. *International Journal of Information Technology and Management, 19*(2–3), 162–180. https://doi.org/10.1504/IJITM.2020.106217.

5. Ringle, C. M., Sarstedt, M., Mitchell, R., & Gudergan, S. P. (2020). Partial least squares structural equation modeling in HRM research. *International Journal of Human Resource Management, 31*(12), 1617–1643. https://doi.org/10.1080/09585192.2017.1416655.
6. Taamneh, A., Alsaad, A. K., & Elrehail, H. (2018). HRM practices and the multifaceted nature of organization performance: The mediation effect of organizational citizenship behavior. *EuroMed Journal of Business, 13*(3), 315–334. https://doi.org/10.1108/EMJB-02-2018-0010.
7. Shrivastava, G., Kataria, N., Chabani, Z., Tongkachok, K., & Salameh, A. A. (2022). Role of E-Hrm practices on business effectiveness in the digital era–an empirical study. *Academy of Entrepreneurship Journal, 28*(S2), 1–9.
8. Shah, N. H., Michael, F., & Chalu, H. (2020). Conceptualizing challenges to electronic human resource management (e-HRM) adoption: A case of small and medium enterprises (SMEs) in Tanzania. *Asian Journal of Business and Management, 8*(4), 37–48.
9. Rahman, A. U., Saeed, M., & Garg, H. (2022). An innovative decisive framework for optimized agri-automobile evaluation and HRM pattern recognition via possibility fuzzy hypersoft setting. *Advances in Mechanical Engineering, 14*(10). https://doi.org/10.1177/16878132221132146.
10. Nowak, K. L., & Rauh, C. (2005). The influence of the avatar on online perceptions of anthropomorphism, androgyny, credibility, homophily, and attraction. *Journal of Computer-Mediated Communication, 11*(1), 153–178. https://doi.org/10.1111/j.1083-6101.2006.tb00308.x.
11. Nambisan, S., Lyytinen, K., Majchrzak, A., & Song, M. (2017). Digital innovation management: Reinventing innovation management research in a digital world. *MIS Quarterly: Management Information Systems, 41*(1), 223–238. https://doi.org/10.25300/MISQ/2017/41:1.03.
12. Mohammad, T., & Darwish, T. K. (2022). Human resource management in Jordan: Challenges and future prospects. In *HRM in the Global South: A Critical Perspective.* https://doi.org/10.1007/978-3-030-98309-3_7.
13. Li, X., & Sung, Y. (2021). Anthropomorphism brings us closer: The mediating role of psychological distance in User–AI assistant interactions. *Computers in Human Behavior, 118.*https://doi.org/10.1016/j.chb.2021.106680.
14. Kreiser, P. M., Kuratko, D. F., Covin, J. G., Ireland, R. D., & Hornsby, J. S. (2021). Corporate entrepreneurship strategy: Extending our knowledge boundaries through configuration theory. *Small Business Economics, 56*(2), 739–758. https://doi.org/10.1007/s11187-019-00198-x.
15. Khajeh, H., Bahari, A., & Rashki, A. (2021). TCF7L2 polymorphisms in type 2 diabetes, insight from HRM and ARMS techniques. *International Journal of Advanced Biological and Biomedical Research, 9*(3), 204–214.
16. Kaushal, N., Kaurav, R. P. S., Sivathanu, B., & Kaushik, N. (2023). Artificial intelligence and HRM: Identifying future research Agenda using systematic literature review and bibliometric analysis. *Management Review Quarterly, 73*(2), 455–493. https://doi.org/10.1007/s11301-021-00249-2.
17. Girisha, M. C., & Nagendrababu, K. (2019). E-Human resource management (E-HRM): A growing role in organizations. *International Journal of Management Studies, 6*(1), 98–104.
18. Farrukh, M., Ansari, N., Raza, A., Wu, Y., & Wang, H. (2022). Fostering employee's pro-environmental behavior through green transformational leadership, green human resource management and environmental knowledge. *Technological Forecasting and Social Change, 179.*https://doi.org/10.1016/j.techfore.2022.121643.
19. Doz, Y. (2020). Fostering strategic agility: How individual executives and human resource practices contribute. *Human Resource Management Review, 30*(1). https://doi.org/10.1016/j.hrmr.2019.100693.
20. Epley, N., Waytz, A., Akalis, S., & Cacioppo, J. T. (2008). When we need a human: Motivational determinants of anthropomorphism. *Social Cognition, 26*(2), 143–155. https://doi.org/10.1521/soco.2008.26.2.143.
21. Chen, S., Jiang, W., Li, X., & Gao, H. (2021). Effect of employees' perceived green hrm on their workplace green behaviors in oil and mining industries: Based on cognitive-affective system theory. *International Journal of Environmental Research and Public Health, 18*(8). https://doi.org/10.3390/ijerph18084056.

22. Chauhan, S., & Singh, A. (2022). E-HRM "electronic human resource management"–its effectiveness in hotels. *PUSA Journal of Hospitality and Applied Sciences, 8*(2), 12–17.
23. Chalvatzis, K. J., Malekpoor, H., Mishra, N., Lettice, F., & Choudhary, S. (2019). Sustainable resource allocation for power generation: The role of big data in enabling interindustry architectural innovation. *Technological Forecasting and Social Change, 144*, 381–393. https://doi.org/10.1016/j.techfore.2018.04.031.
24. Caligiuri, P., De Cieri, H., Minbaeva, D., Verbeke, A., & Zimmermann, A. (2020). International HRM insights for navigating the COVID-19 pandemic: Implications for future research and practice. *Journal of International Business Studies, 51*(5), 697–713. https://doi.org/10.1057/s41267-020-00335-9.
25. Budhwar, P., Malik, A., De Silva, M. T. T., & Thevisuthan, P. (2022). Artificial intelligence–challenges and opportunities for international HRM: A review and research agenda. *International Journal of Human Resource Management, 33*(6), 1065–1097. https://doi.org/10.1080/09585192.2022.2035161.
26. Botelho, C. (2020). The influence of organizational culture and HRM on building innovative capability. *International Journal of Productivity and Performance Management, 69*(7), 1373–1393. https://doi.org/10.1108/IJPPM-05-2019-0228.
27. Austin, A. A., Carpenter, T. D., Christ, M. H., & Nielson, C. S. (2021). The data analytics journey: Interactions among auditors, managers, regulation, and technology*. *Contemporary Accounting Research, 38*(3), 1888–1924. https://doi.org/10.1111/1911-3846.12680.
28. Al-Qudah, S., Obeidat, A. M., Shrouf, H., & Abusweilem, M. A. (2020). The impact of strategic human resources planning on the organizational performance of public shareholding companies in Jordan. *Problems and Perspectives in Management, 18*(1), 219–230. https://doi.org/10.21511/ppm.18(1).2020.19.
29. Aboalsamh, H. M., Khrais, L. T., & Albahussain, S. A. (2023). Pioneering perception of green Fintech in promoting sustainable digital services application within smart cities. *Sustainability (Switzerland), 15*(14). https://doi.org/10.3390/su151411440.

Chapter 25
Financial Information Quality: Analysis of Cloud Accounting Adoption on UAE Firms

Nora Azima Noordin, Ahmad Hayek, Mirjana Sejdini, Aysha Humaid, Nouf Sultan, Bashair Abdulla, and Mariam Yousif

Abstract This research investigates the impact of cloud accounting on the quality of financial information in the UAE through a quantitative method. A questionnaire was employed as the primary data collection tool, with 233 participants from various industries providing responses. The analysis yielded key findings supporting three hypotheses. H01, aligning with previous research confirmed the benefits of cloud accounting for small and medium-sized businesses, emphasizing its similarity to traditional accounting methods while ensuring secure financial information storage. Similarly, H02 indicated a significant effect of cloud accounting on the data efficiency of financial information quality, supported by studies that emphasizing the direct influence of information technology on data quality and efficiency. Furthermore, H03 revealed a notable impact on the data mining of financial information quality, corroborated by research highlighting the facilitative role of cloud computing in accessing and managing data. Overall, the study concludes that cloud accounting has a statistically significant and positive effect on data storage, data efficiency, and

N. A. Noordin (✉) · A. Hayek · M. Sejdini · A. Humaid · N. Sultan · B. Abdulla · M. Yousif
Faculty of Business, Higher Colleges of Technology, Sharjah Women's Campus, PO Box 7947, Sharjah, UAE
e-mail: nazima@hct.ac.ae

A. Hayek
e-mail: ahayek@hct.ac.ae

M. Sejdini
e-mail: msejdini@hct.ac.ae

A. Humaid
e-mail: H00417675@hct.ac.ae

N. Sultan
e-mail: H00388176@hct.ac.ae

B. Abdulla
e-mail: H00369664@hct.ac.ae

M. Yousif
e-mail: H00388667@hct.ac.ae

© The Author(s), under exclusive license to Springer Nature Switzerland AG 2024
A. Emrouznejad et al. (eds.), *Business Analytics and Decision Making in Practice*, Lecture Notes in Operations Research, https://doi.org/10.1007/978-3-031-61589-4_25

data mining in the selected UAE firms. These findings align with previous research on cloud computing's impact on accounting information systems and data mining tools, emphasizing the simplification of data access and central management of software and data storage. This research contributes to the understanding of cloud accounting's transformative potential, providing valuable insights for businesses and policymakers in the UAE.

Keywords Cloud accounting · Data storage · Data mining · Data efficiency · Financial information quality · UAE

Introduction

Cloud accounting is the practice of performing basic accounting tasks, such as managing and balancing the books, using software that resides in a remote location and is frequently delivered in an as-a-service model, such as software as a service (Saas), internet as a service (Iaas), platform as a service (Paas), and network as a service [1]. Cloud accounting is similar to self-introduced or conventional on-site accounting programming, except that it is enabled on remote servers, similar to the SaaS (Software as a Service) approach. When data is transferred to "the cloud," it is processed there and then delivered back to the client. The cloud-based accounting is the practice of storing and managing accounting data in the cloud. This means that the data is stored on remote servers that are accessed over the internet. Cloud-based accounting software can be accessed from anywhere with an internet connection, which makes it ideal for businesses with remote employees or those that need to access their financial data on the go. SaaS is a software distribution model in which a cloud provider hosts programs and makes them available to end customers via the internet. It is a method of delivering programs over the internet. Instead of installing and maintaining software, you simply access it via the internet, releasing yourself from the complexities of software and device maintenance. In this model, an independent software vendor (ISV) can contract with a third-party cloud provider to host the application. In the case of larger corporations such as Microsoft, the cloud provider may also be the software vendor. IaaS is utilized by businesses that want to outsource their data centers and computer resources to a cloud provider. Infrastructure components such as servers, storage, networking hardware, and virtualization resources are hosted by IaaS providers.

According to Parlińska and Petrovska [2] cloud accounting systems have an advantage over traditional accounting software in that they provide a somewhat safe (and often significantly more secure) means of keeping financial data. Cloud accounting has both advantages and disadvantages. Accessing customer accounts is simple because cloud accounting is accessible online. The reports for the clients are now easily accessible from anywhere at any time. Consequently, cloud accounting made information management and access easier for businesses. Data breaches may happen as a drawback of cloud accounting if an organization's workstation or PC containing basic financial information is lost or stolen. In any case, cloud accounting removes all traces of financial data from company PCs and encrypts cloud data access using a private key. By adopting cloud computing platforms, information sharing is also less difficult. Two users simply need to share their special credentials in order to access the same platform when using cloud accounting. Conventional techniques typically depend on transmitting data across computer drives, which are vulnerable to loss or theft.

The quality of financial reporting is defined as how accurately and honestly a financial statement describes the situation and performance of the finances of an organization [3]. As a result, precise information regarding the company's financial condition, activities, and cash flow creation is a necessity for a financial statement to be considered high quality. Relevance and reliability are two crucial qualitative aspects of financial data, according to CertDa Homepage [4]. To be useful, financial information needs to be accurate and relevant at the same time. Improving other aspects of quality will also raise the quality of financial data. Comparability, verifiability, timeliness, and understandability are the enhancing characteristics. Financial information quality is related to how accurately and fairly a financial statement represents the situation and performance of a company's finances. This indicates that the firm's overall performance in the market is reflected in the quality of its financial information [3].

The quality of financial data can be enhanced by cloud-based accounting in a number of ways. First, built-in controls in cloud-based accounting systems can aid in error prevention. These systems might, for instance, be equipped with functions that automatically identify possible mistakes or mandate certain actions from users before allowing them to save data [5]. Second, tracking and managing financial data may be simpler with cloud-based accounting systems. This is due to the fact that these systems frequently offer up-to-date information on financial performance. Errors can be found and fixed faster with the use of this data. Third, sharing financial information with stakeholders may be simpler with cloud-based accounting systems. This is due to the fact that most users of these systems can access financial data from any location with an internet connection. This can facilitate better financial reporting transparency and help stakeholders comprehend a company's economic condition. Cloud accounting may help organizations increase the quality of their financial information by lowering the risk of errors, enhancing the effectiveness of accounting procedures, and giving them access to real-time financial data, according to a different study by Dimitriu and Matei [6]. Thus, the primary goal of this research is to examine the impact of cloud accounting on the quality of financial information in the UAE. Data mining, data efficiency, and data storage are all taken into consideration in this study. Furthermore, this research signifies a progression in the ongoing exploration by researchers into the potential advantages brought about by data science in the realms of accounting, auditing, and the quality of financial information. The main impulse was referred to Noordin et al. [7] to identify variations in the perception of how AI impacts audit quality in the UAE. Additionally, in accordance with the findings of Hayek et al. [8], an additional investigation was conducted to scrutinize the perspectives of external auditors regarding the practicality and usefulness of Machine

Learning in the auditing domain within the UAE. The current study was motivated by Hayek [9] prior research on the application of Big Data Analytics in external audit, prompting an in-depth exploration of the intricate relationship between the data science domain and its consequences for financial accounting and auditing.

In light of the aforementioned, this study could assist accounting professionals by providing them with a more thorough understanding of such topics and assisting cloud accounting developers in making appropriate adjustments in response to accountants' needs. Additionally, it will boost the decision-making process of the firms by enhancing the quality of the findings while using these technologies.

Literature Review

Quality of Financial Information and Decision Making

When making decisions, the primary factor is the quality of the financial information and reports available. Financial statements may aid management in enhancing the way the company is run, according to Georgen Scarborough Associates homepage [10]. It will be easier for management to handle day-to-day operations more effectively if they have a thorough understanding of revenue and cost issues. Financial statements can be used to more easily identify development opportunities that are achievable.

In addition to that, the financial statements provide the management with an overview of the financial situation and assist them in managing the company. Also, it will increase the decision-making capacity of lenders and investors by supplying information through a range of statements, such as an income statement and balance sheet. Investors and analysts can use that information from financial statements to assess a company's creditworthiness and worth. They will therefore be able to decide whether or not it is beneficial to invest in the company. By giving them thorough financial statements, the company may give potential investors all the information they need to assess the viability of the venture. High-quality reporting is valuable for decision-making because it captures the economic reality of the company's operations during the reporting period and accurately reflects the financial situation of the business after the reporting period, according to the [11].

Cloud Accounting and Information Quality

Since the beginning of the COVID-19 pandemic, there have been significant changes in cloud computing and the quality of financial information. During the pandemic, cloud computing technology gained popularity as businesses sought to increase scalability, enable remote work, and ensure business continuity. According to Prajapati

[12], the UAE government has established a rule that all businesses in the UAE must abide by, which is that all businesses in the UAE must comply with the international financial reporting standards (IFRS) in all financial processes. This rule is intended to help manage all businesses in the UAE and prevent any issues with maintaining the quality of financial information. This demonstrates that the government of the UAE recognizes the value of maintaining and improving the quality of the accounting records. The use of cloud computing has resulted in significant consequences for the quality of financial information, as companies are increasingly depending on machine learning, artificial intelligence, and data analytics to obtain insights into their financial operations. Real-time data access and analysis made possible by the integration of cloud-based financial management systems enhanced decision-making procedures and the accuracy of financial reporting [13].

Cloud-based accounting can improve the quality of financial information [5]. The study mention that cloud-based accounting can enhance data timeliness, accuracy, and completeness. Additionally, this study showed that cloud-based accounting can aid in lowering the fraud risk. This is due to the fact that most cloud-based accounting solutions come with integrated security measures that can aid in preventing illegal access to financial data. According to research conducted in Lebanon to examine how using cloud accounting may enhance the qualitative characteristics of financial reports and information. Skafi et al. [14] mentioned that adopting cloud systems has many benefits, including the software's ability to be accessed from any data center that offers different levels of security for better protection of company data and a reduction of the need to purchase costly and effective tools and equipment because all processing, applications, and account setting take place within the software, which is the cloud. These benefits may lead to a reduction in the company's operating and finance expenses, which benefits the company overall. One of the benefits of cloud computing systems for businesses, is by employing these services produces accurate and high-quality accounting information and reports, which may be used to make rational decisions. Mohammadi and Mohammadi [15] reported in his article that using cloud computing enhances the quality and reliability of accounting information and reports. The research also mentioned that using cloud computing improperly involves a number of risks. These risks include misperceptions regarding the human risks associated with using cloud computing, security issues, and contract difficulties. The results of this study, however, do not raise any serious concerns about the possibility that adopting cloud computing could negatively impact the standards of international financial reports. A study was carried out in Palestine by Alhelo [16] to investigate the effects of cloud computing adoption on enhancing the effectiveness of accounting information systems. In this study, the researcher concentrated on the Palestine Stock Exchange during the COVID-19 pandemic. According to this study, the use of cloud computing directly improves the effectiveness, reliability, and security of accounting information systems for service companies listed on the Palestine Exchange.

Hypotheses Development

According to Pawar et al. [17], cloud computing is the delivery of computing services via the Internet, including servers, storage, database systems, networking, applications, and analytics, with the goal of accelerating innovation, increasing resource flexibility, and reducing costs. Instead of investing in their own data centers or computing infrastructure, businesses can rent anything from software to storage from a cloud provider. The range of options available for cloud computing services has expanded to encompass artificial intelligence, natural language processing, basic office applications, networking, storage, and processing power. The same study asserts that storage is an essential part of cloud computing. Storage can be used for an enterprise database or for simple data storage, similar to what is done with a local hard drive. Cloud storage stores data across multiple third-party services, as opposed to traditional networked data storage, which keeps data on dedicated servers.

Cloud storage has a number of benefits. First of all, by offering a variety of file-sharing choices, it streamlines data sharing and makes information exchange easier for clients. Although many cloud service providers allow two users to share data on the same platform, very few provide inter-file sharing capabilities. Second, automation is a major advantage of cloud storage since it functions like a hard drive on your device, preventing interruptions to continuing tasks while saving data to the cloud. The service provider facilitates this autonomy, making sure that the actions of one user don't impact those of other users. Additionally, cloud storage companies use redundant servers to protect data and place a high priority on data preservation. Data is securely kept, even in the unlikely case of a database center failure. Finally, because cloud storage enables businesses to outsource their storage requirements and so lowers the need for internal resources, it is a cost-effective solution. Long-term storage is a popular choice for both individuals and enterprises because so many companies offer cost-effective options. According to Owolabi and Izang [18] research, cloud accounting software is used to store accounting data in the cloud. Data pertaining to accounting is accessible to owners and employees alike from any location with an Internet connection. Distributed computing is becoming more and more common place among enterprises. From networked machinery to online training programs, people all over the world are using the cloud to communicate with clients and create increasingly alluring strategic policies. Zebua and Widuri [19] mentioned that the desktop accounting system used by businesses to handle transaction accounting processes is very inefficient or cannot be integrated with other systems, such as those that run on a single computer and where data can only be transferred via USB drive or hard drive. This study discusses this issue, making the adoption of cloud accounting crucial. This kind of storage is unreliable and unsafe. Users are concerned by insecure data storage since, although it is stored on a local server, only one person has access to it, meaning that critical customer information and financial details are not visible to others. Consequently, the first hypothesis is represented:

H1: Cloud Accounting does not have a significant effect on data storage of financial information quality in selected firms in the UAE.

According to Smith [20] the process of making data easy to use, manage, and access is known as data efficiency. Larger scale firms are often concerned about it since their broad networks and records might make it difficult to locate and use a particular piece of data. Also, he stated that data efficiency has a considerable hardware component, even though what is most challenging in the configuration and setup process is organizing data in a way that makes it easier to find and access. The process of retrieving data from a hard drive or network might become much more difficult than it needs to be due to an outdated and ineffective system. This indicates that using a cloud computing system will improve data efficiency by storing and organizing data efficiently and conveniently relocating data. Arranging information across internal data centers, virtual environments, and cloud repositories to make it easier to manage and access is a challenge for these larger enterprises. This demonstrates that data efficiency is linked mainly to the cloud systems, if the cloud system is utilized the data efficiency level will be improved.

The most effective utilization of data by organizations is achieved through the arrangement of information storage systems and the application of diverse procedures to data. This is also known as data efficiency. When data is efficient, people who need it may find it more quickly and easily, depending on the specific use case. The most important advantages of data efficiency for firms include improved analytics quality, increased output, and reduced costs. Consequently, these advantages are considered objectives to be achieved for the businesses in the market. Thus, the second hypothesis was developed:

H2: Cloud Accounting does not have a significant effect on data efficiency of financial information quality in selected firms in the UAE.

Li et al. [21] delve into the realm of data mining to enhance the quality of financial information. The authors argue that data mining techniques can play a pivotal role in identifying and rectifying errors, inconsistencies, and anomalies in financial datasets, thereby improving the reliability and accuracy of financial information. One of the key advantages of data mining in financial data quality enhancement is its ability to uncover hidden patterns and relationships within large datasets. Through the application of advanced algorithms, data mining can identify outliers and anomalies that may indicate data entry errors, fraudulent activities, or other irregularities in financial records. By detecting these issues early on, organizations can take proactive measures to address them, preventing potentially costly errors and financial mismanagement. Furthermore, the authors emphasize the importance of data cleansing and preprocessing techniques in data mining for financial information quality. They argue that by cleaning and preparing the data effectively, organizations can ensure that the information fed into their financial systems is accurate and reliable from the outset. This, in turn, reduces the likelihood of downstream errors and discrepancies.

[8] Hayek et al. also highlight the role of predictive modelling in improving financial data quality. By using historical financial data and advanced machine learning algorithms, organizations can forecast future financial trends and identify potential

risks. This predictive capability enables better decision-making and risk management, ultimately contributing to the overall quality of financial information. Data mining techniques offer a powerful means of enhancing the quality of financial information. By leveraging advanced algorithms, data cleansing processes, and predictive modelling, organizations can detect and rectify errors, improve data accuracy, and make more informed financial decisions. Ali et al. [22] explore the application of data mining in the realm of financial fraud detection. The authors argue that data mining techniques are crucial for identifying fraudulent activities, anomalies, and patterns within financial data, aiding organizations in safeguarding their financial assets and maintaining trust in financial systems. One of the key advantages of data mining in detecting financial fraud is its ability to analyze vast amounts of data in real-time. Financial transactions generate enormous datasets, making it impossible for humans to manually review every transaction for potential fraud. Data mining algorithms can sift through these datasets rapidly, flagging suspicious activities for further investigation. The authors also emphasize the importance of anomaly detection in financial fraud prevention. Data mining techniques can identify unusual patterns or deviations from normal transaction behaviour, which may be indicative of fraud. By setting up alert systems based on these anomalies, organizations can respond swiftly to potential threats and minimize financial losses. Furthermore, Hayek et al. [8] highlight the role of machine learning in building predictive models for fraud detection. These models can learn from historical data to recognize new and evolving fraud patterns. This adaptive capability is crucial in an environment where fraudsters continually adapt their tactics to bypass traditional security measures. By leveraging real-time analysis, anomaly detection, and machine learning, organizations can proactively identify and mitigate fraudulent activities, safeguarding their financial integrity and the trust of their stakeholders. As a result, the third hypothesis is represented;

H3: Cloud Accounting does not have a significant effect on data mining of financial information quality in selected firms in the UAE.

Methodology

The study used a cross-sectional survey approach to look into the relationship of financial information and cloud accounting. Professional accountants, auditors, and information technology specialists from a range of industries, including financial services, technology, healthcare, education, and construction, were the study's target demographic. The survey instrument was created using Microsoft Forms and consists of 35 questions on an online survey with a Likert-type scale. Emails containing this questionnaire were sent to companies that are listed on the Dubai Financial Market.

Figure 25.1 illustrates the collection of 233 responses, of which 38% came from the financial services sector, 23% from the healthcare, education, and other professional services sector, 20% from manufacturing, 9% from the technology sector, and 5% from each of the other sectors. Using the Cronbach's Alpha test, the validity

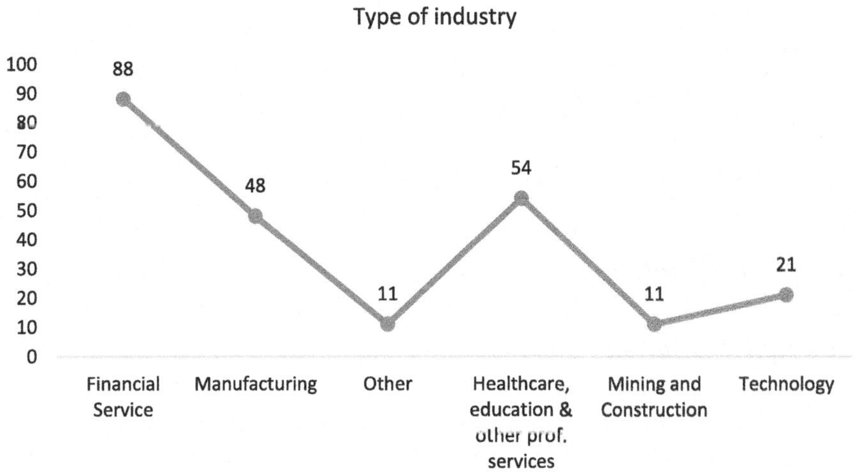

Fig. 25.1 Responses rate based on type of industry

and reliability of the research instrument were evaluated. Data analysis involved descriptive statistics and linear regression to test the hypotheses formulated in the study.

Reliability Test

The research instrument's reliability was assessed using Cronbach's Alpha, yielding a result of 0.91, indicating a high level of consistency as in Table 25.1. According to the established literature, a Cronbach's Alpha of 0.70 and above is considered acceptable. Consequently, this study adopts 0.70 as the minimum threshold for Cronbach's Alpha. As all study variables exhibit a Cronbach's Alpha exceeding 0.70, their reliability is affirmed.

Table 25.1 Reliability test result

Cronbach's α	N of items	Mean	Std. deviation
0.91	232	3.99	0.58

Table 25.2 Descriptive statistics

Variable	N	Mean	Std. deviation
Cloud accounting	232	4.08	0.66
Data storage	232	3.84	0.71
Data efficiency	232	3.91	0.73
Data mining	232	4.15	0.54

Results and Finding

Descriptive Analysis

Table 25.2 displays the mean and standard deviation of questionnaire items corresponding to the study variables (CA, DS, DE, and DM). Descriptive analysis reveals that all questionnaire items scored a mean scale higher than 3.00, indicating a positive statistical perspective. This same analysis was applied to the study variables as a whole, with calculated mean and standard deviation values. All variables exhibited scores exceeding the mean scale of 3.00, affirming their statistical positivity. The presented data suggests that a majority of accountants believe in the influential role of Financial Information Quality metrics (Data Storage, Data Efficiency, Data Mining) through cloud accounting in shaping changes in financial information quality.

Hypothesis Testing Using Linear Regression

H01: Cloud Accounting does not have a significant effect on Data Storage of financial information quality in selected firms in the UAE

In examining Tables 25.3, 25.4, and 25.5, a linear regression analysis was employed to assess the hypothesis. The calculated t-value of 9.28, significant at the 0.05 level (DS = 0.65, P = 0.0000 < 0.05), leads to the rejection of the null hypothesis as indicated in Table 25.5. This implies a substantial and positive impact of cloud accounting on the data storage quality of financial information for the chosen firms in the UAE. This finding aligns with the perspectives presented by Akpan et al. [5] and Al-Zoubi [23]. Additionally, Table 25.3 indicates a positive and strong relationship (R = 0.70) between the variables, affirming that Cloud Accounting has the potential to enhance the efficiency of data storage for the selected firms in the UAE. The analysis of variance (ANOVA) in Table 25.4 gauges the model's significance and fitness. The computed F-value of 223.54 surpasses the tabulated F-value of 0.00 at a 5% significance level, leading to the rejection of the null hypothesis and the acceptance of the alternative hypothesis.

H02: Cloud Accounting does not have a significant effect on Data Efficiency of financial information quality in selected firms in the UAE.

Table 25.3 Model Summary for H1

Model	R	R^2	Adjusted R^2	RMSE
1	0.70	0.49	0.49	0.47

Table 25.4 ANOVA analysis result for H1

Model	Sum of Squares	df	Mean Square	F	p
Regression	49.34	1	49.34	223.54	0.0000
Residual	50.77	230	0.22		
Total	100.11	231			

Table 25.5 Coefficient analysis result for H1

Variable	Unstandardized coefficient	Std. error	t	p
Intercept	1.58	0.17	9.28	0.0000
DS	0.65	0.04	14.95	0.0000

To investigate H2, linear regression analysis was utilized, revealing a significant t-value of 10.21 at the 0.05 level (DE = 0.56, P = < 0.00001 < 0.05), resulting in the rejection of the null hypothesis, as detailed in Table 25.8. This suggests a substantial and positive influence of cloud accounting on the data efficiency of financial information within the selected firms in the UAE. This observation aligns with the viewpoints presented by Akpan et al. [5] and Mugyenyi [24]. Furthermore, Table 25.6 illustrates a positive and robust relationship (R = 0.62) between the variables, underscoring Cloud Accounting's potential to enhance data efficiency for the chosen firms in the UAE. The analysis of variance (ANOVA) in Table 25.7 evaluates the model's significance and fitness. With a computed F-value of 145.96 surpassing the tabulated F-value of 0.00 at a 5% significance level, the null hypothesis is rejected, and the alternative hypothesis is accepted (Table 25.8).

Table 25.6 Model Summary for H2

Model	R	R^2	Adjusted R^2	RMSE
1	0.62	0.39	0.39	0.52

Table 25.7 ANOVA analysis result for H2

Model	Sum of squares	df	Mean square	F	p
Regression	38.87	1	38.87	145.96	<0.00001
Residual	61.24	230	0.27		
Total	100.11	231			

Table 25.8 Coefficient analysis result for H2

Variable	Unstandardized coefficient	Std. error	t	p
Intercept	1.89	0.18	10.21	<0.00001
DE	0.56	0.05	12.08	<0.00001

H3: Cloud Accounting does not have a significant effect on Data Mining of financial information quality in selected firms in the UAE.

To explore H3, a linear regression analysis was applied, revealing a noteworthy t-value of 3.65 at the 0.05 level (DM = 0.75, P = $< 0.00001 < 0.05$), leading to the rejection of the null hypothesis, as outlined in Table 25.11. This implies a substantial and positive impact of cloud accounting on the data mining aspect of financial information within the chosen firms in the UAE. This finding resonates with the perspectives presented by Akpan et al. [5] and Ogunsola [25]. Furthermore, Table 25.9 illustrates a positive and robust relationship (R = 0.62) between the variables, emphasizing Cloud Accounting's capacity to enhance data mining for the selected firms in the UAE. The analysis of variance (ANOVA) in Table 25.10 assesses the model's significance and fitness. With a computed F-value of 144.99 surpassing the tabulated F-value of 0.00 at a 5% significance level, the null hypothesis is rejected, and the alternative hypothesis is accepted (Table 25.11).

Table 25.9 Model Summary for H3

Model	R	R^2	Adjusted R^2	RMSE
1	0.62	0.39	0.38	0.52

Table 25.10 ANOVA analysis result for H3

Model	Sum of squares	df	Mean square	F	p
Regression	38.87	1	38.71	144.99	<0.00001
Residual	61.40	230	0.27		
Total	100.11	231			

Table 25.11 Coefficient analysis result for H3

Variable	Unstandardized coefficient	Std. error	t	p
Intercept	0.96	0.26	3.65	<0.00001
DM	0.75	0.06	12.04	<0.00001

Discussion

The objective of this research study was to investigate the impact of Cloud accounting on the quality of financial information in the UAE. Utilizing a quantitative survey approach, the study employed a questionnaire as the primary data collection tool. A total of 233 participants from various industries provided responses to the questionnaire. The subsequent analysis of the gathered data led to the following key findings. H01 result agreed with [26] who conducted research on cloud accounting in Nigeria. This research confirmed that cloud accounting is beneficial for small and medium-sized businesses in Nigeria and shows that the advantage of utilizing this system is that it functions similarly to accounting by using the web and storing financial information safely. This result also agreed with [27], they performed research in Iran to test The Impact of Information Technology Methods on Accounting Information Quality. Their test's findings indicated that there is a substantial correlation between the use of information technology techniques and the quality of accounting information.

Similarly, the result for H02 indicates a significant effect of Cloud Accounting on the Data Efficiency of financial information quality in the selected firms. The very low p-value reinforces the statistical significance of the observed effect. The findings of H02 agree with the study of Abdelraheem et al. [28] where they performed a study to determine how information technology at the Nile Bank in Sudan affected the quality of accounting data. According to their study, information technology's ability to gather, process, store, and share data has a direct influence on the quality and efficiency of financial data. The results also agreed with [5] where they demonstrated that Cloud accounting has beneficial effects on the data efficiency and quality of financial information of selected firms in Nigeria. Based on the examination of data efficiency, it was shown that cloud accounting practices have an effect on the efficiency of financial information.

H03 also shows a significant effect, suggesting that Cloud Accounting has a notable impact on the Data Mining of financial information quality in the chosen firms. In summary, all three hypotheses are supported by the data, indicating that Cloud Accounting has a statistically significant and positive effect on Data Storage, Data Efficiency, and Data Mining of financial information quality in the selected firms in the UAE. Based on Al-Zoubi [23] research that has been conducted to determine how cloud computing affects the components of the accounting information system and its quality. They demonstrated that the users of cloud computing will find it simpler to get the data they need, since they may access all financial reports offered by the established system at any time, by the authority granted by the organization to its customers. These study's results support our findings regarding the effect of cloud accounting on the data mining of financial information since we discovered that cloud systems do not affect data mining. Sharma et al. [29] research align with our findings as will regarding the data mining since they discovered that through the use of data mining in cloud computing, businesses can centrally manage their software and data storage while guaranteeing their consumers receive effective, dependable,

Table 25.12 Hypothesis summary finding

Hypothesis	t	p	Result
H1: Cloud Accounting does not have a significant effect on Data Storage of financial information quality in selected firms in the UAE	14.95	0.0000	Significant effect
H2: Cloud Accounting does not have a significant effect on Data Efficiency of financial information quality in selected firms in the UAE	12.08	<0.00001	Significant effect
H3: Cloud Accounting does not have a significant effect on Data Mining of financial information quality in selected firms in the UAE	12.04	<0.00001	Significant effect

and secure services, which means that there is a relationship between data mining and cloud accounting as cloud computing's data mining tools are a crucial component of today's enterprises. Table 25.12 summarize the hypothesis results.

Conclusion and Recommendation

In conclusion, the UAE government stands at a pivotal role offering a remarkable opportunity to empower businesses through the establishment of state-of-the-art data centers. This initiative creates a positive environment conducive to the widespread adoption of cloud accounting, instilling confidence in companies and fostering a more optimistic outlook. Additionally, actively addressing concerns about data center quality will further contribute to the success of this technological transition. Furthermore, the government's proactive efforts to reduce the cost of data and associated accessories serve as a compelling incentive for organizations to enthusiastically embrace new technologies. This strategic move not only makes technological advancements more financially accessible but also encourages a broader and more positive uptake, fostering an environment of innovation and progress.

Encouraging organizations to invest proactively in automated accounting systems and prioritize comprehensive training programs for accounting staff represents a positive and forward-thinking approach. This not only enhances overall efficiency in financial reporting and other accounting functions but also positions businesses to embrace technological advancements with enthusiasm and confidence. In parallel, UAE universities play a crucial role in this positive trajectory by empowering accounting and finance students with valuable training in adopting cloud systems. This proactive preparation ensures that graduates are well-equipped for future job opportunities in the country, contributing to the overall success of the technological transformation in the financial landscape. Together, these concerted efforts create a promising landscape for the integration of cloud accounting systems, fostering a culture of innovation and excellence in the UAE business environment.

References

1. Al Okaily, M., Al Khawaldi, A., Abdulrahman, A., & Al Qudah, H. (2022). Cloud-based accounting information systems usage and its impact on Jordanian SMEs' performance: The post-COVID-19 perspective. *Journal of Financial Reporting and Accounting, 21*(4), 126–155.
2. Parlińska, M., & Petrovska, I. (2017). Cloud computing and its benefits. *Information Systems in Management, 6*(4), 309–317.
3. Tang, Q., Chen, H., & Lin, Z. (2016). How to measure country-level financial reporting quality? *Journal of Financial Reporting and Accounting, 14*(2), 230–265.
4. CertDa Homepage, Retrieved November 13, 2023 from https://accounting.binus.ac.id/2021/10/01/qualitative-characteristics-of-financial-reports/
5. Akpan, J. U., Igbekoyi, O. E., Ogungbade, O. I., & Osaloni, B. O. (2023). Effect of cloud accounting on financial information quality of selected firms in Nigeria. *International Journal of Research and Innovation in Social Science, 7*(1), 1175–1193.
6. Dimitriu, O., & Matei, M. (2014). A new paradigm for accounting through cloud computing. *Alexandru Ioan Cuza University of Iasi,* 840–846
7. Noordin, N. A., Hussainey, K., & Hayek, A. F. (2022). The use of artificial intelligence and audit quality: An analysis from the perspectives of external auditors in the UAE. *Journal of Risk and Financial Management, 15*(8), 339.
8. Hayek, A. F., Noordin, N. A., & Hussainey, K. (2022). Machine learning and external auditor perception: An analysis for UAE external auditors using technology acceptance model. *Accounting and Management Information Systems, 21*(4), 475–500.
9. Hayek, A. F. (2022). Data science and external audit. In *Sustainable development through data analytics and innovation: Techniques, processes, models, tools, and practices.* Cham: Springer International Publishing.
10. Georgen Scarborough Associates homepage. Retrieved November 12, 2023 from https://www.gsacpa.com/2021/07/13/how-financial-statements-can-be-helpful-in-decisionmaking/.
11. CFA Institute homepage. Retrieved November 12, 2023 from https://www.cfainstitute.org/en/membership/professional-development/refresher-readings/evaluating-quality-financial-reports
12. Prajapati, J. Accounting and financial reporting in Dubai–guide. *Avyanco.* Retrieved November 12, 2023 from https://avyanco.com/news/accounting-financial-reporting-in-dubai/
13. Yuan, Y., Zhao, L., & Zhang, Y. (2020). Cloud computing: A review of information security and financial implication. *Journal of Cloud Computing: Advances, Systems and Applications, 9*(1), 1–14.
14. Skafi, M., Yunis, M. M., & Zekri, A. (2020). Factors influencing SMEs' adoption of cloud computing services in Lebanon: An empirical analysis using TOE and contextual theory. *IEEE Access, 8*, 79169–79181.
15. Mohammadi, S., & Mohammadi, A. (2014). Effect of cloud computing in accounting and comparison with the traditional model. *Research Journal of Finance and Accounting, 5*(23), 104–114.
16. Alhelo, E. (2022). The impact of the adoption of cloud computing on improving the efficiency of accounting information systems during the COVID-19 pandemic (a field study on the service companies listed on the palestine stock exchange). *Arab Journals Platform*
17. Pawar, A. B., Ghumbre, S. U., & Jogdand, R. M. (2023). Study and analysis of various cloud security, authentication, and data storage models: A challenging overview. *International Journal of Decision Support System Technology (IJDSST), 15*(1), 1–16.

18. Owolabi, S. A., & Izang, J. U. (2020). Cloud accounting and financial reporting. *International Journal of Research Publication, 60*(1), 21–28.
19. Zebua, S. U. L. I. N. A., & Widuri, R. I. N. D. A. N. G. (2023). Analysis of factors affecting adoption of cloud accounting in Indonesia. *Journal of Theoretical and Applied Information Technology, 101*(1), 89–105.
20. Smith, P. Improving data efficiency to produce better quality insights and improved productivity. RecordPoint. Retrieved November 15, 2023 from https://www.recordpoint.com/blog/improving-data-efficiency-to-produce-better-quality-insights-and-improved-productivity
21. Li, W., Zhou, Q., Ren, J., & Spector, S. (2019). Data mining optimization model for financial management information system based on improved genetic algorithm. *Information Systems and eBusiness Management,* 1–19
22. Ali, A., Abd Razak, S., Othman, S. H., Eisa, T. A. E., Al-Dhaqm, A., Nasser, M., & Saif, A. (2022). Financial fraud detection based on machine learning: A systematic literature review. *Applied Sciences, 12*(19), 9637.
23. Al-Zoubi, A. M. (2017). The effect of cloud computing on elements of accounting information system. *Global Journal of Management and Business Research: Accounting and Auditing, 17*(3), 1–8.
24. Mugyenyi, R. (2018). Adoption of cloud computing services for sustainable development of commercial banks in Uganda. *Global Journal of Computer Science and Technology: B Cloud and Distributed, 18*(1), 1–9.
25. Ogunsola, E. A. (2021). Effect of cloud accounting on the financial reporting quality of SMEs in Nigeria. *Bingham University Journal of Accounting and Business (BUJAB), 6*(2)
26. Owolabi, S. A., & Izang, J. U. (2020). Cloud accounting and financial reporting qualities of Smes in Nigeria: An overview. *Cloud Accounting and Financial Reporting Qualities of SMEs in Nigeria: An Overview, 60*(1), 8–8.
27. Najafi, A., Soleimanpur, S., & Morady, Z. (2022). The impact of information technology methods on accounting information quality. *Journal of Information and Organizational Sciences, 46*(1), 63–77.
28. Abdelraheem, A. a. E., Hussaien, A. M., Mohammed, M. a. A., & Elbokhari, Y. a. E. (2021). The effect of information technology on the quality of accounting information. *Accounting,* 191–196
29. Sharma, R., Sharma, R., & Bhati, J. P. (2017). Impact of cloud computing datamining in digitalization. *International Journal of Applied Engineering Research, 12*(22), 12716–12720.

Chapter 26
Occupational Health and Safety Hazards Analysis by AHP Among Agriculture Workers in Sharjah, UAE

Mohammad Miftaur Khadem and Sultan Yosuif AlAli

Abstract The study investigates workplace safety concerns in the city's agricultural sector through interviews with occupational health and safety experts. A cross-sectional design is used to enhance data collection from a large group of respondents selected through the convenience sampling approach. A semi-structured interview with occupational safety experts and a structured survey with workers in Sharjah emirate's agriculture sector is done in this research. Respondents' feedback is collected through semi-structured interviews through the analytic hierarchy process (AHP) and evaluated against current health and safety practices in the agricultural sector of Sharjah emirate. The research results indicate that the workers prioritize exposure to heat as the most significant sub-factor of the physical occupational health and safety hazards in the sector. Additionally, roadside crashes are considered the most severe sub-factor of physical threats. The results also indicate that unclean irrigation water is the most significant sub-factor for the biological and chemical hazards. At the same time, poisonous plants are the most severe sub-factors in this category. It is important to provide sufficient awareness and training programs for workers in health and safety topics in order to identify the effects of different hazards within the agriculture sector in Sharjah emirate. Additionally, organizations must be aware of how to implement proper health and safety systems within their workplace in order to protect their workers from the effects of these hazards. Proper Occupational Health and Safety Systems that are specifically designed for the agricultural sector containing details on how to address relevant hazards must be developed and implemented within Sharjah emirate.

Keywords Safety risks · Occupational safety · Workplace · Safety procedures · AHP · Workers' health

M. M. Khadem (✉) · S. Y. AlAli
Department of Industrial Engineering and Engineering Management, University of Sharjah, Sharjah, UAE
e-mail: mkhadem@sharjah.ac.ae

© The Author(s), under exclusive license to Springer Nature Switzerland AG 2024
A. Emrouznejad et al. (eds.), *Business Analytics and Decision Making in Practice*, Lecture Notes in Operations Research, https://doi.org/10.1007/978-3-031-61589-4_26

Introduction

Occupational safety and health risks exist in every industry, including the agricultural sector. These risks can range from severe physical and mental dangers to mild hazards. Employers are legally and ethically obliged to create safe work settings for their employees. In the UAE, the government requires companies to implement effective measures for safeguarding employees' safety in the workplace. The National Crisis Emergency and Disasters Authority (NCEMA) created the UAE's occupational health and safety management system (OHSMS) as a part of the initiative to align organizational safety efforts with national standards. The system provides guidelines for all companies in the UAE to maintain high occupational safety standards. The agricultural sector in the UAE has transformed over the years, creating opportunities for firms to hire more workers and adopt new technologies to maximize productivity. However, these improvements also highlight the importance of research on occupational health and safety risks in the industry.

Protecting employees from work-related safety risks is a global public health concern. Every year, millions of people are killed or injured in workplace accidents. This is a particularly pressing issue in small- and medium-sized firms with poor technical capability and insufficient resources to implement appropriate occupational health and safety programs. In the UAE, economic growth has led to the creation of many new jobs. However, the government's capacity to train occupational health inspectors has not kept pace with this growth. This means that many workers in the UAE may be exposed to safety risks without adequate protection. Research on occupational health risks in the agricultural sector of the UAE is essential to raise awareness of these risks and develop effective programs to protect workers. Recent Progress on Research on Occupational Health Hazards in the UAE.

A recent study on occupational health risks in the UAE examined the availability of PPE and the level of burnout, stress, and anxiety among frontline health workers caring for COVID-19 patients. The study found that most healthcare workers were satisfied with the quality and availability of PPE, but some workers experienced depression, anxiety, and burnout. The study recommended that health organizations should develop better occupational health plans to protect workers' mental well-being. Research on occupational health and safety in the UAE shows a high rate of work-related safety hazards in the construction sector. Employees in nearly 70% of construction firms in the country have insufficient knowledge of occupational health and safety policies, and many organizations lag in implementing safety plans [1]. Only 18% of construction firms conduct continuous safety and health training, and the percentage of organizations without occupational safety training plans is even higher in some regions. Accidents in the construction sector severely affect workers, their families, and society.

A study by Al Faqeeh, Khalid, and Osman [3] examined the connection between workplace conditions and employees' safety behaviors and attitudes in the healthcare sector in the UAE. The study found that organizational managers significantly influence employees' safety behaviors and attitudes. Hospital leaders' commitment to safety programs motivates workers to respond to institutional changes for preventing occupational health risks, which increases compliance with established regulations. Employees' safety attitude also depends on organizational culture and influences individuals' conduct. A study by Al Mazrouei et al. [4] examined the relationship between employees' perceptions of process safety and top managers' safety behaviors in an oil refining firm in Abu Dhabi, UAE. The study found that employees' engagement in occupational health programs and safety performance is positively correlated with organizational leaders' safety practices. This suggests that company leaders play a vital role in promoting a safety culture in workplaces. A study by Kambris, Khan, and Al Falasi [7] examined workers' concerns in the automotive repair industry in Dubai, UAE. The study found that workers in this sector have insufficient knowledge of occupational health and safety risks and protective tools. Safety training was not conducted at the automotive firm involved in the research, and inadequate PPE was provided to the workers. The study highlights the lack of occupational safety and training programs in small automotive firms in the UAE.

The identification of these hazards are essential for the analytical hierarchy process (AHP) analysis, acting as alternatives. Their various sub-factors are treated as the criteria in the AHP analysis. Overall, the AHP values are weighted to determine the priorities and severity of the different sub-factors of physical, biological, and chemical occupational health and safety hazards in Sharjah emirate's agricultural sector.

Methods

In order to achieve the objectives of this study, a multicriteria decision making (MCDM) approach using an analytical hierarchy process (AHP) was adopted. The analytical hierarchy process (AHP) is a multi-criteria decision-making method that organizes complex problems in a hierarchical order by simplifying the evaluation of all applicable criteria. In a decision-making situation, it assists the decision-maker in determining the relative importance of the attributes. A pair-wise comparison made among the factors; this step is an initial process for the analytical hierarchy process (AHP). A nine-point scale is used to rank the importance of the two attributes, ranging from similarly important to one attribute being significantly more important than the other [8]. The Saaty scale is a scale that has been proposed to describe the relative value of various factors.

To calculate the weights for various criteria, the analytical hierarchy process (AHP) method creates a pairwise comparison matrix $[A]$ as follows:

$$A = \left[a_{ij}\right] = \begin{bmatrix} a_{11} & a_{12} & \cdots & a_{1n} \\ a_{21} & a_{22} & \cdots & a_{2n} \\ \vdots & \vdots & \ddots & \vdots \\ a_{n1} & a_{n2} & \cdots & a_{nn} \end{bmatrix} = \begin{bmatrix} 1 & a_{12} & \cdots & a_{1n} \\ 1/a_{12} & 1 & \cdots & a_{2n} \\ \vdots & \vdots & \ddots & \vdots \\ 1/a_{1n} & 1/a_{2n} & \cdots & 1 \end{bmatrix} \tag{26.1}$$

The matrix A is a reciprocal matrix since $a_{ij} = 1/a_{ji}$, where, $a_{ij} = w_i/w_j$ and n is number of criteria. If a_{ij} represents the importance of criteria i over criteria j and a_{jk} represents the importance of j over k, then, a_{ik}, the importance of criteria i over criteria k, must be equal to $(a_{ij}.a_{jk})$ for the judgements to be consistent for all i, j, k. The pairwise comparison result is placed in the upper triangle of the pairwise matrix. The main diagonal is the comparison of the elements, with a value of 1. The lower triangle is the reciprocal of the upper triangle and is the pairwise comparison matrix.

(1) **Step 3: Structuring the normalized matrix [A₁]**

In the above matrix, calculate the sum of every column.

$$y_j = \sum_{i=1}^{n} a_{ij} \tag{26.2}$$

Then, divide every member of the matrix respectively by its obtained column sum. After that, take the average for each row of the normalize matrix $[A_1]$ to obtain the relative importance weight of ith criterion. So normalized matrix $[A_1]$ as follows:

$$A_1 = \begin{bmatrix} a'_{11} \; a'_{12} \; \ldots \; a'_{1n} \\ a'_{21} \; a'_{22} \; \ldots \; a'_{2n} \\ \vdots \quad \vdots \quad \ddots \quad \vdots \\ a'_{n1} \; a'_{n2} \; \ldots \; a'_{nn} \end{bmatrix} \; and\; a'ij = \frac{a_{ij}}{\sum_{i=1}^{n} a_{ij}} \text{for i}, j = 1, 2, 3, \ldots, n \tag{26.3}$$

(2) **Step 4: Finding eigenvalue and eigenvector**

The common eigenvalue solution in numerical analysis is used to determine the eigenvector or dominant vector and the maximum eigen value of the pairwise comparison matrix. So eigen vector W calculated as the following:

$$W = \begin{bmatrix} W1 \\ W2 \\ \vdots \\ Wn \end{bmatrix} \text{ and } W_i = \frac{\sum_{i=1}^{n} a\prime ij}{n} \tag{26.4}$$

Then to calculate the eigenvalue W', the pairwise comparison matrix A is first multiplied by the calculated eigenvector W to acquire a new vector which is shown as follows:

$$W' = A \times W = \begin{bmatrix} 1 & a_{12} & \cdots & a_{1n} \\ 1/a_{12} & 1 & \cdots & a_{2n} \\ \vdots & \vdots & \ddots & \vdots \\ 1/a_{1n} & 1/a_{2n} & \cdots & 1 \end{bmatrix} \times \begin{bmatrix} W1 \\ W2 \\ \vdots \\ Wn \end{bmatrix} = \begin{bmatrix} W'1 \\ W'2 \\ \vdots \\ W'n \end{bmatrix} \tag{26.5}$$

And

$$\lambda\max = \frac{1}{n}(\frac{W'1}{W1} + \frac{W'2}{W2} + \cdots + \frac{W'n}{Wn}) \tag{26.6}$$

where:

W: Eigenvector.

W': Eigenvalue of given matrix.

λmax : Maximum eigenvalue of the pair-wise comparison matrix.

(3) **Step 5: Check the reliability of the results through the Consistency Index (CI) and the Consistency Ratio (CR)**

Saaty [9] suggested that the highest eigen value of the matrix A is λmax could be used to measure (CI) and (CR) as the following:

The consistency index (CI) is calculated as follows:

$$(CI) = \frac{\lambda_{max} - n}{n - 1} \tag{26.7}$$

where:

λmax : Maximum eigenvalue.

n: Number of criteria.

Now, some randomly generated consistency index (RI) values are given as follows (Table 26.1).

Finally, the consistency ratio (CR) is:

$$(CR) = \frac{CI}{RI} \tag{26.8}$$

where RI is the random consistency index (RI).

If the CR < 10%, then the pair-wise comparison/judgement matrix is acceptable. Otherwise, it is suggested to update the judgement matrix to eliminate the inconsistency. The progression must be a cyclic process until the consistency condition is made satisfactory.

Table 26.1 A reasonable assumption RI value based on number of attributes

n	1	2	3	4	5	6	7	8	9	10	11	12	13	14	15
RI	0	0	0.58	0.9	1.12	1.24	1.32	1.41	1.45	1.49	1.51	1.54	1.56	1.57	1.58

Source Saaty and Ozdemir (2003)

Results and Discussion

This study evaluates occupational health and safety hazards among workers in the agricultural sector in Sharjah emirate. Accordingly, the research achieved three specific objectives. First, to understand the current occupational health and safety regulations and practices in Sharjah emirate's agricultural sector. Second, to examine the prioritized occupational health and safety hazards in the agricultural sector in the city. Third, to assess and evaluate current occupational health and safety practices in the city's agricultural sector against the discovered hazards. Consequently, this section presents the research results illustrating the occupational health and safety factors that are most and least significant for physical and biological, and chemical alternatives.

Physical Factors

The first step was to identify the most dominant physical occupational health and safety hazards in the agricultural sector within the emirate of Sharjah through conducting a semi-structured interview with health and safety experts. Additionally, the AHP method was utilized as the main method of analysis. The result of the AHP method can be seen in Table 26.2.

The results in Table 26.2 show that exposure to heat (average = 26%) is the most significant sub-factor of the physical occupational health and safety hazards in the city's agricultural sector, followed by exposure to radiation (20%). In other words, most organizations in the agricultural sector in Sharjah emirate prioritize exposure to heat and radiation in their occupational health and safety management plans. However, the results suggest that electrocution (2%) and roadside crashes (2%) are the lowest sub-factor of the physical occupational health and safety hazards considered by the agricultural sector's management.

Use of hazardous tools, Dim lights, and loud noise are identified as the top four most frequent hazards in the sector. Thus, the next part is to validate the results through the consistency measures of the AHP analysis. Table 26.3 presents the consistency measures of the frequencies and averages of the sub-factors of the physical occupational health and safety hazards shown in Table 26.2.

As evident in Table 26.3, the calculated value of CR (0.0952) is less than 0.1 ($CR < 0.1$) or 10%, implying that the inconsistency is acceptable. Likewise, a CR of less than 0.1 means that the matrix is suitable for consistency. Thus, the subjective evaluation of the sub-factors of the physical occupational health and safety hazards in Sharjah emirate's agricultural sector considered in this study is significantly consistent.

Table 26.4 summarizes the results of the severity of the sub-factor of the physical occupational health and safety hazards in Sharjah emirate's agricultural sector considered in this study. The findings indicate that roadside crashes (29%) are the most severe sub-factor, followed by electrocution (23%) and falls from height (19%).

Table 26.2 Prioritization of physical occupational health and safety hazards in the agricultural sector

Physical/ frequency	Exposure to heat	Exposure to radiation	Use of hazardous tools	Trips and slips	Dim lights and loud noise	Falls from height	Strikes by moving machinery or a piece of moving machinery	Electrocution	Roadside crashes	Average
Exposure to heat	1/3	3/7	3/7	0.19	0.23	0.20	0.22	0.18	0.18	26%
Exposure to radiation	0.16	0.21	0.32	0.19	0.23	0.18	0.19	0.16	0.18	20%
Use of hazardous tools	0.08	0.07	0.11	0.25	0.34	0.15	0.15	0.14	0.13	16%
Trips and slips	0.11	0.07	0.03	0.06	0.03	0.15	0.15	0.14	0.13	10%
Dim lights and loud noise	0.16	0.11	0.04	0.25	0.11	0.18	0.22	0.18	0.20	16%
Falls from height	0.04	0.03	0.02	0.01	0.02	0.03	0.01	0.04	0.07	3%
Strikes by moving machinery or a piece of moving machinery	0.05	0.04	0.03	0.02	0.02	0.10	0.04	0.10	0.09	5%
Electrocution	0.04	0.03	0.02	0.01	0.01	0.01	0.01	0.02	0.01	2%
Roadside crashes	0.04	0.03	0.02	0.01	0.01	0.01	0.01	0.04	0.02	2%
Total	1	1	1	1	1	1	1	1	1	1.000

Table 26.3 Consistency measures physical occupational health and safety hazards

	Physical/frequency consistency measure	
Exposure to heat	2.7764	10.538
Exposure to radiation	2.1864	10.869
Use of hazardous tools	1.7865	11.275
Trips and slips	0.9577	9.9261
Dim lights and loud noise	1.6900	10.504
Falls from heights	0.2647	9.3299
Strikes by moving machinery or a piece of moving machinery	0.5084	9.4730
Electrocution	0.1607	9.6374
Roadside crashes	0.1954	9.3932

· Consistency ration (CR) = 0.0952
· Consistency index (CI) = 0.1381
· Random consistency index (RI) = 1.45

However, trips and slips (2%) and dim lights and loud noise (2%) are the least severe sub-factors of occupational health and safety risk, followed by exposure to heat (4%), use of hazardous tools (5%), exposure to radiation (8%), and strikes by moving machinery or a piece of moving machinery (9%).

The results in Table 26.5 indicate that the calculated value of CR (0.096866157) is less than 10%, implying that the subjective evaluation of the sub-factors of the physical occupational health and safety hazards considered in this study is significantly consistent.

Biological and Chemical Factors

Table 26.6 indicates the results of the normalized matrix for the frequencies and averages of the biological and chemical occupational health and safety hazards in the agricultural sector in Sharjah emirate analyzed using the analytical hierarchy process. Based on these findings, using unclean irrigation water (38%) is the most significant sub-factor of biological & chemical occupational health and safety hazard prioritized in Sharjah emirate's agricultural sector, followed by non-dissolving fertilizers (24%). However, the analysis reveals that contact with poisonous plants is the least significant sub-factor of biological & chemical occupational health and safety hazards considered in the sector.

As evident in Table 26.7, the calculated value of the Consistency Ratio (0.017821967) is less than 10%, implying that the subjective evaluation of the sub-factors of physical and chemical occupational health and safety hazards considered in this study is significantly consistent.

Table 26.4 Severity of physical occupational health and safety hazards in the agricultural sector

Physical/ severity	Exposure to heat	Exposure to radiation	Use of hazardous tools	Trips and slips	Dim lights and loud noise	Falls from height	Strikes by moving machinery or a piece of moving machinery	Electrocution	Roadside crashes	Average
Exposure to heat	0.03	0.01	0.01	0.09	0.11	0.02	0.01	0.03	0.04	4%
Exposure to radiation	0.13	0.05	0.12	0.11	0.13	0.03	0.03	0.04	0.07	8%
Use of hazardous tools	0.08	0.02	0.04	0.07	0.09	0.02	0.02	0.04	0.06	5%
Trips and slips	0.01	0.01	0.01	0.02	0.05	0.02	0.01	0.02	0.04	2%
Dim lights and loud noise	0.004	0.01	0.01	0.01	0.02	0.02	0.01	0.02	0.04	2%
Falls from height	0.12	0.26	0.23	0.18	0.16	0.14	0.28	0.11	0.12	19%
Strikes by moving machinery or a piece of moving machinery	0.13	0.11	0.12	0.11	0.13	0.04	0.07	0.06	0.09	9%
Electrocution	0.21	0.26	0.23	0.20	0.16	0.29	0.28	0.22	0.18	23%
Roadside crashes	0.21	0.26	0.23	0.20	0.16	0.43	0.28	0.45	0.36	29%
Total	1.00	1.00	1.00	1.00	1.00	1.00	1.00	1.00	1.00	1.000

Table 26.5 Consistency measures severity of physical occupational health and safety hazards

Physical/frequency	Consistency measure	
Exposure to heat	0.3617	9.18076
Exposure to radiation	0.8278	10.3476
Use of hazardous tools	0.4843	9.96692
Trips and slips	0.2089	9.27827
Dim lights and loud noise	0.1415	9.39614
Falls from heights	2.0538	11.0195
Strikes by moving machinery or a piece of moving machinery	0.9897	10.5542
Electrocution	2.4634	10.8614
Roadside crashes	3.0203	10.5077

· CR = 0.096866157
· CI = 0.140455928
· RI = 1.45

Table 26.6 Prioritization of biological and chemical occupational health and safety hazards in the agricultural sector

Biological & chemical/ frequency	Exposure to pesticides	Inhalation of organic dust	Use of non-dissolving fertilizers	Inhalation of fumes	Poisonous plants	Use of unclean irrigation water	Average
Exposure to pesticides	0.08	0.07	0.08	0.05	0.11	0.10	8%
Inhalation of organic dust	0.16	0.14	0.11	0.18	0.17	0.13	15%
Use of non-dissolving fertilizers	0.24	0.274	0.23	0.27	0.22	0.20	24%
Inhalation of fumes	0.16	0.07	0.07	0.09	0.11	0.10	10%
Poisonous plants	0.04	0.05	0.06	0.05	0.06	0.07	5%
Use of unclean irrigation water	0.32	0.41	0.45	0.36	0.33	0.40	38%
Total	1	1	1	1	1	1	1.000

Table 26.8 summarizes the results of the AHP analysis of the severity of the various sub-factors of biological and chemical occupational health and safety hazards in Sharjah emirate's agricultural sector, as considered in this study. The findings suggest that contact with poisonous plants (36%) is the most severe sub-factor of the biological and chemical occupational health and safety hazards in the city's agricultural sector, followed by exposure to pesticides (23%) and the use of unclean irrigation water (15%). Nevertheless, the analysis reveals that the inhalation of fumes

Table 26.7 Consistency measures biological and chemical occupational health and safety hazards

Physical/frequency	Consistency measure	
Exposure to pesticides	0.4826	6.030
Inhalation of organic dust	0.9115	6.135
Use of non-dissolving fertilizers	1.4754	6.173
Inhalation of fumes	0.6131	6.074
Contact with poisonous plants	0.3146	6.096
Use of unclean irrigation water	2.33727	6.153

- CR = 0.017821967
- CI = 0.022099239
- RI = 1.24

(5%) is the lowest sub-factor, followed by using non-dissolving fertilizers (8%) and inhalation of organic dust (12%). Thus, the sector believes that contact with poisonous plants is the most severe sub-factor of the biological and chemical occupational health and safety hazards, while inhalation of fumes is the least dangerous.

Table 26.9 indicates the calculated values of the consistency ratio, consistency index, and average random index. The CR (0.02066154) value is less than 10%, implying that the subjective evaluation of the sub-factors of the biological and chemical occupational health and safety hazards considered in this study is significantly consistent. In other words, the consistency of the matrix is statistically acceptable in

Table 26.8 Severity of biologics and chemical occupational health and safety hazards in the agricultural sector

Biological and chemical / severity	Exposure to pesticides	Inhalation of organic dust	Use of non-dissolving fertilizers	Inhalation of fumes	Poisonous plants	Use of unclean irrigation water	Average
Exposure to pesticides	0.22	0.23	0.24	0.22	0.19	0.27	23%
Inhalation of organic dust	0.11	0.11	0.16	0.17	0.13	0.07	12%
Use of non-dissolving fertilizers	0.07	0.06	0.08	0.11	0.10	0.07	8%
Inhalation of fumes	0.05	0.04	0.04	0.06	0.08	0.05	5%
Poisonous plants	0.44	0.34	0.32	0.28	0.38	0.41	36%
Use of unclean irrigation water	0.11	0.23	0.16	0.17	0.13	0.14	15%
Total	1	1	1	1	1	1	1.000

Table 26.9 Consistency measures for severity

Physical/frequency	Consistency measure	
Exposure to pesticides	1.414	6.191
Inhalation of organic dust	0.752	6.060
Use of non-dissolving fertilizers	0.489	6.066
Inhalation of fumes	0.314	6.084
Contact with poisonous plants	2.234	6.191
Use of unclean irrigation water	0.953	6.177
· CR = 0.0204 · CI = 0.0256 · RI = 1.24		

evaluating the various sub-factors of the biological and chemical occupational health and safety risks in Sharjah emirate's agricultural sector.

As illustrated in Fig. 26.1, the study's results indicate that the health and safety experts in the sector prioritize exposure to heat as the most significant sub-factor of the physical occupational health and safety hazards.

However, the stakeholders believe that roadside crashes are the most severe sub-factor in the sector. The results suggest that trips and slips are the least severe hazards presently posing risks to Sharjah emirate's agricultural sector employees. The increased risk of exposure to heat is influenced by multiple factors, including an agricultural firm's work schedule, shift rules, remuneration schemes, and management approaches.

As evident in Fig. 26.2, most of the participants in this study identify the use of unclean irrigation water as the most significant sub-factor of the biological and chemical occupational health and safety hazards in Sharjah emirate's agricultural sector. Interestingly, poisonous plants are the least significant sub-factors despite being deemed the most severe hazard. Therefore, farmworkers are often exposed to the risk of using unclean irrigation water. The study's findings in this regard are in congruity with Dickin et al.'s [6] finding that contaminated water poses a

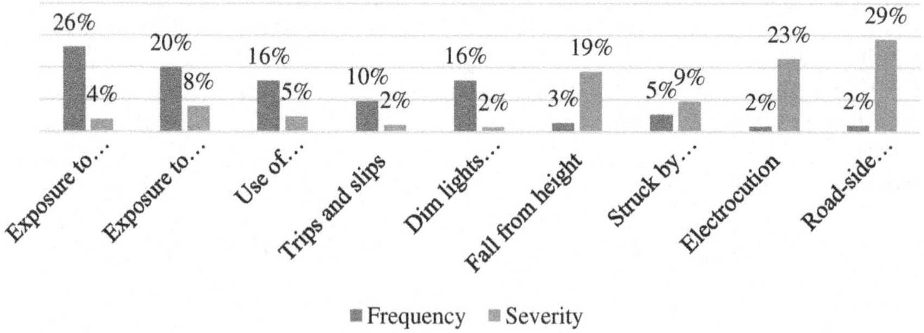

Fig. 26.1 Frequency and severity of sub-factors of physical occupational health and safety hazards in Sharjah Emirate's Agricultural Sector

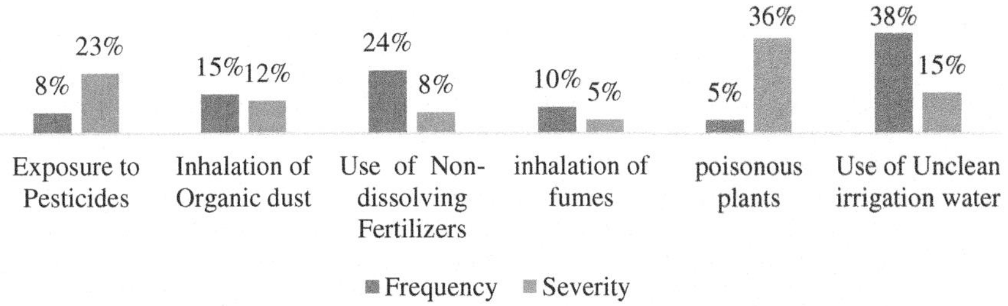

Fig. 26.2 Frequency and severity of sub-factors of biological and chemical occupational health and safety hazards in Sharjah Emirate's Agricultural Sector

significant health risk for farmworkers as most farmworkers resort to using such water due to depleting freshwater resources and the water scarcity associated with global warming and climate change. Further, Ali et al.'s [5] study in Bangladesh demonstrates that about 90% of farmworkers suffer various illnesses, including skin rashes and breathing difficulties, due to contact with poisonous plants. These revelations are in keeping with the results of the present study, indicating that many (approximately 36%) of the participants perceive toxic plants as the most severe sub-factor of the biological and chemical occupational health and safety hazards present in Sharjah emirate's agricultural sector. However, the Sharjah emirate government needs a specific policy guideline for managing the occupational health and safety risks associated with poisonous plants and the use of unclean irrigation water.

Conclusion

This study aims to assess the state of occupation health and safety hazards of Sharjah emirate's agricultural sector from the experts perspective. The research employs a qualitative survey approach, collecting data from 28 participants using semi-structured interviews. The data is analyzed using the analytical hierarchy process. The research results indicate that the workers prioritize exposure to heat as the most significant sub-factor of the physical occupational health and safety hazards in the sector. Additionally, roadside crashes are considered the most severe sub-factor of physical threats. The results also indicate that unclean irrigation water is the most significant sub-factor for the biological and chemical hazards. At the same time, poisonous plants are the most severe sub-factors in this category. In answering the research questions, the study reveals that there is a lack of specific occupational health and safety regulations for the agricultural sector within the emirate of Sharjah. Although some occupational health and safety guidelines do exist within the emirate, the existing regulatory guidelines are too general to address the sector workers' unique and distinct health and safety needs.

References

1. Abukhashabah, E., Summan, A., & Balkhyour, M. (2019). Causes of occupational accidents and injuries in the construction industry in Jeddah city. *Journal of King Abdulaziz University-Meteorology Environment and Arid Land Agriculture Sciences, 28*(1), 105–116.
2. Ajab, S., Ádam, B., Al Hammadi, M., Al Bastaki, N., Al Junaibi, M., Al Zubaidi, A., Hegazi, M., Grivna, M., Kady, S., Koornneef, E., Neves, R., Uva, A. S., Sheek-Hussein, M., Loney, T., Serranheira, F., & Paulo, M. S. (2020). Occupational health of frontline healthcare workers in the United Arab Emirates during the COVID-19 Pandemic: A snapshot of summer 2020. *International Journal of Environmental Research and Public Health, 18*(21), 1–15.
3. Al Faqeeh, F., Khalid, K., & Osman, A. (2019). Integrating safety attitudes and safety stressors into safety climate and safety behaviour relations: The case of healthcare professionals in Abu Dhabi. *Oman Medical Journal, 34*(6), 504–513.
4. Al Mazrouei, M. A., Khalid, K., Davidson, R., & Sundarakani, B. (2019). 'Modeling the impact of safety climate on process safety in a modern process industry: The case of the UAE's oil-refining industry. *Cogent Business and Management, 6*(1), 1–21.
5. Ali, Y., Al-Kafy, A., Rahaman, Z. A., Islam, F., Rahman, R., Ara, I., Akhtar, R., & Javed, A. (2022). Comparative occupational health risk between tobacco and paddy farming people in Bangladesh. *SSM-Mental Health, 2*(1), 1–6.
6. Dickin, S. K., Schuster-Wallace, C. J., Qadir, M., & Pizzacalla, K. (2017). A review of health risks and pathways for exposure to wastewater use in agriculture. *Environmental Health Perspectives, 124*(7), 900–909.
7. Kambris, M. K., Khan, S., & Al Falasi, S. N. (2019). Perceptions of health and safety among workers in the automotive repair industry in Dubai (United Arab Emirates): A cross-sectional exploratory study. *Journal of Ecophysiology and Occupational Health, 19*(3&4), 1–11.
8. Saaty, R. W. (1987). The analytic hierarchy process—what it is and how it is used. *Mathematical Modelling, 9*(3), 161–176. https://doi.org/10.1016/0270-0255(87)90473-8.
9. Saaty, T. L. (1980). *Analytic hierarchy process McGraw- Hill.*

Chapter 27
The Future of Construction: Investigating the Latest Trends and Innovations in Off-Site Construction Solutions for Worker Safety in the United Arab Emirates

Maryam Omar AlZarooni and Hamad S. J. Rashid

Abstract One of the riskiest industries for workers is the construction industry, which has the highest fatal accidents death rate when compared to other industries. With the potential for falls, being struck by things, electrocution, and getting trapped between objects, safety is a major issue for those working in the construction business. Therefore, improving workers safety in controlled environments is a fundamental requirement within the construction industry that is growing in the United Arab Emirates. Although prefabrication and modular construction are now continuing to have increasing trends in the construction industry and have a significant impact in increasing the efficiency, nevertheless, it is feared that this new construction approach would bring new threats to worker's safety or can have its positive effect on safety. Accordingly, this research aims to evaluate the influence introduced by the new prefabrication technology on workers safety within construction sites. A questionnaire was used to gather data from the relevant parties, which included project managers, project engineers, safety engineers, site engineers and workers. The questionnaire was divided into three primary sections: Training and Education, Safety Compliance, and Incident Rates. 106 participants from various construction companies in the United Arab Emirates provided the pertinent data, which was then analyzed using three-way ANOVA (Analysis of Variance) technique to examine the relationship between the three mentioned variables. The gained results showed that the Incident Rates, Safety Compliance, Training and Education were all statistically significant. The interaction between all three variables showed no combined effect on the safety score.

M. O. AlZarooni · H. S. J. Rashid (✉)
Industrial Engineering and Engineering Management Department, University of Sharjah, Sharjah, UAE
e-mail: hamad.rashid@sharjah.ac.ae

M. O. AlZarooni
e-mail: U21100902@sharjah.ac.ae

© The Author(s), under exclusive license to Springer Nature Switzerland AG 2024

A. Emrouznejad et al. (eds.), *Business Analytics and Decision Making in Practice*, Lecture Notes in Operations Research, https://doi.org/10.1007/978-3-031-61589-4_27

Keywords Construction management · Prefabricated construction · Risk management · Safety compliance · Worker safety

Introduction

The construction industry is one of the most dangerous for workers, having the highest number of deaths than other industries, and for this reason, construction safety has always been the first priority for any contractor on all job sites [1]. In recent years, there has been an accelerated shift towards the use of prefabrication in new construction.

Prefabrication is the assembly of parts that are manufactured off-site in a factory and then transported to their final assembly or destination. Furthermore, the process involves three stages which are production, transportation, and assembly. On the other hand, modularisation is the process by creating structures which can be broken up into separate parts for an easier assembly and transportation [2].

Understanding the many benefits and limitations of the off-site construction solutions can help the industry professionals as well as homeowners and consumers to make more informed decisions regarding the construction, design, management of buildings, and increase workers safety.

The thesis will investigate the difference in construction methods, both Prefabricated approach and Traditional Construction to see which one has a higher impact towards worker safety. The thesis will also focus on optimizing the assembly process, a critical aspect of modern construction.

The assembly phase also brings unique difficulties and risks that call for careful attention. This choice is essential as it will ensure smooth coordination and lower the possibilities of mistakes or injuries.

Studying the need for enhancing worker's safety in the UAE through a prefabricated approach is important for several reasons:

- Growing Construction Industry: The United Arab Emirates has witnessed a swift urbanization and infrastructure growth, resulting in a flourishing building sector. The likelihood of workplace accidents increases with an increase in construction activities. Having a better understanding of and using prefabricated techniques can help create safer working environments.
- Technological Advancements: Prefabricated and modular pieces in construction projects are becoming easier to design and coordinate because of the widespread use of modern technologies like Building Information Modeling (BIM). Efficiency and Productivity: Prefabrication techniques can increase the productivity and efficiency of construction. Traditional construction methods have the potential to reduce the number of accidents by using controlled conditions for component production, optimizing procedures, and decreasing on-site activity.

However, research gap in this area may include:

- Limited Localized Research: Research on worker safety and prefabrication specifically pertaining to the construction industry in the United Arab Emirates is lacking. Studies that are specific to a given area can shed light on its prospects and problems.
- Comparative Studies: Studies comparing prefabrication and conventional building techniques in the United Arab Emirates might shed light on the safety consequences of various construction strategies.
- Worker Perception and Training: Research could be necessary to understand the way employees view prefabrication in terms of safety and what kind of training is required to make sure safety precautions are successfully implemented.

The construction industry is constantly evolving and as it moves forward into the future, it is important to stay up to date on the innovations and latest trends in construction [2].

Through the implementation of prefabricated techniques, stakeholders in the UAE construction industry can improve worker safety by making well-informed decisions by filling in these research gaps.

Literature Review

The Prefabrication Process

Prefabrication refers to the method or practice of assembling components of structures or buildings at a location different from the construction site such as a factory or any other manufacturing site. Moreover, the process involves transporting either sub-assemblies or complete assemblies to the construction site [3].

The Growth of the Prefabricated Approach in the United Arab Emirates

The growing adoption of advanced construction methods to reduce the rising number of unskilled labors and the time represents one of the key factors which can positively influence the sales of prefabricated building components in the United Arab Emirates (UAE) [4].

Furthermore, the usage of Prefabricated off-site solutions helps in decreasing construction debris and pollutants and minimizing the overall cost of the project as well as human errors. Prefabricated Buildings components are manufactured using advanced technologies with controlled quality specifications which can extend the average lifespan of buildings in the UAE [4]. Moreover, IMARC Group expects the prefabricated market will grow by 7% within the year 2021–2026 [4].

These benefits along with the increasing preferences of contractors for off-site construction are strengthening the growth of the market in the country [4]. Moreover, the UAE Government began numerous development programs which are a long-term plan for the economic development of the emirate.

Benefits of Prefabrication for the Safety of Workers

Prefabrication has been acclaimed for a variety of benefits such as cost savings, time savings, better site security, higher quality, and improvement for site safety. Moreover, there are several benefits of applying prefabrication into the construction industry.

Controlled Environment: Prefabrication can put workers into a controlled environment where many risks of the traditional construction sites can be avoided. For example, offsite construction eliminates risks from weather conditions in which there are no slippery surfaces from rain that helps in cutting down slips and falls and avoiding risk of electrocution to exposed electrical lines resulting from moisture [1].

Less On-Site Workers: Less workers onsite means less workers who can potentially be killed, injured, or involved in any safety accident. Furthermore, the workers that install the prefabricated pieces are more experienced allowing sites to leverage the expertise of workers knowledge while minimizing the need for the less trained workers in proper safety procedures [1]. Moreover, a less workers onsite results in a less crowded jobsite which can lead to lessen the potential of workers to run into issues due to a crowded working environment or lack of space.

Safer Scheduling: Offsite construction can help in safer installation practices and better planned schedules for installations which can help in minimizing the need for rushed work or jammed spaces [1].

Less Waste and Clutter: Offsite construction enables a better material flow where less materials that are stored on-site can reduce debris and clutter to achieve a more organized, cleaner, and safer project site [1].

Minimize Risky Site Elements and Heights—Since falls are the number one killer of the Fatal Four, maximizing the work done at the ground level can have a vast impact [1]. Furthermore, with off-site construction, permanent scaffolding is used, which presents less opportunities for error than the frequently assembled and reassembled scaffolding which is used on traditional sites.

Traditional Construction Versus Prefabricated Approach

Over the years, the building sector has undergone substantial change, with two main approaches that stand out: traditional construction and the prefabricated approach.

In the building industry, worker safety is of utmost importance and plays a major role in determining whether to use prefabricated materials or traditional construction techniques. We will look at how each strategy affects worker safety in this introduction, as well as the steps taken to guarantee the safety and welfare of construction workers.

Traditional Construction and Worker Safety

Traditional Construction and Worker Safety has a long history and well-established procedures; thus, it prioritizes worker safety, but it also presents a unique set of safety risks:

On-Site Hazards: Traditional construction sites expose workers to a variety of risks associated with their employment, such as the possibility of falls, accidents involving heavy machinery, and exposure to inclement weather.

Scaffolding and Heights: Workers in traditional construction frequently perform work at heights that need the use of scaffolding, ladders, and other equipment, all of which raise the possibility of accidents.

Safety Protocols: Rules and procedures pertaining to safety, such as personal protective equipment (PPE), are essential to mitigate on-site risks in traditional construction.

Skilled Labor: The need for skilled labor in traditional construction means that in addition to the general safety precautions, personnel need to receive intensive training in safety procedures and practices.

Prefabricated Approach and Worker Safety

Prefabricated Approach and Worker Safety prioritizes worker safety and offers unique advantages in this regard.

Controlled Environment: Prefabrication occurs in a controlled factory environment, minimizing potential safety dangers and limiting exposure to external variables, such as harsh weather.

Efficient Assembly: Prefabricated components are made to assemble quickly and safely on the job site, reducing the need for labor-intensive and dangerous construction work.

Worker Training: Prefabrication workers can need specific industrial work training that focuses on safe machine operation and assembly procedures.

Reduced On-Site Risks: There can be a significant reduction in the number of laborers on-site and the related dangers when a large percentage of the construction is done off-site.

Precision and Quality Control: Precision and strict quality control are made possible by prefabrication, which produces building components that are safer and more dependable.

Construction Safety Study in the UAE

This section will focus on the identification of factors influencing safety in prefabrication technology, and the confirmation of the research methodology.

This study was necessary because there were very few studies about the safety performance of the UAE's construction industry and none about the Emirate of Dubai up to that point.

A study under the title "Health and Safety Influence on the Construction Project Performance in United Arab Emirates (UAE)" has been conducted with a total of 130 companies participating. Furthermore, the study showcased the following in UAE's construction companies.

71% have no training of workers [5] 0.86% of respondents admitted that they did not adhere to reporting accidents procedures [5] 0.69% have a serious lack of understanding of the importance of Health and Safety policy importance [5].

Prefabricated Construction as a Tool for Good Mental Health

The prefabrication approach can deliver a quality of projects in a shorter time while increasing the productivity which can positively impact the mental health of workers by potentially weakening the impacts of industry-related stressors that causes other stressors [6].

Safety risks are easier to identify at an early stage of prefabricated construction and can facilitate a safer construction than the traditional approach.

Through the reduction in on-site overlap, project duration and unambiguous design, and construction information, work pressure and all the other associated work-related stressors such as long working hours, work overload and unsafe work speed could be reduced.

Moreover, to a better psychological work environment, prefabricated construction also enhances a cleaner construction with the ease of housekeeping due to the disaggregation of tasks through standardization which ensures that workers carry out their duties in a good physical work environment and reduces the chances of safety incidents [6].

Industry Related Stressors

Industry Related Stressors dictate the level of physical, technological, mental, and other requirements of workers to fulfill their workplace obligations. Furthermore, excessive work pressure can lead to work overload which is a function of heterogeneity of every project in the industry.

The technical instructions in the construction industry are continually amended to suit every project with an increased opportunity of experiencing ambiguity in work-related instructions and procedures. Moreover, other implications of time pressure on construction workers are long working hours which can play a major role in the poor work-life balance of construction workers [6].

Management and Organizational Stressors

Management and Organizational Stressors come from management responses and impacts from the industry related stressors. Furthermore, interpersonal conflicts for example can arise from improper communication from new and ever-changing ideas and instructions.

Personal Stressors

Personal Stressors can be grouped under the personal classification such as marital status, gender discrimination, age discrimination, gender discrimination, harassment, and financial difficulties which results in low socioeconomic status among construction workers.

The construction workplace attracts workers from all over the world with different backgrounds and this has promoted culturally motivated barriers such as cultural conflicts, language barriers and racial discriminations.

Assembly Stage in Prefabrication

Prefabrication, a method of building that involves producing building components off-site before they are installed on the construction site, has many benefits, including lower costs, better quality control, and quicker project completion.

The assembly of these pre-made components at the construction site is a crucial step in the prefabrication process. While prefabrication can increase the efficiency and security of construction, the assembly phase also brings unique difficulties and risks that call for careful attention.

Prefabrication's assembly phase entails inherent hazards because of the nature of the work required and the difficulty in coordinating numerous components. The following are some of the main reasons why prefabrication assembly can be risky:

Handling Heavy Components: Prefabricated components can be heavy and unwieldy, including concrete panels, steel frames, or modular pieces. Workers may experience strain because of lifting, placing, and aligning these components, which raises the possibility of musculoskeletal injury.

Work at Heights: For a structure to be stable, prefabricated components must be connected, fastened, and aligned correctly. Safety risks might result from assembly mistakes or inadequacies that affect the building's structural integrity.

Equipment Hazards: When heavy machinery, such as cranes and forklifts, is used during assembly, there is a risk of accidents caused by the equipment if it is not operated by trained workers or if it malfunctions.

What Could Go Wrong?

As prefabrication becomes more popular in the construction sector, it is important to distinguish between its various uses and forms and investigate the effects on worker welfare to avoid hidden dangers becoming a statistic [7]. Off-site methods and processes can only present chances for better work conditions if the kinds and sources of risk are recognized by people engaged in their conception and implementation. It is imperative to acknowledge that mishaps that occur during prefabrication component assembly are not intrinsic to the prefabrication process; rather, they are the consequence of human mistake, insufficient training, or other site-specific elements. Clear communication, meticulous planning, stringent adherence to safety procedures, and appropriate training can all help to reduce these hazards and guarantee the safe assembly of prefabricated components. Prefabrication's safety benefits, like regulated manufacturing conditions and less work on-site, can frequently result in a safer building process overall.

Methodology

For conducting this research, Mixed Research Methods will be used to collect the appropriate amount of qualitative and quantitative data. By utilizing the advantages of each method and minimizing its drawbacks, a mixed methods approach which integrates quantitative and qualitative research methods such as questionnaires, interviews, the Delphi method, and statistical analyses like 3-way ANOVA improves the validity and reliability of research.

1. Comprehensive Understanding: In-depth investigation and comprehension of complicated phenomena are made possible by qualitative approaches like interviews, which also offer insightful information about the participants' varying experiences and viewpoints. However, quantitative techniques such as statistical analyses provide a methodical and controlled way to collect numerical data, making it possible to spot patterns and trends in a bigger sample.
2. Triangulation: Triangulating results by combining qualitative and quantitative data using techniques like the Delphi approach enhances the research's overall validity. Through the process of cross-verification, triangulation reduces the influence of individual approach biases and increases the study's resilience.
3. Quantitative Validation: To quantify variables and establish statistical associations, questionnaires, and statistical analyses like 3-way ANOVA are used. This gives the research more objectivity and generalizability. A more rigorous basis for making conclusions is provided by the application of statistical tools, which enable the detection of statistically significant differences or correlations.
4. Qualitative Enrichment: When combined with quantitative data, qualitative techniques, especially interviews can reveal subtle details and provide rich contextual information that may not be revealed by using standardized measurements

alone. This qualitative enrichment contributes to a more thorough and integrated comprehension of the research context.

5. Delphi Method for Consensus: The Delphi approach helps experts come to a consensus by using controlled feedback and repetitive rounds of surveys. The act of reaching a consensus improves the validity of the study results, particularly when discussing difficult or ambiguous subjects.

Nonetheless, it is imperative to acknowledge possible constraints:

- Validity Threats where the likelihood of validity concerns, such as questions about the application of qualitative results or the possibility of confounding variables in quantitative assessments, increases when a variety of methods are used. To support the study's overall validity, there is a need to put these concerns to rest through the application of rigorous methodologies.
- Resource Intensiveness such as when it comes to time, money, and experience, using blended methodologies can be resource intensive. To guarantee that both the qualitative and quantitative components are completed successfully, plan and assign resources carefully.
- Integration Challenges to extract valuable insights from the combination of qualitative and quantitative data, integration must be done carefully. To effectively combine data, the use of right approaches, making sure that the mixed methods approach provides value rather than creating discrepancies.

Questionnaire Development

To assess the worker safety of the prefabricated approach in the construction site, a structured questionnaire was created and sent to various site managers, project engineers, safety officials, and employees.

The questionnaire was developed after a careful analysis of the body of research on the impact of prefabrication on worker safety. Uncovering important themes like incident rates, safety compliance, and the significance of training and education in guaranteeing worker safety in prefabrication settings by thoroughly reviewing pertinent papers. Based on these discoveries, focused inquiries to measure and investigate these factors to gather complex viewpoints and advance knowledge of the safety dynamics related to prefabrication in the building sector.

The questionnaire consisted of the following segments:

Segment 1: Demographics

The first section of our questionnaire goes into demographics, a critical first step in completely comprehending our audience's different backgrounds. Creating a detailed profile of our responders by gathering important demographic information such as job title, organization and construction sector, years of experience, and most experience from construction site projects in the UAE.

This fundamental data will not only allow us to adjust following questions to the individual needs and viewpoints of our audience, but it will also allow us to evaluate and interpret survey results through a lens that considers the many contexts in which our participants live.

Segment 2: Workplace Safety Assessment Survey

The sections for Segment 2 included the following three variables:

- Incident Rates: A variety of occupational mishaps and occurrences that endanger worker safety in the context of prefabricated building are covered such as slips, trips and falls, struck by objects, caught-in/between objects, electrical incidents, collapse or structural failures and material handling incidents.
- Safety Compliance: When using a prefabricated approach to improve worker safety in construction, safety compliance refers to following established rules and procedures. Examples of safety measures include emergency response plans, frequent safety audits, safety communication protocols, incident reports, and investigations. When taken as a whole, these safety precautions help to promote a culture of safety compliance, guaranteeing that the well-being of construction workers is given top priority and preserved throughout the prefabrication phase.
- Training and Education: To improve worker safety in the construction industry, a prefabricated method calls for extensive training and educational programs that provide employees the information, abilities, and awareness they need to deal with the difficulties that come with prefabrication. Prefabrication procedures, equipment operation, material handling, fall prevention, regulatory compliance, and continuous learning are all included in the training and educational components.

The second section of our Workplace Safety Assessment Survey offers a thorough approach to assessing various aspects of workplace safety, splitting into three areas, and enriched with 5–6 well prepared questions using a deliberate blend of Likert-scale and multiple-choice questions.

The section on Incident Rates delves into the frequency and nature of workplace incidents, adopting a complex scale to describe the severity and impact of these occurrences. Safety Compliance examines adherence to specified safety protocols, as well as qualitative and quantitative dimensions of compliance.

The Training and Education section investigates the efficacy of safety education programs to gain a comprehensive grasp of employee experiences and opinions. This dual-method approach not only offers a more detailed investigation of respondents' attitudes, but it also delivers a fuller, more comprehensive picture.

Segment 3: Pilot Study

The third and most important section of our questionnaire is a thorough pilot study aimed at refining and optimizing the survey instrument for wider use. Furthermore, the questionnaire went out to 11 carefully selected responders, whose different viewpoints and experiences provided significant insights regarding the questions' clarity, relevance, and efficacy.

Their helpful feedback served as a compass, allowing us to spot potential ambiguities, tweak phrasing, and improve the survey's overall structure. Based on this essential feedback, changes were made to ensure that the questionnaire is not only understandable but also completely captures the delicate aspects of workplace safety.

Sampling Design

The majority of the study's target group are project and site engineers who operate on building sites. Project and site engineers, who play a crucial role in supervising construction activities, were chosen to carry out the survey on improving worker safety in the United Arab Emirates using a prefabricated method. They are in a unique position to evaluate safety procedures and spot potential hazards related to prefabrication because of their direct knowledge and proficiency in overseeing on-site operations. As important participants in the project's implementation, project and site engineers are well-versed in the unique difficulties presented by prefabricated construction techniques. As such, they are in a good position to learn important lessons about safety precautions and to help develop focused interventions that put the health and safety of workers in the UAE construction industry first.

There are two ways that the questionnaire was created and designed:

- Electronic Questionnaire: One hundred safety officers, engineers, managers, and laborers from the construction industry in the United Arab Emirates filled out the survey.
- Hard Copy Questionnaire: Six papers are distributed to prefabricated professionals to solicit their opinions on the various construction techniques.

The total sample obtained from the questionnaire was 106 respondents. After getting all the responses from the respondents, both Linkert scale and multiple-choice questions were further quantified out of 5.

Each section regarding the three variables was summed up with a total of 25, 25 and 30. When adding all the independent variables score, the dependent score for worker safety turned out to be out of 80%.

Moreover, inserted ranges from high to low labeled from (1 to 3) with 1 being very low, 2 being moderate, and 3 being very high for all the summed up independent variables to further insert it in the SPSS model to generate the data of how effective and significant the independent variables are on the dependent variable.

Sample Size

Sample size is used to study the number of individuals or observations in a current study. After getting the data generated by Dubai Municipality's Statistic Centre, the following population size for buildings under construction by 2022 is obtained to be 13,236. However, the targeted population are only experts in the construction industry such as site engineers and project managers having a population size of 125.

Therefore, inputting the target population number into the equation of the sample size will yield a sample size of 95 as shown below.

$$n = \frac{\frac{z^2 * p(1-p)}{e^2}}{1 + (\frac{z^2 * p(1-p)}{e^2 N})} = \frac{\frac{(1.96)^2 * (0.5)(1-0.5)}{(0.05)^2}}{1 + (\frac{(1.96)^2 * (0.5)(1-0.5)}{(0.05)^2 (125)})} = 95 \quad (27.1)$$

- Population Size (N) = 125.
- Sample Size (n) = 95.
- Z-Score = 1.96 at 95% Confidence Interval.
- Sample Proportion (p) = 0.5.
- Margin of Error (e) = 0.05.

Accordingly, the number of respondents yielded from the questionnaire was (106), which is higher than our sample size (95). To improve the robustness of this study, further research with a larger sample size could be beneficial in the future.

Data Analysis Techniques

A three-way ANOVA was utilized to examine the association between the variables (Incident Rates, Safety Compliance, Training and Education) on Worker's Safety to meet the study's objectives and ensure its validity, followed by interview with experts and Delphi method to confirm and rank the results.

Results and Discussion

This chapter provides a thorough explanation of the data analysis process and the outcomes that were produced. The results that are displayed follow the order in which the tests were conducted. The demographics of the individuals who responded to the questionnaire were displayed through a descriptive analysis. Subsequently, a preliminary analysis was conducted to examine the test assumptions, including homogeneity and normality. The study's second section exhibits the findings of the 3-way ANOVA that was employed, goes into great depth about how the independent variables affected the dependent variable, and illustrates how the prefabricated technique has affected overall worker safety in UAE construction. Performed all our analyses using a 95% confidence interval. The last section of the discussion showcases the results of interviews conducted alongside ranking them through the Delphi Method to give further insights, depth, credibility and identify future directions within this study.

Results from the Questionnaire (Descriptive Analysis)

The questionnaire was delivered both online and in physical form to project managers and site engineers working on various building projects around the United Arab Emirates. A total of 106 replies were collected, and all pertinent data for this study was produced using the SPSS Statistics Software.

- Job Title: Most respondents to the study were site and safety engineers with a total of 19.81% and 16.04% which suggests that they are an important group with extensive understanding of worker safety in the UAE, especially in relation to prefabricated construction methods.
- Years of Experience in the Construction Field: According to the demographic analysis, 35.85% of the respondents had six to ten years of experience in the construction industry, which represents a significant pool of seasoned experts qualified to offer advice on how to improve worker safety using a prefabricated approach. Following that, experience levels of 1–5 years and 11–15 years had yields of 21.70% and 21.70%, respectively. The remaining age groups, 16–20 years old, over 20 years old, and younger than a year, contributed 10.38%, 5.66%, and 4.72%, in that order.
- Construction Sector: According to the demographic split, most respondents (40.57%) were from the residential construction industry, while 21.70% were from the commercial sector. Both results are consistent with our thesis, which focuses on improving worker safety in both industries. Moreover, Infrastructure/Transportation at 13.21% and Energy/Utilities at 12.26%. Other construction sectors had a uniform percentage of 0.94%.
- Organization Sector: According to the survey results, most participants are employed in the private sector (54.72%), with the government sector coming in second (23.58%). This finding is consistent with our research thesis, which focuses on improving worker safety, especially in these two organizational sectors.
- Size of the Organization with most experience: Most respondents work in organizations with more than 200 employees having a percentage of 63.21%. Following that with 33.02% working in organizations with 50–200 employees and 3.77% working in organizations with less than 50 employees.
- Most experience from construction site projects in the UAE: With a total of 60 responses, most respondents with experience working on construction sites in the UAE were based in Dubai, making up 56.6% of the sample. This finding is crucial for determining the sample size because it ensures representation from a significant geographic context within the UAE.

Preliminary Analysis

- Levene's Test: Levene's Test is the Homogeneity of Variances for each combination of the groups of the three independent variables. The homogeneity test is one of the primary presumptions that must be fulfilled to perform an ANOVA test. "Levene's Test of Equality of Error Variance" is used in our study to determine which population locations for our variables had the same variance. Based on the observed results, accepting the null hypothesis of homogeneity of variances since the p value of 0.115 is greater than our α value of 0.05. The results of the Levene's test indicated that each group's variance was equal. ($F(10,88) = 1.616$, $p = 0.115$).
- Normality of the Sample: Looking into the Skewness and Kurtosis Table and by dividing the statistic by the standard error most pairs for the z value are between −3.29 and 3.29, therefore concluding that the data is normally distributed.3-Way ANOVA only requires approximate normal data because it is quite robust to violations of normality, meaning that the assumption can be a little violated and still provide valid results. Since our data is large ($N = 106$), using analytical tests such as "Shapiro-Wilk test of normality" will have a major drawback since the p-value decreases with the increase of sample size, meaning that it will be less than 0.05. To avoid this problem, graphical methods should be used to demonstrate normality.

Reporting the Main Effects

The primary conclusions of the 3-Way ANOVA analyses will be compiled in this part, along with a discussion of all the pertinent data that can be deduced from the results.

A factorial ANOVA [3-Way ANOVA] was conducted to compare the main effects of Incident Rates, Safety Compliance, and Training and Education [III] as well as their interaction effects on the Worker Safety Score [DV].

All three independent variables were statistically significant at $p < 0.001$, meaning that there is a substantial difference across all levels, according to the data. Additionally, the table allows us to draw the following conclusions.

- Incident Rates, Safety Compliance, and Training and Education effects were all statistically significant at $p < 0.001$ (This indicates that there is a significant difference across all levels).
- The main effect of Incident Rates yielded an effect size of 0.282, indicating that 28.2% of the variance in the Worker Safety was explained by Incident Rates. ($F(2,88) = 17.240$, $p < 0.001$).
- The main effect of Safety Compliance yielded an effect size of 0.324, indicating that 32.4% of the variance in the Worker Safety was explained by Safety Compliance.

(F(2,88) = 21.087, p < 0.001).

- The main effect of Training and Education yielded an effect size of 0.423, indicating that 42.3% of the variance in the Worker Safety was explained by Training and Education
(F(2,88) = 32.304, p < 0.001).

The value for small sample size ranges between 0.01 and 0.06, for medium sample size, between 0.06 and 0.014, and for high sample size, 0.14 or above, according to Cohen's rule of thumb. The partial eta squared value is over 0.14, as can be seen in the above table, indicating that it had a significant impact on the dependent variable.

This suggests that the frequency of incidents, adherence to safety regulations, and the provision of training and education all have a notable and substantial influence on the safety results at construction sites.

Reporting the Main Interactions

Accepting the Null Hypothesis where P Value of (0.001) is less than our alpha, therefore there was no combined effect between all pairs.

- The interaction effect between Incident Rates and Safety Compliance was not significant (F(4,88) = 0.509, p = 0.729), indicating that there was no combined effect for Incident Rates and Safety Compliance on the Worker Safety.
- The interaction effect between Incident Rates and Training and Education was not significant (F(3,88) = 0.393, p = 0.758), indicating that there was no combined effect for Incident Rates and Training and Education on the Worker Safety.
- The interaction effect between Safety Compliance and Training and Education was not significant (F(2,88) = 0.091, p = 0.913), indicating that there was no combined effect for Safety Compliance and Training and Education on the Worker Safety.
- The interaction effect between Incident Rates, Safety Compliance and Training and Education was not significant (F(1,88) = 0.074, p = 0.786), indicating that there was no combined effect Incident Rates, Safety Compliance and Training and Education on the Worker Safety.

Results from Interview with Experts

In summary, overall observations point to a dynamic future for the construction industry in the United Arab Emirates, where employee safety during prefabrication will be greatly enhanced by a blend of sustainable practices, technical breakthroughs, and environmental concerns.

Accepting these developments will improve worker wellbeing while setting up the sector for long-term resilience and sustainability.

It's evident that incident rates, safety compliance, training, and education have a major impact on worker safety in the UAE construction industry based on the three-way ANOVA results and the considerable effects they have on worker safety.

Insights and ANOVA results match as follows:

- Incident Rates (Explaining 28.2% of Variance): The importance of incident rates is supported by the insights that highlight safety precautions like proper site training, PPE kits, cutting-edge technologies, and off-site assembly. The qualitative observations imply that PPE provision, training initiatives, and cutting-edge technologies can all help lower the number of occurrences.
- Safety Compliance (Explaining 32.4% of Variance): The following factors, drainage systems, supporting structures, appropriate sanitization, and off-site assembly, all help ensure safety compliance: sustainable construction materials. The ANOVA results show that these procedures are in line with the requirement for adherence to safety standards and laws.
- Training and Education (Explaining 42.3% of Variance): The strong impact of training and education on worker safety is consistent with the insights centered around skill development, modern technologies, off-site assembly, and training programs. The greater proportion of variance accounted for by education and training highlights the need for continuous learning and skill development.

To sum things up, the conclusions drawn from the qualitative data are consistent with the statistical outcomes of the three-way ANOVA.

The ANOVA's findings about the significant effects of incident rates, safety compliance, and training and education on worker safety are consistent with the insights' emphasis on training, safety measures, and compliance practices.

This connection reinforces the notion that worker safety in the UAE construction sector requires a comprehensive strategy that includes a range of safety measures and educational programs.

Results from Delphi Method

Round 1 Discussion:

The first Delphi method round effectively commenced an extensive investigation into improving worker safety in the United Arab Emirates through a prefabricated strategy. The replies from the experts shed light on a number of issues and offered insightful analysis of the 3-way ANOVA results.

Prefabricated constraints, such as issues with customisation and adaptability, stand out among these worries. Environmental factors were also emphasized, including problems like resource depletion and waste production related to prefabrication. Experts were upbeat about prefabrication's ability to lower hazardous emissions because of its well-managed and effective construction processes.

Round 2 Discussion:

Participants in the second Delphi round offered their perspectives on a range of topics pertaining to prefabrication in the building industry. Prefabrication is thought to have favorable environmental effects, especially when it comes to lowering greenhouse gas (GHG) emissions, which is important for maintaining worker health standards. Feedback on the efficacy of training programs was divided; while it was generally acknowledged that personnel needed educational training, there was doubt regarding the practical sessions' ability to raise safety awareness.

Positive acknowledgement of cultural diversity was shown, highlighting the necessity of inclusive training programs to successfully engage multicultural staff. Prefabrication constraints, on the other hand, like joint leaks and problems with product quality, were called out as potential problems that can negatively impact consumer experiences.

Round 3 Discussion:

At the end of the third round of the Delphi process, a consensus was reached among experts, indicating a significant consensus regarding the benefits of prefabrication in terms of lowering dangerous emissions and addressing environmental issues.

By decreasing manual labour and automating procedures, the use of computerized programs and artificial intelligence was considered as a major factor in improving workplace safety. Prominent specialists underscored the significance of prefabrication in mitigating waste, underscoring its capacity to enhance sustainability through optimal material utilization and mitigation of adverse environmental impacts.

The consensus also extended to training efficacy, where responses to the use of AI in virtual sessions and the combination of theoretical and practical instruction were largely positive. The conversation about cultural diversity also highlighted how crucial inclusive training initiatives and development guidelines are to improving workers' abilities.

There was broad consensus regarding the limitations of prefabrication, with experts arguing that traditional construction methods should be abandoned in order to prevent manufacturing flaws and enhance product quality. Finally, the Delphi approach produced a strong consensus about prefabrication's positive environmental effects, especially when it comes to lowering greenhouse gas emissions via cutting-edge technologies like "Waste to Energy," such as biomass energy, gasification, incineration which turns waste materials into energy sources.

The thorough insights gathered from this round offer a solid basis for forming practices and policies.

Conclusion:

By providing a sophisticated and consensus-driven understanding of important aspects related to worker safety, environmental sustainability, and technological advancements in the context of prefabrication in the UAE construction industry, the Delphi method's findings significantly advance the research as a whole. The knowledge acquired has applications for stakeholders, industry professionals, and policymakers in addition to influencing scholarly discourse.

A roadmap for creating focused interventions, educational initiatives, and legislative frameworks that can support the secure and long-term incorporation of prefabrication techniques is provided by the issues that have been identified and their possible solutions.

In general, the expert opinions obtained using the Delphi technique provide a strong basis for directing the future course of construction practices in the United Arab Emirates, bringing them into compliance with international safety, efficiency, and environmental requirements.

Conclusion and Recommendations

In this chapter, the summary of findings and recommendations for future studies will be outlined. This chapter presents the study's key conclusions and provides research questions based on the information gathered. In addition, the study's limitations and recommendations are provided to improve on future research on the safety of building sites.

Research Result Discussion

- Question 1: What are the current safety practices and regulations in the construction industry in the UAE with regards to the use of prefabricated technology?
  - To improve productivity and safety in the building sector, the United Arab Emirates has been actively pushing the use of prefabricated technology in current construction processes.
  - Construction methods, including the use of prefabricated technologies, are governed by a comprehensive set of building norms and standards in the United Arab Emirates. The purpose of these standards and guidelines is to guarantee the excellence and safety of building projects such as the following:
  - Regulatory Authorities: Dubai Municipality, which has set rules and regulations for the use of prefabricated technology, is the regulatory body in charge of regulating building in the United Arab Emirates. There's a chance that other UAE emirates have their own regulatory agencies and specifications.
  - Quality Control: In the construction sector, control procedures are crucial. Prefabricated components must meet specific quality standards to ensure the safety of the final construction.
- Question 2: How does the use of prefabricated technology compare to traditional construction techniques in terms of their impact on the safety of workers in the construction industry in the UAE?

- When compared with traditional construction methods, the use of prefabricated technology in construction can have several effects on worker safety in the UAE's construction sector such as:
  Lower On-Site Risks: Prefabrication is the process of producing individual building modules or even full assemblies in a regulated factory setting. As a result, less labor needs to be done on-site, lowering the chance of encountering risks including inclement weather, uneven terrain, and falling from a height. Reduced heat-related dangers are advantageous for workers in the United Arab Emirates, especially during the hot summer months.
- Managed Work Environment: Factory workers enjoy regulated working conditions that include enough lighting, ventilation, and safety precautions. This may result in a reduction in the likelihood of mishaps caused by overheating, dim lighting, or insufficient ventilation, all of which are common in the construction industry in the United Arab Emirates.
- Quality Assurance: Strict quality control procedures can be implemented during the manufacturing process thanks to prefabrication. This implies that structural flaws in building materials are less likely to occur, which lowers the risk of mishaps or failures while building.
- Speed and Efficiency: Prefabricated building frequently results in quicker project completion, lowering the total amount of time that dangers are present. Due to extended building activity, shorter construction timetables may lead to fewer accidents.
- Fewer Trades Made On-Site: Traditional building techniques frequently necessitate the simultaneous employment of several crafts on the same site, raising the possibility of mishaps and disputes. The construction process can be streamlined by prefabrication, requiring fewer trades to work concurrently on the site.
- Training and Experience: Workers with a diverse set of abilities, such as manufacturing, assembly, and quality control, may be needed for prefabrication. This may influence the kind and degree of training needed, thereby lowering some risks associated with building.
- To simply explain the difference between the use of traditional and prefabricated approach in terms of workers safety in construction, Table 27.1 is provided down below.

• Question 3: What are the most significant safety concerns related to the use of prefabricated technology in the construction industry in the UAE, and how can they be addressed?

  - The use of prefabricated technology in the UAE construction sector has various advantages, but it also raises some safety problems that must be addressed. Some of the most serious safety concerns about prefabricated technologies in the UAE are as follows:
  - Training and Adaptation: Workers might have to adjust to new working practices, which could mean more training and time for adjustment.

Table 27.1 Comparison between the use of traditional and prefabricated approach in terms of workers safety in the construction industry of UAE

Aspect	Prefabricated technology	Traditional construction techniques
On-site risks	A reduced likelihood of on-site accidents as a large amount of work is completed off-site	Higher on-site risks because of substantial on-site construction activity and exposure to weather conditions
Controlled working conditions	The factory environment provides controlled working conditions with enough lighting, ventilation, and safety precautions	On-site circumstances may be less controlled, exposing workers to changing weather and terrain
Quality control	In a production setting, strict quality control minimizes the risk of structural problems	On-site quality control is frequently more difficult, resulting in potential faults and safety hazards
Efficiency & speed	Faster project completion minimizes total risk exposure time	Longer building schedules may prolong exposure to potential hazards
On-site trades	Fewer trades working on-site at the same time, limiting potential disputes and accidents	On-site, multiple trades are frequently working concurrently, increasing the potential of accidents and coordination issues
Transportation risks	The dangers of delivering prefabricated components to the job site. Proper logistics planning is required	Transportation concerns are limited because most materials are delivered directly to the job site
Factory-related hazards	In a factory setting, workers are exposed to production processes and machinery	On-site workers may be exposed to a variety of hazards associated with construction activities
Training & skill requirements for safe operations	Workers may require specialized skills in production, assembly, and quality control	Traditional construction abilities are generally relevant; however, specialist training may be necessary for some tasks
Ergonomics & worker-well being	Due to repetitive work in a production setting, ergonomics may be required	On-site ergonomic difficulties may still exist, particularly in labor-intensive occupations

(continued)

Table 27.1 (continued)

Aspect	Prefabricated technology	Traditional construction techniques
Adaption & training	Workers may need to adjust to new working procedures and get prefabrication process training	Traditional construction workers may require training for specialized tasks, but they often use methods that are known to them
Emergency response	Specific emergency action plans are required in the event of an accident during shipment, offloading, or installation	On-site accidents and incidents are expected at higher rates, thus more emergency response strategies and plans are required

- Factory-Associated Risks: There may still be certain hazards for workers in prefabrication companies because of the machinery and production procedures involved. Ensuring their safety in a factory setting is crucial.
- Risks in Transportation and Assembly: Large prefabricated components may be hazardous to transport to the construction site, especially when it comes to dumping at the site, road safety, and logistics.

- Question 4: What are the most effective solutions for evaluating and improving the impact of prefabricated technology on the safety of workers in the construction industry in the UAE?
  - Evaluating and enhancing the impact of prefabricated technology on worker safety in the UAE construction sector requires a combination of proactive and continuous actions.

Recommendations for Future Research

Respondents have consistently underlined the necessity of incorporating technical improvements into prefabrication procedures, especially given the thesis's limited time frame. The urgency is a result of the data analysis, which highlights how technology can improve productivity, as well as thorough literature research that identifies common health risks, including musculoskeletal illnesses among workers. To limit health risks, a viable approach to addressing these concerns is the investigation and application of ergonomic design concepts into the prefabrication production stage.

Furthermore, the Delphi method's insights highlight respondents' increasing knowledge of environmental issues related to conventional construction. Numerous participants emphasized that prefabrication can greatly lower environmental impact, including energy and trash production. This supports prefabrication's feasibility as an environmentally friendly substitute and provides opportunities for further research and recommendations, highlighting the building industry's potential for sustainable practices.

There are various ways to improve the study's findings that can be addressed for future research:

- Technological Integration Assessment—Investigate the use of emerging technologies such as artificial intelligence, virtual reality, and robotics in off-site construction to determine their potential for improving worker safety. Examine how these technologies can be integrated and optimized within the construction procedure.
- Collaborative Industry Initiatives—Encourage building sector collaboration by involving important stakeholders such as contractors, manufacturers, regulators, and workers. Create channels for exchanging best practices, learnings, and encouraging innovation in off-site construction safety.
- Global Benchmarking—Conduct benchmarking studies by contrasting UAE off-site construction safety practices and innovations with worldwide counterparts. Identify effective tactics and lessons learned from other regions, allowing for the implementation of best practices that are customized to the UAE setting.
- Environmental Impact Assessment—Consider the broader impact and examine the environmental sustainability and worker safety aspects of off-site building alternatives. Examine the ecological footprint of these approaches and look for ways to improve both safety and environmental consequences.
- Ergonomic Design Evaluation—Examine the ergonomic design aspects of off-site building solutions like modular components and prefabricated structures to see how they affect worker safety and comfort. Examine how ergonomics are included in the planning and manufacturing stages.
- Worker Perception and Behavior—Investigate worker safety perceptions in prefabricated buildings and how these perceptions influence safety behavior. Conduct surveys or interviews to learn about worker attitudes regarding safety standards and prefabrication ergonomics.

Limitations of the Study

Several limits were encountered over the course of this investigation due to various constraints, including:

- Time Limitations—The research was conducted in a limited timeframe, which may have limited the depth and breadth of the investigation.
- Online Surveys—There was no control over the behavior of respondents when answering the survey because of employing online questionnaires. Furthermore, there was no control over whether the correspondent was the study's designated respondent.
- Limited Availability of Comprehensive Data—Due to the topic's infancy in the region, finding relevant and detailed information proved difficult, limiting the depth of the investigation, and likely leaving key parts unaddressed.

- Continuation of the Delphi Method—In order to acquire more extensive and comprehensive results, the third round of the applicable Delphi analysis in this research should be conducted with diverse groups of professional experts.

References

1. Johnson, S. J. (2019). *Improving construction safety with prefabrication and modular construction technology*. BIM 360.
2. Bagatsing, R. (2023). *17 Construction trends in 2023*. Manila: Gineers Now.
3. Gausch, K., & Guevara, P. (2022). *An introductory guide to prefabrication*. Safety Culture.
4. Imarc. (2021). UAE prefabricated building and structural steel market.
5. Shibani, A., Saidani, M., & Alhajeri, M. (2013). Health and safety influence on the construction project performance in United Arab Emirates (UAE). *Journal of Civil Engineering and Construction Technology, 4*(2), 32–44.
6. Fagberno, R. K., Sunindijo, R. Y., Illankoon, C., & Frimpong, S. (2023). Influence of prefabricated construction on the mental health of workers: systematic review. *European Journal of Investigation in Health Psychology and Education, 19*.
7. Franks, E. D. (2018). *Safety and health in prefabricated construction: A new framework for analysis*.

Chapter 28
Enhancing Safety Management in UAE Construction Sites Through Site Manager Performance Evaluation

Abdulla Omar Al Zarooni and Hamad S. J. Rashid

Abstract The construction industry is one of the most high-risk domains, having the highest number of occupational accidents compared to other industries globally. One of the main reasons for worker injuries and deaths is the site managers' unsafe behaviour on the construction sites, which creates an environment that normalizes further unsafe acts and leads to the personnel's inability to recognize and identify various hazards within the construction site. This research aims to investigate and evaluate the site manager's contribution towards safety through specific indicators that include their safety-oriented performance, provision of safety equipment, and conduction of safety training. Data was collected from the workers and managers on construction sites through a comprehensive survey that constituted the three main mentioned indicators. The surveyed sample included 174 participants from different construction organizations across the UAE. The data was then analyzed using the 3-way ANOVA technique to study the relationship between the three named variables. Monte Carlo Simulation was used to study the uncertainty and variability of the interactions of the three factors in the construction site. The obtained results showed that the three factors were all statistically significant, indicating that the three independent variables had a positive effect on the safety score of the site manager. The interaction effect between the safety-oriented performance & provision of safety equipment was significant indicating that there was a combined effect between them on the safety score. A set of recommendations are provided to ensure workplace and human well-being that will lead to more safety methods being implemented across the entire UAE construction sector.

Keywords Construction manager · Construction safety · Construction site · Safety management · Site manager · Worker safety

A. O. Al Zarooni · H. S. J. Rashid (✉)
Industrial Engineering and Engineering Management Department, University of Sharjah, Sharjah, UAE
e-mail: hamad.rashid@sharjah.ac.ae

A. O. Al Zarooni
e-mail: U21100904@sharjah.ac.ae

© The Author(s), under exclusive license to Springer Nature Switzerland AG 2024
A. Emrouznejad et al. (eds.), *Business Analytics and Decision Making in Practice*, Lecture Notes in Operations Research, https://doi.org/10.1007/978-3-031-61589-4_28

Introduction

The construction industry in one of the deadliest industries for people to work in if they are not careful. Since 2012, the construction industry has experienced the most workplace deaths than any other industry [1]. In 2021, there has been 986 fatalities recorded, and an average of 949 deaths from 2012–2021 [1]. To minimize the number of fatalities and maintain the safety of the personnel in the workplace, it is recommended to have a site manager who oversees the workers and prioritizes their safety by creating a comprehensive site safety plan, conducting regular safety training, providing new and high-quality equipment, communicating, documenting, and tracking safety incidents [2].

A construction site manager is regarded to have a significant performance impact on the safety of the construction sites. Moreover, site managers are regarded as critical figures; therefore, they should lead based on influence and not by authority, and this can be achieved by sharing the project vision, setting the direction and the pace of the work to be done, empowering and coaching the team, easing the communication between all stakeholders, and demonstrating good character and attitude [3].

Site safety training is an essential requirement to be done within any industry. It is a structured educational program focused on ensuring safety and minimizing risk within the workplace [4]. Since most fatalities are preventable, comprehensive safety training will reduce the number of deaths and, in addition, will ensure the workers' safety, boost operational productivity, and increase cost-savings within a company [4].

Having new and high-quality equipment in the workplace has twofold benefits compared to old equipment. Site managers who provide new and modern safety equipment can help productivity and reduce body strain that can lead to musculoskeletal injuries. Moreover, it can defend the two common hazards in the workplace: "Struck-by" & "Caught-in" Incidents [2].

The UAE strives to enhance safety in the construction site by implementing rules and regulations to consider workers' safety. Recently, a new law was implemented that doesn't permit personnel to work from 12:30 PM—3:00 PM during summer (Hot weather). A fine of 50,000 AED will be imposed on anyone who violates this law. This will, in return, decrease the chance for workers to suffer from heat strokes and avoid heat stress, which is a reason they don't wear their Personal Protective Equipment (PPE) regularly.

The study assesses the safety score of the site managers by looking into their current overall performance and seeing how it can be elevated further to sustain an injury-free environment in the UAE.

Literature Review

Effect of Site Managers' Safety Violations on Workers in the Construction Site

Site Managers are liable to ensure that the daily operations in the workplace are going as planned while providing the personnel with the relevant instructions to ensure their safety and well-being. Since the site managers are directly in charge of the workers, they will tend to communicate with them more often, and depending on how they communicate, the worker's behaviors will be affected in either a positive or a negative way.

Most studies focus on the positive effects of a site manager on safety-related behavior, while studies of the negative social influence of site managers on workers are rare. When workers observe the bad safety regulations of the supervisor, they will most likely breach safety rules and regulations.

The two terms, social learning, and social information are used to explain the work-related attitude of individuals [5]. Social learning is the process through which individuals get influenced by the attitudes of others, while social information uses the social or physical environment characteristics, which results in individuals reacting based on the situation that is related to the environment of the place, which will guide their behavior [5].

The Human Factors Analysis and Classification System (HFACS) is a broad human error framework that provides a tool to assist in the investigation process, target training, and prevention efforts [6]. Active and Latent failures within a workplace that led to the accident can be systematically identified by investigators using this framework. [6]. HFACS is used to investigate the hidden factors that lead to an accident and not to blame other people for their mistakes [6]. Level 1 is about unsafe acts, which are divided into two categories—errors and violations. Errors are unintentional behaviors, while violations are the willingness to breach the rules and regulations. Level 2 mentions the preconditions that cause unsafe acts to occur, such as fatigue, stress, inadequate training, poor equipment, and bad weather conditions. Level 3 states the impact of the supervisor's safety commitment that can result in unsafe acts. Level 4 explores how higher management and organizational culture can affect safety on the frontline. To further emphasize the importance of the supervision level, HFACS placed unsafe supervision on Level 3 in the framework. The safety of the project and people is greatly influenced by many factors under this umbrella, and one of the most critical factors is affected by the supervisory violations towards safety.

A Better Contribution to Safety as a Site Manager

Occupational safety research indicates that the way managerial leadership is executed may significantly affect how occupational risks are managed, and occupational safety is advanced [7]. Both direct and indirect leadership behaviors affect safety culture.

Indirect leadership practices, such as coordinating work by establishing a frequent meeting structure, planning the physical layout of the site, managing safety aids and resources, and structuring the work by establishing and following rules, guidelines, and plans, have a positive impact on safety [8]. Direct management practices, such as proactively monitoring and correcting risks, managing safety risks and violations, and acting as role models for professional workers and subcontractors, have a favorable effect on safety [8].

Leadership Behaviors that are characterized by a lack of active leadership responsibility, unclear communication, and not making risk analyses result in reduced safety at the construction site. The core leadership behaviors that positively influence safety are continuous planning, consistently coordinating between the team, being a role model to the workers, and monitoring and correcting defects [7]. Safety could be improved further by offering site managers leadership development training and education that focuses on communicating clear expectations, carrying out collective risk analyses, and executing proactive monitoring and feedback procedures [7].

How Safety Leadership Styles Impact Employee Safety Behaviour

There are many challenges being faced at multiple levels by management when it comes to the safety of the workplace [9]. At an individual level, workers are urged to follow safety rules and regulations while interacting with each other in a way that fosters safety on the construction site [9]. At a team level, site managers should try to allocate some of their time to creating goals, observing the behaviors of the workers, providing feedback on safety concerns, executing corrective actions when necessary, and giving incentives to promote excellent safety behaviors [9]. At the organizational level, managers are in charge of assisting workers in realizing that their safety in the construction sites is considered to be of utmost priority and is being valued by the higher management in order to strengthen a safety culture that focuses on human well-being and trust both inside and beyond the organization [9].

Leadership plays a vital role in creating a positive work atmosphere for the workers in regard to safety and health since workers are insightful about the people above them and value their safety within an organization [9].

Employees are more likely to adhere to safety policies in the workplace when they perceive management to be devoted to safety. When leaders make it clear via their actions and practices that they are committed to ensuring the integrity of safety in daily operations, employees can clearly understand where leadership stands in regard to safety. Some examples of this can be:

- Management ensures that personal protective equipment (PPE) is available and easily accessible on a daily basis [9].
- Management invests on passive controls to protect workers from any kinds of incidents [9].
- Motivating the workers to continuously meet goals for key performance indicators (KPI) that leads to safety [9].
- Ensuring a safety culture where employees feel comfortable in reporting workplace hazards by providing them with sufficient training and resources [9].

Factors Influencing Safety at the Construction Site

The construction industry is known as the most hazardous and dangerous industry, which scored the highest number of deaths over more than ten years in a row [10]. The incidents and accidents that occurred at the construction site have caused delays in project completion, increase in project costs, reduction of productivity, and created negative impressions of the organization [10]. It is critical to ensure safety and health at the workplace by considering different safety factors. The most studied factors that affect safety performance of construction projects are the construction site manager's performance toward safety, the provision of safety equipment, and the conduction of safety training [10].

Construction Site Manager's Performance Towards Safety. The site manager's performance towards safety can positively lead to a better and safer environment at the construction site, which leads to establishing a safety culture among the workers on the site. Lack of action from the site manager creates workplace hazards, and this can be avoided by spreading safety knowledge among all personnel, having solid leadership that puts health and safety as a top priority, and ensuring that all safe work practices and procedures are being followed.

Provision of Safety Equipment. The most common safety equipment found in the construction site are personal protective equipment (PPE) which are protective clothing that protects most body parts and serves as a final line of defense when engineering controls are not enough to reducing or eliminating the risk [11]. It is utilized to keep an individual free of risk from physical hazards to avoid injuries and fatal accidents. There are many major benefits of using safety equipment such as: It prevents unnecessary injury in the workplace, protects employees from chemical exposure, helps businesses comply with regulatory requirements, and improve employee productivity and efficiency [11].

Conduction of Safety Training. Conducting safety training in the workplace can help employees gain the necessary knowledge and skills to identify hazards, and use proactive measures to prevent accidents and incidents from occurring [12]. For example, in the construction industry, including good safety training will ensure everyone in the team uses the correct tools, make employees aware of potential hazards and know how to respond to them, which will ensure a safe environment with no injuries. Moreover, conducting safety training will decrease employee turnover rate, costs, and illness/injury, while increasing productivity morale of the workers [13].

Methodology

Academic research designs can be divided into four main categories that guide researchers in their exploration of new frontiers with a systematic approach. These categories include Action Research or Applied Research, which involves using scientific methods to analyze and solve real-life problems. Qualitative Research is another type that involves collecting and analyzing data from observations in natural and social settings, while Quantitative Research focuses on using statistical, mathematical models, and logical techniques to generate numerical data and facts. Finally, Mixed Methods combines both Qualitative and Quantitative methods to reduce the weaknesses associated with using a single research design while maximizing strengths, validity, and reliability. The framework that showcases the systematic approach of the applied methodology is presented below:

A. **Main Literature Review**: The main literature review will focus on the effect of the site manager and their impact on the employees in regard to safety. It showcases how safety violations can affect safety and how different safety leadership styles can enhance the site manager's safety performance. This information provides beneficial knowledge and details to get an idea of how different acts can lead to different outcomes when it comes to the safety and well-being of the workers in the workplace, and based on that, the most common factors that affect safety in construction projects were chosen such as construction site manager's performance towards safety (results in having a safe work environment), provision of safety equipment, and conduction of safety training.
B. **Methodology Finalization**: The final framework constructed that defines the systematic approach and techniques used by showcasing the roadmap that will be implemented in order to achieve the study's research objectives. The roadmap was developed based on the site visits conducted in the UAE, which generated our research objectives/questions, and questionnaire itself was developed based on the current observed safety practices.
C. **Data Collection**: A structured questionnaire was distributed to different site managers, safety officers, and workers, which focused on the site managers' performance towards safety, provision of safety equipment, and conduction of safety training. The questionnaire consisted of two main sections. The first section consisted of demographics to learn about the respondent's background. The second section consisted of questions about the three variables mentioned

above, and for each variable, five multiple-choice questions were provided for them answer to learn about the safety of the current safety practices in the workplace. In order to begin the data collection, a pilot study was conducted, and ten people answered the questionnaire in order to check any weaknesses within the questions and check if all questions were comprehensive and valid to the research's objectives. After the pilot study was done, the questionnaire was sent to the rest of the workers, and a total of 174 responses were collected. The responses were analyzed using "Cronbach's Alpa" to test the reliability and validity of the test. Finally, the responses were analyzed using 3-way ANOVA Analysis to study the main and interaction effect of the three variables on the safety of the workplace based on the site manager.

D. **Data Analysis: 3-Way ANOVA**: A 3-way ANOVA approach was conducted after collecting the relevant data from workers in order to study the site managers' performance towards safety, provision of safety equipment, and conduction of safety training. The reason behind choosing those three variables is because, based on past literature reviews, and a most recent one titled: "Factors Affecting Safety Performance at Construction Projects: A Literature Review," they were perceived to be the most important factors that affect the safety of construction projects, which therefore can affect the well-being of the workers. This data analysis tool was used to examine the main and interaction effects to get a comprehensive understanding of the main and combined impact on the overall safety of the workplace based on the overall performance of the site managers.

E. **Data Analysis: Monte Carlo Simulation**: Monte Carlo Simulation is a risk simulation method that is used to study uncertainty by generating multiple different scenarios. In our study, using Microsoft Excel, a model was developed that generated 10,000 different scenarios on the site managers' performance towards safety, provision of safety equipment, and conduction of safety training to see the site manager's overall score within the UAE. The probability for each score was influenced based on the site visits conducted, and the site manager's score was categorized into good safety behavior, neutral safety behavior, and bad safety behavior. Although Monte Carlo Simulation and ANOVA Analysis are two distinct statistical tools that serve different purposes in the context of research, these kinds of simulation can be integrated with each other to further enhance the study and make it more robust. The answers obtained from the simulation were used later to compare the results obtained from the questionnaire itself.

F. **Conclusions**: The study's findings contribute to enhancing construction site management practices, minimizing the risk of accidents, and improving safety outcomes that benefit the people working in the construction industry. The main solutions that were concluded are related to enhancing the work environment to make it more safety-friendly, implementing more developed and tailored safety training programs to be organized depending on the current challenges facing the workers in the construction site, implementing better PPE and safety equipment to be used in the construction site, and enhancement of current safety protocols.

Questionnaire Development

A structured questionnaire was developed and distributed among different site managers, safety officers, and workers, which focuses on the site managers' performance towards safety, provision of safety equipment, and conduction of safety training. The questionnaire consisted of two main sections. The first section consisted of demographics to learn about the respondent's background. The second section consisted of questions about the three variables mentioned above, and for each variable, five multiple-choice questions were provided for them to answer to learn about the safety of the current safety practices in the workplace. Those questions were developed based on the site visits that were conducted in the UAE.

In order to begin the data collection, a pilot study was conducted, and ten people answered the questionnaire in order to check any weaknesses within the questions and check if all questions were comprehensive and valid to the research's objectives. After the pilot study was done, the questionnaire was sent to the rest of the workers, and a total of 174 responses were collected. The responses were analyzed using "Cronbach's Alpha" to test the reliability and validity of the test. Finally, the responses were quantified and used in the 3-Way ANOVA Analysis to study the main and interaction effect of the three variables on the safety of the workplace based on the site manager.

Sampling Design and Sample Size

The target population of this study consists mostly of buildings that are under construction, with the main target audience to answer the questionnaire being the workers on the construction site since they are directly under the site manager's supervision and know the most about his interactions with them. Since most of the workers don't speak a specific language and can't read, the questionnaire was designed and distributed in different ways:

- **Electronic Questionnaire**: The Online questionnaire was designed and distributed to members of the construction industry to provide their feedback.
- **Hard Copy Questionnaire in 3 Different Languages**: As mentioned, since most workers in the construction site can only read in a single language, the questionnaire was designed to have 2 versions, Arabic & English, and distributed upon the workers. Since most workers there were Indians who speak Urdu, a foreman who can read the questionnaire sat with each worker individually and explained to them in Urdu to get the relevant data from them during their break sessions.

After the data collection period was done, the total amount of respondents who answered the questionnaire was obtained to be 174. For the sample size, the targeted population was assumed to be around 50,000 buildings under construction. This number was assumed because the only data that was provided for the public was from Abu Dhabi, Dubai, and Sharjah, in which the following statistics were shared on the Internet:

- Dubai Statistical Center showcased that in Dubai, there were 13,236 buildings were under construction and 2559 building permits issued in 2022.
- Federal Competitiveness and Statistics Centre showcased that there were 4145 building permits issued in Abu Dhabi City, 1710 building permits issued in Al Ain City, and 438 permits issued in Al Dhafra City, making that a total of 6293 building permits in 2015 [Latest data in their website].
- Sharjah Municipality announced that they issued 1,265 permits in 2022.

Those three Emirates cover around 95% of the UAE, so Ajman, Umm Al Quwain, Ras Al Khaimah, and Fujairah will have less data compared to them. Building permits are different than buildings under construction, and our targeted population will be based on the buildings that are under construction. Inputting the targeted population number to following sample size equation will yield a sample size of 382 as seen below.

$$n = \frac{\frac{z^2 * p(1-p)}{e^2}}{1 + (\frac{z^2 * p(1-p)}{e^2 N})} = \frac{\frac{(1.96)^2 * (0.5)(1-0.5)}{(0.05)^2}}{1 + (\frac{(1.96)^2 * (0.5)(1-0.5)}{(0.05)^2 (50{,}000)})} = 382 \quad (28.1)$$

where:

n sample size.
z z-score at 95% confidence interval.
p sample proportion.
e error of margin.
N population size.

Data Analysis Techniques

To fulfill the objectives and validity of the study, a 3-Way ANOVA was used to examine the main and interaction effects of the variables to get a comprehensive understanding of the main and combined impact on the safety of the workplace based on the overall performance of the site managers. Moreover, a Monte Carlo Simulation was used to generate multiple scenarios based on the site manager's performance towards safety, how effectively they conduct safety training programs, and their provision of safety equipment in the workplace to assess their impact on safety outcomes. The results were categorized into good safety behavior, neutral safety behavior, and bad safety behavior, which was used later to compare to the results obtained from the questionnaires.

Results and Discussion

In this chapter, a detailed description of the analysis of the data and the obtained results is explained. The results presented are in sequence with the performance of the tests. A descriptive analysis was used to showcase the demographics of the people who answered the questionnaire. Following that, a preliminary analysis was used to study the assumptions of the tests used, such as homogeneity and normality. The final section presents the results obtained from the 3-Way ANOVA and Monte Carlo Simulation that was used in this study, discusses in detail the effect of the independent variables on the dependent variable, and displays the overall safety score of the construction site managers in the UAE. All analyses were conducted on a 95% confidence interval.

Descriptive Analysis

The online and hard copy version of the questionnaire were distributed among the workers in different construction sites across all the UAE. A total of 174 responses was obtained overall, and was used in the SPSS Statistics program to generate all the relevant results for this study. The following section displayed the demographics of the respondents who contributed in adding their inputs for this study.

- **Job Title**: The Majority of the respondents were workers in the construction field being 67.24% of the total amount of respondents.
- **Years of Experience in the Construction Field**: People between 6–10 years of construction experience yielded 32.18% of the total amount of respondents, making them the major group. Following that in, 1–5 years of experience yielded 27.01%, 11–15 years of experience yielded 25.29%. Other age groups ranging between 16–20 years, more than 20 years, less than a year contributed 8.05%, 4.02%, and 3.45% respectively.
- **Construction Sector**: Most companies are specialized in Residential Construction with 125 responses. Following it is the commercial construction with 62 response, industrial construction with 45 responses, energy and utilities with 6 responses, and infrastructure & transportation with 4 responses.
- **Organization Sector**: The Majority of the respondents work in the private sector, consisting of 90.23% of the total amount of respondents, while 6.90% consisted of workers from the government sector, and 2.87% consisted of workers from the semi-government sector.
- **Size of the Organization with Most Experience**: In terms of size of the organization, the majority of the companies had more than 200 employees being 76.24% of the total amount of respondents. Meanwhile, 15.52% consisted of companies with 50–200 employees, and 8.05% consisted of employees with less than 50 employees.

- **Most Experience from Construction Site Projects in the UAE**: Most workers had experience in Dubai with 72 responses, following that is Sharjah with 60 responses, Umm Al Quwain with 52 responses, Abu Dhabi with 19 responses, Ajman with 16 responses, Fujairah with 8 responses, and Ras Al Khaimah with 2 responses. In addition to that, 10 responses were obtained stating that they got experience for all of the seven Emirates equally.

Preliminary Analysis

- **Homogeneity Test**: One of the main assumptions to be done in order to conduct an ANOVA test, is the homogeneity test. In our study we implemented a quick "Levene's Test of Equality of Error Variance" to check where our variables have the same variance in the population. Based on the obtained results, since the p value is less than 0.001 which is less than our α value (0.05), we reject the null hypothesis of homogeneity of variances. The Assumption of homogeneity is violated, which means that the Levene's test showed that the variances in the group were not equal ($F(10,163) = 8.703$, $p < 0.001$). There is a school of thought that suggests that ANOVA is Robust to violations of homogeneity of variance if we have a sample size of greater than 30 and equal sample sizes, so this assumption can be assumed to be valid.
- **Normality Test of the Variables and Residuals** The normality of the variables and residuals can be derived from the Skewness and Kurtosis of our data. By dividing the statistic value by the standard error in the Skewness and Kurtosis Table, we can derive that the z value for most pairs is between −3.29 and 3.29, concluding that the data is normally distributed.
- **Reliability and Validity of the Test**: In order to test the internal consistency of the items in the study, Cronbach's Alpha reliability test was conducted. The minimum acceptable Cronbach's Alpha value should be no less than 0.60, and the greater the value, the more reliable the result. Moreover, in order to check the validity of the tests, the Corrected Item-Total Correlation should be between 0.30 and 0.70 for a good scale [14]. The following table summarizes the results obtained from the Cronbach's Alpha test for each variable (Table 28.1).

Based on the table above, the results obtained is reliable and valid, since the corrected item-total correlation ranged between 0.30 to 0.70, and Cronbach's Alpha reliability test is above 0.60.

Reporting the Effects of 3-Way ANOVA

A factorial ANOVA [3-Way ANOVA] was conducted to compare the main effects of **Construction Site Manager's Performance towards Safety, Safety Equipment,**

Table 28.1 Reliability and validity test

Variable	Item	Corrected item-total correlation	Cronbach's alpha coefficient	Cronbach's alpha if item deleted
Safety performance	SP01	0.550	0.814	0.795
	SP02	0.743		0.731
	SP03	0.510		0.804
	SP04	0.622		0.773
	SP05	0.622		0.774
Safety equipment	SE01	0.606	0.784	0.727
	SE02	0.626		0.724
	SE03	0.514		0.761
	SE04	0.508		0.760
	SE05	0.563		0.742
Safety training	ST01	0.540	0.793	0.765
	ST02	0.664		0.722
	ST03	0.595		0.747
	ST04	0.624		0.738
	ST05	0.483		0.785

and Safety Training as well as their interaction effects on the **Safety Scores.** A summary of the results is as follows:

- **The main effect**: **Construction Site Manager's Performance Towards Safety** was statistically significant, meaning that it has an effect on the safety score of the site manager.
- **The main effect**: **Safety Equipment** was statistically significant, meaning that it has an effect on the safety score of the site manager.
- **The main effect**: **Safety Training** was statistically significant, meaning that it has an effect on the safety score of the site manager.
- **The interaction effect**: **Construction Site Manager's Performance Towards Safety and Safety Equipment** was significant, indicating that there was a combined effect for Construction Site Manager's Performance Towards Safety and Safety Equipment on the Safety Scores.

The rest of the interaction effects were not significant, meaning that it had negligible combined effect on the study.

Reporting the Effects of Monte Carlo Simulation

The Monte Carlo Simulation is used in this study to generate multiple random scenarios of site manager performance and access their impact on safety outcomes.

The worker's behaviors were observed during the site visits around the UAE, and based on that, the overall behavior of all worker was assumed and turned into a probability in order to test the Monte Carlo Simulation. The overall score for each factor was rated out of 25, and the probability assumed of each overall score can be seen below (Table 28.2).

Based on the probability, random scores out of 25 were being generated for 10,000 different scenarios to measure how effective the site managers are in terms enhancing safety in the workplace. The random numbers generated yielded the following results:

- 42% of the site managers have good safety performance.
- 53% of the site managers have neutral safety performance.
- 5% of the site managers have bad safety performance (Table 28.3).

Comparing the Monte Carlo result to the results obtained from the questionnaire we can see that its approximately the same result (Table 28.4).

Table 28.2 Probability out of 25 for the monte carlo simulation

Number	Probability (%)	CDF
5	0.82	0.00
6	0.82	0.01
7	0.82	0.02
8	0.82	0.02
9	0.82	0.03
10	0.82	0.04
11	0.82	0.05
12	0.82	0.06
13	1.26	0.07
14	1.26	0.08
15	8.26	0.09
16	8.26	0.17
17	8.26	0.26
18	8.26	0.34
19	8.26	0.42
20	8.26	0.50
21	8.26	0.59
22	8.26	0.67
23	8.26	0.75
24	8.26	0.83
25	8.26	0.92

Table 28.3 Monte carlo simulation results

Construction site manager safety rank		Probability (%)
Good	4203	42.03
Neutral	5263	52.63
Bad	534	5.34

Table 28.4 Questionnaire results

Construction site manager safety rank		Probability (%)
Good	84	48.28
Neutral	87	50.00
Bad	3	1.72

Recommendations

Recommendations for Current Research

- The current practices have workers wear standard PPE, which is not always recommended, especially during the extreme heat conditions in the UAE. The temperature can rise to 50 degrees Celsius, and workers cannot work under those conditions at all. Therefore, there should be a need to start implementing different variations of PPE for different scenarios, such as cooling PPE, to avoid workers getting heat stress, enhance their comfort, and ensure the workers do their respective duties without compromising their well-being. This will, in return, result in increased productivity in the workplace. Moreover, even though the UAE doesn't reach freezing temperatures, it is recommended to still design insulating PPE for the workers to use just in case some are not comfortable during the winter, which can avoid cold-health issues that some workers can suffer from. Hybrid PPE can be further explored and tested to fit all weather conditions by making use of effective modular design to customize PPE at will and have adjustable features to be used that can be modified freely.
- Since there are different local rules and regulations in all seven emirates when it comes to the safety in the construction site, it is recommended that all Emirates agree upon a unified safety compliance to follow in order to ensure consistency when performing safety practices and avoid all sorts of miscommunication in the future.
- Implementing more developed and tailored safety training programs to be organized depending on the current challenges facing the workers in the construction site. Some of the challenges that need to be addressed and have a customizable training problem could be safety practices in extreme heat conditions, the use of better mechanism to avoid diverse workforce language barriers, high-rise

construction training on safety harnesses and fall prevention, and efficient emergency response procedures. Doing efficient safety training for the workers will ensure and promote their well-being and ensure their safety is set as a top priority.

Recommendations for Future Research

- Other complex independent variables to be studied, such as:
  - Investigating how site managers ensure adherence to safety regulations to maintain safety in the workplace.
  - Analyzing the communications channels between the higher management, site managers, workers, and the relevant authorities regarding safety protocols and emergencies.
  - Study of site managers involve workers in enhancing the safety protocols by providing their feedback towards safety in the workplace.
  - Check if site managers implement risk and crisis mitigation plans to see if they are taking into consideration proactive measures in their daily operations to reduce the probability of accidents, and reactive measures to make sure site managers are prepared for all sorts of incidents/accidents that may arise in the workplace.
- Benchmarking the current safety management practices in the UAE with global standards provides a roadmap for improvement towards safety. This technique will identify gaps and areas of necessary improvement to get a better understanding of the strengths and weaknesses of the current safety management system. Moreover, the UAE construction industry can adopt proven strategies to enhance safety management by learning from the best practices that are done worldwide.
- A more detailed study could be conducted to study the performance of the site managers in the least developed emirates (Ajman, Umm Al Quwain, Ras Al Khaimah, Fujairah). Each emirate may face different and unique challenges in the construction site, and it is good to understand how site managers cope in the least developed emirates in order to identify gaps and get better improvements. Tailored safety programs could be implemented for each of the emirates based on the shortcomings in safety practices of the site managers, which in return will enhance safety and ensure the well-being of the workers. Even though this study had many inputs from the emirate of Umm Al Quwain, the other three named emirates need to be further explored.

References

1. NSC. (2023). Most dangerous industries. Retrieved from National Safety Council: https://injuryfacts.nsc.org/work/industry-incidence-rates/most-dangerous-industries/
2. Tracy, T. (2023). 7 ways to improve construction site safety. Retrieved from Procore: https://www.procore.com/library/improve-construction-site-safety
3. Rowe, S. F. (2007). Managing and leading small projects. Retrieved from pmi: https://www.pmi.org/learning/library/managing-leading-small-projects-7245
4. Caburao, E. A. (2023). Strengthening workplace safety through site safety training. Retrieved from Safety Culture: https://safetyculture.com/topics/workplace-safety/site-safety-training/
5. Liang, H., & Zhang, S. (2019). Impact of supervisor's safety violation on an individual worker within a construction crew. *Safety Science*, 679–691.
6. Skybrary. (2022). Human factors analysis and classification system (HFACS). Retrieved from Skybrary: https://www.skybrary.aero/articles/human-factors-analysis-and-classification-system-hfacs
7. Grill, M., & Nielsen, K. (2019). Promoting and impeding safety–a qualitative study into direct and indirect safety leadership practices of constructions site managers. *Safety Science,* 148–159.
8. Cramogroup. (2019). How site managers can contribute to better safety. Retrieved from Cramogroup: https://www.cramogroup.com/en/how-site-managers-can-contribute-to-better-safety/
9. McRae, M. (2023). How safety leadership styles impact employee safety behaviors. *Safety Science,* 1–138.
10. Abas, N. H., Yusuf, N., Suhaini, N. A., Kariya, N., Mohammad, H., & Hasmori, M. F. (2020). Factors affecting safety performance of construction projects: A literature review. *Journal of Materials Science & Engineering.*
11. Tarlengco, J. (2023). Everything you need to know about PPE safety. Retrieved from Safety Culture: https://safetyculture.com/topics/ppe-safety/
12. HSENetwork. (2023). The role of health and safety training in preventing accidents and injuries. Retrieved from HSE Network: https://www.hse-network.com/the-role-of-health-and-safety-training-in-preventing-accidents-and-injuries/
13. Regan, R. (2023). What is workplace safety training and why do you need it? Retrieved from Connecteam: https://connecteam.com/workplace-safety-training-need/
14. Ferketich. (1991). Focus on psychometrics: Aspects of item analysis. *Research in Nursing & Health,* 165–168.

Chapter 29
Exploring the Influence of Financial Technologies on Asset Price Dynamics: An Analytical Blueprint

Panagiotis G. Artikis, Evaggelia Kopanaki, and Polyxeni G. Tsitsiri

Abstract The complicated link between financial technology (FinTech) and asset pricing processes are examined in this article. It is presented an analytical framework combining foundational theories from the fields of Information Economics, Behavioral Finance, and Market Efficiency with the most recent FinTech advances like blockchain, data analytics, and automated trading systems. More specifically, it is used a combination of financial, and behavioral economics with technological adoption models. The incorporation of actual evidence strengthens theoretical principles. Data analytics, blockchain, automated trading, peer-to-peer lending, and regulatory technologies are the key five elements in the context of Fintech that affect asset value and are examined on this paper. The implications of how these results affect a variety of participants, such as investors and regulators, are discussed in this paper. Finally, the research focuses on the importance of testing the theory using empirical data and analyzing the ethical concerns of employing these technologies in asset evaluation.

Keywords Asset pricing · Financial technologies · Conceptual framework

Introduction

Asset pricing is a key financial science topic. Some financial innovative works are the Capital Asset Pricing Model (CAPM) [1, 2] and the Efficient Market Hypothesis [3]. These theoretical principles have influenced academic research and financial market investments. In contrast, financial technology, mainly known as FinTech, is forcing a reevaluation of current models, and generating concerns about the possible impact

P. G. Artikis · E. Kopanaki · P. G. Tsitsiri (✉)
University of Piraeus, 80, M. Karaoli & A. Dimitriou St., 18534 Piraeus, Greece
e-mail: polyxenits@unipi.gr

P. G. Artikis
e-mail: partikis@unipi.gr

E. Kopanaki
e-mail: evik@unipi.gr

© The Author(s), under exclusive license to Springer Nature Switzerland AG 2024
A. Emrouznejad et al. (eds.), *Business Analytics and Decision Making in Practice*,
Lecture Notes in Operations Research, https://doi.org/10.1007/978-3-031-61589-4_29

on asset pricing frameworks. FinTech advancements like blockchain and machine learning have changed financial institutions, however the immediate influence on asset price is still unknown.

Despite vast research on FinTech's effects on risk management [4], and financial inclusion [5], the investigation of FinTech's effects on asset pricing has received little attention. Asset pricing affects market efficiency and capital allocation; thus academics, investors, and regulators must grasp its consequences. FinTech technologies like robo-advisors, which manage assets that worth hundreds of billions of dollars, support our investigation [6].

The paper seeks to explain in a conceptual framework how financial technology may affect asset price. This framework categorizes the different FinTech instruments important to asset pricing, explores the theoretical concepts that explain their impacts, and offers the dimensions and channels via which FinTech might increase its influence. This research also investigates the actual effects on investors and regulators.

This study will focus on financial technology advances including machine learning, blockchain, and data analytics. This research does not analyze how financial technology like mobile banking and payment systems impact asset prices, but these technologies affect financial markets. Since this research is theoretical, future research will concentrate on experimentally confirming of this conceptual framework.

The key parts of this paper are listed below: Sect. Literature Review reviews current research on traditional asset pricing models and FinTech integration. Section Key Factors Affecting Asset Pricing Through Financial Technologies develops the conceptual framework, whereas Sect. Conceptual Framework examines the theoretical concepts. Section Theoretical Underpinnings explains the methodology that used, while Sect. Methodology examines how financial technology affects asset price. Section Implications offers the study's main implications, while the Sect. Conclusions summarizes the key findings and suggests additional research.

Literature Review

With the Capital Asset Pricing Model (CAPM) as a foundation, academic interest in asset pricing theories has grown since the early 1900s. CAPM was created by Sharpe [2] and Lintner [1]. According to this concept, systematic risk, defined by the asset's beta coefficient, is the major factor affecting future returns. French and Fama [7] established the Three-Factor Model, and CAPM was enhanced to incorporate size and book-to-market value.

The well-known Capital Asset Pricing Model (CAPM) [1, 2], and Arbitrage Pricing Theory (APT) [8] are traditional asset pricing frameworks that determine asset prices. Most pricing algorithms in these paradigms presume rational participants and efficient markets. By explaining financial derivative pricing methodologies, the Black–Scholes-Merton model [9, 10] added complexity. These classic methods may be used to evaluate FinTech's influence on asset pricing frameworks.

The rise of FinTech has spurred academic research on its numerous financial market effects. Gomber et al. [6] examine how FinTech affects trading, lending, and risk management in the financial industry. These developments have occurred, but prior research has not examined FinTech's direct influence on asset pricing.

Asset pricing research has increasingly used machine learning, a specialized FinTech subset. According to Gu et al. [11], machine learning algorithms can predict stock price movements better than traditional asset pricing models.

Asset pricing and market dynamics are being studied in relation to the effects of algorithmic trading [12] and robo-advice services [13]. According to Tapscott and Tapscott [14], blockchain technology is anticipated to have a substantial influence on asset values by addressing financial market information asymmetry.

Despite these advances, the influence of financial technology on asset prices are particularly not examined. To fill this gap, our research developed a conceptual framework to explain how financial technology may affect financial asset price.

Key Factors Affecting Asset Pricing Through Financial Technologies

Technology in financial markets has made asset pricing systems more complicated in various ways. This section will concentrate on the five main factors in the context of financial technology that affect asset prices; data analytics, blockchain, automated trading, peer-to-peer lending, and regulatory technologies, which are emerging technologies.

Recently, data analytics has been a major effect on financial markets, especially asset pricing. Advanced analytical methods provide more accurate and dynamic asset valuations, according to Chen et al. [15]. Market activity, customer preferences, and risk concerns are better understood with these tactics. Machine learning and AI systems are highly beneficial to forecast short-term asset value changes [16].

Blockchain technology can transform asset pricing landscape. Maurer et al. [17] believe that it creates a transparent and immutable record of all financial transactions, minimizing market manipulation and information asymmetry.

Due to its dual nature, high-frequency trading and other automated trading platforms affect asset values in several ways. While flash crashes are commonly connected to them, high-frequency trading increases market liquidity and lowers the bid-ask spread [12, 18]. Market flash collapses are temporary. According to [19], peer-to-peer (P2P) lending and crowdfunding platforms allow users to get capital access but employ unusual traditional variables into the asset pricing equation. These include social proof and network effects.

RegTech, or regulatory technology, affect asset values indirectly by affecting compliance costs and the requirement to adapt to new rules. RegTech automates conformance procedures, improving market efficiency. This may affect asset values [5].

Conceptual Framework

This section provides a theoretical framework for examining the complicated relationship between FinTech and asset price. This section explains financial technology and describes the relationship between FinTech and asset pricing. According to Arner et al. [5], "financial technologies" (or "FinTech") are technical advancements that improve financial services. Peer-to-peer lending, robo-advisors, blockchain, and algorithmic trading are examples [20].

We use a conceptual method to explain how FinTech may affect asset price. This stratified model is based on the Capital Asset Pricing Model (CAPM) [1, 2] asset pricing assumptions. These tiers are layered utilizing FinTech methods including machine learning, blockchain, and data analytics. Market sentiment, regulatory frameworks, and information efficiency affect the interplay between these levels.

FinTech affects asset values in numerous ways. Blockchain technology may improve asset values by increasing transparency and reduce information asymmetry, according to Tapscott and Tapscott [14]. This might improve financial transactions. Algorithmic and high-frequency trading may affect asset prices by changing liquidity, transaction volume, and market depth [12, 21] observed that behavioral factors like robo-advisors may affect investors' asset valuation and trading choices.

The existing literature provides insights into the individual effects of financial technology on asset price, but comprehensive study is few. Most studies focus on specific issues like market efficiency or investor behavior rather than unifying them into a conceptual framework. This research addresses this problem by offering a comprehensive conceptual framework that incorporates the different ways financial technology may affect asset pricing [22].

The diagram below illustrates asset price and FinTech theory and their relationship. Financial technology and asset pricing models are defined at the top of this illustration. These models demonstrate fundamentals. Arrows show impact and interaction. FinTech, for example, has an impact on asset pricing models and is affected by external factors. The bottom half addresses how investment technology affects asset price, including machine learning, blockchain, and data analytics. The sideboard arrow shows how this interplay affects behavioral elements and decision-making (Fig. 29.1).

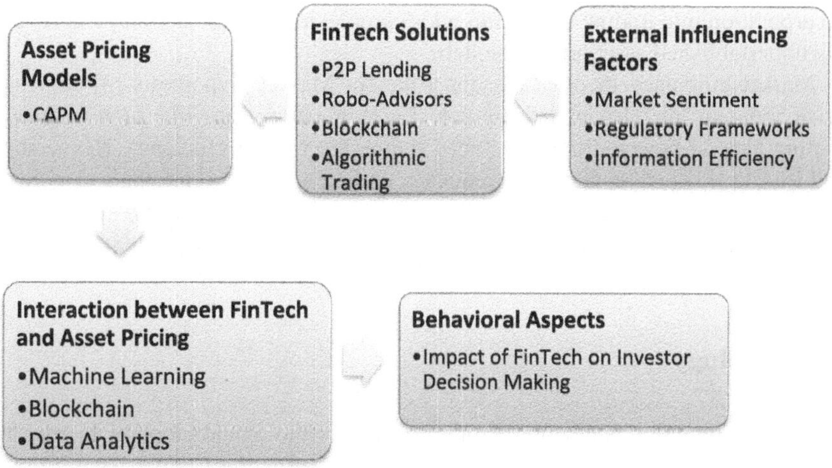

Fig. 29.1 FinTech and asset pricing interaction

Theoretical Underpinnings

This research relies on Information Economics, Behavioral Finance, and Market Efficiency Theories. These theoretical approaches provide light on how financial innovations affect asset values.

Our conceptual framework uses the Capital Asset Pricing Model (CAPM) [1, 2], and the Fama–French Three-Factor Model [7]. These models provide traditional metrics for assessing the anticipated profits and risks of financial investments. However, they emerged in a technological setting that is quite different from financial technology today.

Akerlof [23] and Stigler [24] show how information asymmetry and diffusion affect market players. According to Pagano and Volpin [25], FinTech approaches like blockchain, and automated trading systems affect financial market information accessibility and usage. These technologies might considerably decrease information asymmetry, affecting asset prices.

Behavioral finance blends psychology with financial market research. Prospect Theory, developed by Kahneman and Tversky [26], explains investor behavior and asset prices via cognitive biases and heuristics. As more financial technology platforms utilize behavioral nudges, including robo-advisors, it is crucial to study how they exploit investor psychology to affect asset prices [27].

The Technology uptake Model (TAM) by Davis [28] and the Diffusion of Innovations Theory by Rogers [29] may also be used to examine the adoption of developing technology trends. Their relevance to our study is significant since FinTech adoption speed and extent may affect asset prices [6].

By incorporating financial technology into asset pricing frameworks, it is the main novelty of our conceptual framework. Gu et al. [11] show that quick machine learning incorporation into trading algorithms adds complexity. Complexity may improve or disrupt established asset pricing models.

Market efficiency theories like the Efficient Market Hypothesis [3] say asset prices contain all relevant information. FinTech innovations like high-frequency trading and AI-powered analytics need rethinking market efficiency. Hendershott and Riordan [12] claim that high-frequency trading may distort pricing and increase market volatility. Due to FinTech technologies, market microstructure—the study of how trading procedures impact price formation [30]—has grown in importance.

Methodology

Financial economics, behavioral economics, and technological adoption theories are used in this multidisciplinary study. The research uses the Capital Asset Pricing Model (CAPM) [1, 2] as its analytical framework. Behavioral economics is used in FinTech to examine how psychological variables affect asset prices. The Technology Acceptance Model [28] measures financial technology adoption pace and effect on asset value.

For this research, basic assumptions were made. First, it supports Fama's [3] Efficient Market Hypothesis, which suggests market efficiency levels. According to Gomber et al. [6], financial technology has penetrated the market enough to affect asset values. As proven in behavioral economics research, financial markets include both rational and irrational players, as indicated by previous behavioral economics studies. This research assumes stable regulation to simplify analysis.

Despite its theoretical approach, this study supports its assertions with actual evidence. More specifically, the theoretical claims are chosen based on peer-reviewed academic publications. Additionally, reputable databases like CRSP or Datastream provide empirical financial measurements like stock prices, trade volumes, and other indicators. The research includes data from respected industry studies like Gartner and Statista to support its FinTech adoption rates claims.

Implications

This research seeks to further asset pricing-financial technology debates. Our results' effects on investors, regulators, and opportunities for academic research are examined in this section.

Understanding how financial technology affects asset values may help investors make big money. Chen et al. [15] found that investors who understood how data analytics affects forecasting models made better decisions. Understanding how blockchain technology reduces information asymmetry may help investors evaluate assets [17].

The regulatory authorities must constantly monitor how financial technology is affecting asset valuation standards. Understanding this is important for managing automated trading systems, which may bring new market vulnerabilities [18]. The growing number of peer-to-peer financial platforms necessitates innovative regulatory frameworks to guarantee equal asset value and investor protection [31].

This study sets the framework for future research. Further research utilizing real-world datasets to confirm this conceptual framework is crucial. Few studies have examined behavioral finance and financial technology [12], which is mostly unexplored.

Conclusions

This research sought to understand how financial technology affects asset values. The study focuses how data analytics, blockchain, automated trading, peer-to-peer finance, and regulatory technologies affect asset value. These improvements may have substantial effects, such as employing blockchain technology to solve information asymmetry [17] and improving asset value prediction models using machine learning algorithms [11].

According to [32], these theoretical implications lead to improve asset pricing theory integrating economic, behavioral, and technical drivers. Chen et al. [15] suggest that these findings should prompt investors to rethink their investing strategy in light of technology advances. According to Kirilenko et al. [18], Policymakers should evaluate and adjust regulatory frameworks to market trends.

According to a key criterion, empirical research employing real-world data is necessary to validate the suggested conceptual framework. Arner et al. [5] also suggest studying the ethical issues surrounding asset pricing using these tools.

Acknowledgements This work has been partly supported by the University of Piraeus Research Center.

References

1. Lintner, J. (1965). The valuation of risk assets and the selection of risky investments in stock portfolios and capital budgets. *The Review of Economics and Statistics, 47*(1), 13–37.
2. Sharpe, W. F. (1964). Capital asset prices: A theory of market equilibrium under conditions of risk. *The Journal of Finance, 19*(3), 425–442.
3. Fama, E. F. (1970). Efficient capital markets: A review of theory and empirical work. *The Journal of Finance, 25*(2), 383–417.
4. Li, C., He, S., Tian, Y., Sun, S., & Ning, L. (2022). Does the bank's FinTech innovation reduce its risk-taking? Evidence from China's banking industry. *Journal of Innovation & Knowledge, 7*(3), 100219. https://doi.org/10.1016/j.jik.2022.100219

5. Arner, D. W., Barberis, J., & Buckley, R. P. (2016). FinTech, RegTech, and the reconceptualization of financial regulation. *Northwestern Journal of International Law & Business, 37*(3), 371–413.
6. Gomber, P., Koch, J. A., & Siering, M. (2018). Digital finance and FinTech: Current research and future research directions. *Journal of Business Economics, 88*(3), 289–341.
7. Fama, E. F., & French, K. R. (1993). Common risk factors in the returns on stocks and bonds. *Journal of Financial Economics, 33*(1), 3–56.
8. Ross, S. A. (1976). The arbitrage theory of capital asset pricing. *Journal of Economic Theory, 13*(3), 341–360.
9. Black, F., & Scholes, M. (1973). The pricing of options and corporate liabilities. *Journal of Political Economy, 81*(3), 637–654.
10. Merton, R. C. (1973). Theory of rational option pricing. *The Bell Journal of Economics and Management Science, 4*(1), 141–183.
11. Gu, S., Kelly, B., & Xiu, D. (2020). Empirical asset pricing via machine learning. *The Review of Financial Studies, 33*(5), 2223–2273.
12. Hendershott, T., & Riordan, R. (2013). Algorithmic trading and the market for liquidity. *Journal of Financial and Quantitative Analysis, 48*(4), 1001–1024.
13. Fisch, J. E., Laboure, M., & Turner, J. A. (2019). The Emergence of the Robo-advisor. *The disruptive impact of FinTech on retirement systems, 13*, 13–37.
14. Tapscott, D., & Tapscott, A. (2016). *Blockchain revolution: how the technology behind bitcoin is changing money, business, and the world*. Penguin.
15. Chen, H., Chiang, R. H., & Storey, V. C. (2014). Business intelligence and analytics: From big data to big impact. *MIS Quarterly, 36*(4), 1165–1188.
16. Fischer, T., & Krauss, C. (2018). Deep learning with long short-term memory networks for financial market predictions. *European journal of operational research, 270*(2), 654–669. https://doi.org/10.1016/j.ejor.2017.11.054
17. Maurer, B., Nelms, T. C., & Swartz, L. (2013). "When perhaps the real problem is money itself!": The practical materiality of Bitcoin. *Social Semiotics, 23*(2), 261–277.
18. Kirilenko, A. A., Kyle, A. S., Samadi, M., & Tuzun, T. (2017). The flash crash: High-frequency trading in an electronic market. *The Journal of Finance, 72*(3), 967–998.
19. Lin, M., Prabhala, N. R., & Viswanathan, S. (2013). Judging Borrowers by the Company They Keep: Friendship Networks and Information Asymmetry in Online Peer-to-Peer Lending. *Management Science, 59*(1), 17–35. https://doi.org/10.1287/mnsc.1120.1560
20. Arner, D. W., Barberis, J., & Buckley, R. P. (2015). The evolution of fintech: A new post-crisis paradigm? *Georgetown Journal of International Law, 47*(4), 1271–1319.
21. D'Acunto, F., Prabhala, N., & Rossi, A. G. (2019). The promises and pitfalls of robo-advising. *The Review of Financial Studies, 32*(5), 1983–2020. https://doi.org/10.1093/rfs/hhz014
22. Varga, D. (2017). Fintech, the new era of financial services. *Vezetéstudomány-Budapest Management Review, 48*(11), 22–32. https://doi.org/10.14267/VEZTUD.2017.11.03
23. Akerlof, G. A. (1970). The market for "lemons": Quality uncertainty and the market mechanism. *Quarterly Journal of Economics, 84*(3), 488–500.
24. Stigler, G. J. (1961). The economics of information. *Journal of Political Economy, 69*(3), 213–225.
25. Pagano, M., & Volpin, P. F. (2012). Securitization, transparency, and liquidity. *The Review of Financial Studies, 25*(8), 2417–2453.
26. Kahneman, D., & Tversky, A. (1979). Prospect theory: An analysis of decision under risk. *Econometrica, 47*(2), 263–292.
27. Bhattacharya, U., Hackethal, A., Kaesler, S., Loos, B., & Meyer, S. (2012). Is unbiased financial advice to retail investors sufficient? Answers from a large field study. *The Review of Financial Studies, 25*(4), 975–1032.
28. Davis, F. D. (1989). Perceived usefulness, perceived ease of use, and user acceptance of information technology. *MIS Quarterly, 13*(3), 319–340.
29. Rogers, E. M. (1962). *Diffusion of Innovations*. Free Press
30. O'Hara, M. (1995). *Market microstructure theory*. Blackwell

31. Navaretti, G. B., Calzolari, G., & Pozzolo, A. F. (2017). FinTech and banks: Friends or Foes? *European Economy*, (2), 9–30.
32. Daniel, K., & Titman, S. (1997). Evidence on the characteristics of cross sectional variation in stock returns. *The Journal of Finance*, *52*(1), 1–33. https://doi.org/10.1111/j.1540-6261.1997.tb03806.x

SPRINGER NATURE

GPSR Compliance

The European Union's (EU) General Product Safety Regulation (GPSR) is a set of rules that requires consumer products to be safe and our obligations to ensure this.

If you have any concerns about our products, you can contact us on ProductSafety@springernature.com

In case Publisher is established outside the EU, the EU authorized representative is:

Springer Nature Customer Service Center GmbH
Europaplatz 3
69115 Heidelberg, Germany

The manufacturer's authorised representative in the EU is Springer Nature Customer Service Centre GmbH, Europaplatz 3, 69115 Heidelberg, Germany. If you have any concerns regarding our products, please contact ProductSafety@springernature.com

Printed and bound by CPI Group (UK) Ltd, Croydon, CR0 4YY

28/09/2026

02224061-0001